The Latin American Studies Book Series

The Latin American Studies Book Series promotes quality scientific research focusing on Latin American countries. The series accepts disciplinary and interdisciplinary titles related to geographical, environmental, cultural, economic, political and urban research dedicated to Latin America. The series publishes comprehensive monographs, edited volumes and textbooks refereed by a region or country expert specialized in Latin American studies.

The series aims to raise the profile of Latin American studies, showcasing important works developed focusing on the region. It is aimed at researchers, students, and everyone interested in Latin American topics.

Submit a proposal: Proposals for the series will be considered by the Series Advisory Board. A book proposal form can be obtained from the Publisher, Juliana Pitanguy (juliana.pitanguy@springer.com).

José Maria Landim Dominguez ·
Ruy Kenji Papa de Kikuchi ·
Moacyr Cunha de Araújo Filho ·
Ralf Schwamborn · Helenice Vital
Editors

Tropical Marine Environments of Brazil

Spatio-Temporal Heterogeneities and Responses to Climate Changes

Springer

Editors
José Maria Landim Dominguez
Department of Oceanography
Universidade Federal da Bahia
Salvador, Bahia, Brazil

Ruy Kenji Papa de Kikuchi
Department of Oceanography
Universidade Federal da Bahia
Salvador, Bahia, Brazil

Moacyr Cunha de Araújo Filho
Department of Oceanography
Universidade Federal de Pernambuco
Recife, Pernambuco, Brazil

Ralf Schwamborn
Department of Oceanography
Universidade Federal de Pernambuco
Recife, Pernambuco, Brazil

Helenice Vital
Universidade Federal do Rio Grande do Norte
Natal, Rio Grande do Norte, Brazil

ISSN 2366-3421 ISSN 2366-343X (electronic)
The Latin American Studies Book Series
ISBN 978-3-031-21331-1 ISBN 978-3-031-21329-8 (eBook)
https://doi.org/10.1007/978-3-031-21329-8

This Springer imprint is published by the registered company Springer Nature Switzerland AG
The registered company address is: Gewerbestrasse 11, 6330 Cham, Switzerland

Preface

This book presents a synthesis of almost a decade of research conducted under the umbrella of the National Institute of Science and Technology in Tropical Marine Environments (inctAmbTropic), supplemented by contributions of our partners (Chaps. 2 and 3).

The National Institutes of Science and Technology Program (INCTs) were started in 2008 and financed by the Brazilian National Fund for Scientific and Technological Development (FNDCT) in partnership with state research support foundations. This program enabled the configuration of interregional networks of collaboration with national coverage and academic, scientific, and technological performance compatible with the best international programs.

This book synthesizes the current knowledge about the major tropical marine environments of Brazil at three spatial scales: (i) the coastal zone, (ii) the continental shelf, and (iii) the Tropical Atlantic basin. Additionally, it discusses the impacts and implications of the ongoing climate changes in these environments.

This volume contains eight chapters written by 68 authors who are leading specialists in their fields of research and provides a state-of-the-art assessment about the Climate Variability and Change in Tropical South America, Mangrove Swamps, Deltas, Coral Reefs, Continental Shelf, Zoo- and Ichthyoplankton Communities, and Circulation, Biogeochemical Cycles, and CO_2 Flux in the Tropical Atlantic Ocean.

The authors have made an effort to present information in a very didactical way so the book can also be used as a textbook for undergraduate and graduate students and managers interested in the subject.

Salvador, Brazil — José Maria Landim Dominguez
Salvador, Brazil — Moacyr Cunha de Araújo Filho
Recife, Brazil — Ralf Schwamborn
Recife, Brazil — Ruy Kenji Papa de Kikuchi
Natal, Brazil — Helenice Vital

Acknowledgements

We are very grateful for the support provided by the Brazilian National Council for Scientific Research and Development (CNPq) (grants 565054/2010-4 and 465634/2014-1), the Coordination for the Improvement of Higher Education Personnel (CAPES), and the Foundation for Research Support of the State of Bahia (FAPESB) (grants 8936/2011 and inc004/2019). Without their support, most of the information presented herein would not have been produced. We also thank all our colleagues for their effort in the production of this volume, in special Dr. Jose Antônio Marengo and Dr. Pedro Walfir Martins e Souza-Filho and their collaborators who promptly accepted our invitation to contribute to this book. At Springer, we especially thank Juliana Pitanguy for her guidance during the elaboration of the initial proposal for this book.

Contents

Editors and Contributors

About the Editors

José Maria Landim Dominguez Full professor at the Department of Oceanography of the Federal University of Bahia, Salvador, Brazil. He holds a Ph.D. in Marine Geology and Geophysics from the University of Miami—RSMAS, USA. He presently coordinates the National Institute of Science and Technology for Tropical Marine Environments (inctAmbTropic). His main research interests include wave-dominated deltas, coastal sedimentation, sea-floor mapping, sea-level change, climate change, coastal sedimentation, and high-resolution seismic. e-mail: landim@ufba.br

Ruy Kenji Papa de Kikuchi Full professor at the Department of Oceanography of the Federal University of Bahia, Salvador, Brazil. He holds a Ph.D. in Geology from the Federal University of Bahia. He presently coordinates the Reefs and Coralline Environments and the Oil Spill Working Groups of the National Institute of Science and Technology for Tropical Marine Environments (inctAmbTropic). His main research interests are coral reef mapping, assessment, and monitoring and coral sclerochronology and geochemistry. e-mail: kikuchi@ufba.br

Moacyr Cunha de Araújo Filho Associate professor at the Department of Oceanography of the Federal University of Pernambuco, Recife, Brazil. He holds a Ph.D. in Physics and Chemistry of the Environment from the Institut National Polytechnique de Toulouse, France. His main research interests are ocean observation, forecasting and monitoring processes, and systems in the tropical Atlantic and has contributed to different initiatives/projects, e.g., PIRATA, H2020 Triatlas, JPI-MicroplastiX, inctAmbTropic, and Proantar-MEPHYSTO. More recently, he has also been involved in developing infrastructure and a framework for sharing ocean observations, devices, and best practices, with a focus on linking offshore and coastal programs. e-mail: moa@ufpe.br

Ralf Schwamborn Associate professor at the Department of Oceanography of the Federal University of Pernambuco, Recife, Brazil. He holds a Ph.D. from the University of Bremen, Germany (Center for Tropical Marine Ecology—ZMT). His main research interests are marine biology, marine ecology, fisheries biology, advanced statistics, and ecological modeling. His current research projects include the use of stable isotopes in food webs, zooplankton ecology, microplastics, decapod crustacean larvae, shrimp fisheries, coral reefs, mangrove ecology, size spectra, robust statistics, bootstrapping, novel statistical models, and development of new software (e.g., fishboot R package). e-mail: ralf.schwamborn@ufpe.br

Helenice Vital Full professor at the Federal University of Rio Grande do Norte, Natal, Brazil. She holds a Ph.D. in Natural Sciences from the University of Christian Albrechts at Kiel, Germany. Her major research interests include coastal erosion, sea-level changes, incised valleys, environmental monitoring, geohazards, high-resolution seismic, sea-floor mapping, coastal dynamics, and the Brazilian equatorial margin. e-mail: helenice.vital@ufrn.br

Contributors

Carlos Eduardo Rocha Duarte Alencar Southwestern Bahia State University (UESB), Jequié, Bahia, Brazil;
Federal University of Rio Grande do Norte (UFRN), Natal, Rio Grande do Norte, Brazil

Lincoln Muniz Alves National Institute for Space Research (INPE), São Jose dos Campos, São Paulo, Brazil

Tercio Ambrizzi University of São Paulo (USP), São Paulo, São Paulo, Brazil

Nils Edvin Asp Federal University of Pará (UFPA), Bragança, Pará, Brazil

Alex Cardoso Bastos Federal University of Espírito Santo (UFES), Vitória, Espírito Santo, Brazil

Paula Nepomuceno Campos Federal Rural University of Amazônia (UFRA), Belém, Pará, Brazil

Rayssa de Lima Cardoso Federal University of Maranhão (UFMA), São Luís, Maranhão, Brazil

Luiz Cortinhas Solved—Solutions in Geoinformation, Belém, Pará, Brazil

Marcos Heil Costa Federal University of Viçosa (UFV), Viçosa, Minas Gerais, Brazil

Ana Paula Cunha National Center for Monitoring and Early Warning of Natural Disasters (CEMADEN), São José dos Campos, São Paulo, Brazil

Laura Rodrigues da Conceição University of São Paulo (USP) São Paulo, São Paulo, Brazil

Delzenira Silva do Nascimento da Costa Federal University of Maranhão (UFMA), São Luís, Maranhão, Brazil

Leticia Cotrim da Cunha Brazilian Research Network on Global Climate Change (Rede CLIMA), São José dos Campos, São Paulo, Brazil;
State University of Rio de Janeiro (UERJ), Rio de Janeiro, Rio de Janeiro, Brazil

Wilson Rocha da Nascimento Júnior Vale Institute of Technology, Belém, Pará, Brazil

Alex Costa da Silva Federal University of Pernambuco (UFPE), Recife, Pernambuco, Brazil

Elton Alex Correa da Silva Federal Rural University of Amazônia (UFRA), Belém, Pará, Brazil

Leiliane Souza da Silva Federal Universidade of Pará (UFPA), Belém, Pará, Brazil

Paula Cilene Alves da Silveira Federal University of Maranhão (UFMA), São Luís, Maranhão, Brazil

Patrícia Silva Basílio Dantas Federal University of Pernambuco (UFPE), Recife, Pernambuco, Brazil

Moacyr Cunha de Araújo Filho Federal University of Pernambuco (UFPE), Recife, Pernambuco, Brazil;
Brazilian Research Network on Global Climate Change (Rede CLIMA), São José dos Campos, São Paulo, Brazil

Julia Martins de Araújo University of São Paulo (USP), São Paulo, São Paulo, Brazil

Maria Elisabeth De Araújo Federal University of Pernambuco (UFPE), Recife, Pernambuco, Brazil

Tereza Cristina Medeiros de Araújo Federal University of Pernambuco (UFPE), Recife, Pernambuco, Brazil

Claudine Dereczynski Federal University of Rio de Janeiro (UFRJ), Rio de Janeiro, Brazil

Helen Michelle de Jesus Affe Brazilian Research Network on Global Climate Change (Rede CLIMA), São José dos Campos, São Paulo, Brazil;
State University of Rio de Janeiro (UERJ), Rio de Janeiro, Rio de Janeiro, Brazil

Ruy Kenji Papa de Kikuchi Federal University of Bahia (UFBA), Salvador, Bahia, Brazil

Francielly Alcântara de Lima Federal Universidade of Pará (UFPA), Belém, Pará, Brazil

Nuno Filipe Alves Correia de Melo Federal Rural University of Amazônia (UFRA), Belém, Pará, Brazil

Marilia de Dirceu Machado de Oliveira Federal University of Bahia (UFBA), Salvador, Bahia, Brazil

Christiane Sampaio de Souza Federal University of Bahia (UFBA), Salvador, Bahia, Brazil

Cesar Guerreiro Diniz Solved—Solutions in Geoinformation, Belém, Pará, Brazil

José Maria Landim Dominguez Federal University of Bahia (UFBA), Salvador, Bahia, Brazil

Xiomara Franchesca García Díaz Federal Rural University of Amazônia (UFRA), Belém, Pará, Brazil

Elisabeth Cabral Silva Falcão Federal University of Rio Grande do Sul (UFRGS), Imbé, Rio Grande Do Sul, Brazil

Caroline Vieira Feitosa Federal University of Ceará (UFC), Fortaleza, Ceará, Brazil

João Lucas Leão Feitosa Federal University of Pernambuco (UFPE), Recife, Pernambuco, Brazil

Marcus Emanuel Barroncas Fernandes Federal University of Pará (UFPA), Bragança, Pará, Brazil

Gabriela Guerra A. A. Figueiredo Federal University of Pernambuco (UFPE), Recife, Pernambuco, Brazil

Fulvio Aurélio Morais Freire Federal University of Rio Grande do Norte (UFRN), Natal, Rio Grande do Norte, Brazil

Gbekpo Aubains Hounsou-Gbo Federal University of Ceará (UFC), Fortaleza, Ceará, Brazil

Zelinda Margarida de Andrade Nery Leão Federal University of Bahia (UFBA), Salvador, Bahia, Brazil

Anna Evelin Coimbra Libório Federal University of Maranhão (UFMA), São Luís, Maranhão, Brazil

Juliana Franco Lima Federal University of Maranhão (UFMA), São Luís, Maranhão, Brazil

Nathália Lins-Silva Federal University of Pernambuco (UFPE), Recife, Pernambuco, Brazil

Simone Maria de Albuquerque Lira Federal University of Pernambuco (UFPE), Recife, Pernambuco, Brazil

João Paulo Nobre Lopes Vale Institute of Technology, Belém, Pará, Brazil

Paulo de Oliveira Mafalda Jr Federal University of Bahia (UFBA), Salvador, Bahia, Brazil

Catarina da Rocha Marcolin Federal University of Sul da Bahia (UFSB), Porto Seguro, Bahia, Brazil

Jose Antônio Marengo National Center for Monitoring and Early Warning of Natural Disasters (CEMADEN), São José dos Campos, São Paulo, Brazil

Jussara Moretto Martinelli-Lemos Federal Universidade of Pará (UFPA), Belém, Pará, Brazil

Pedro Walfir Martins e Souza-Filho Vale Institute of Technology, Belém, Pará, Brazil;
Federal University of Pará (UFPA), Belém, Pará, Brazil

Sigrid Neumann-Leitão Federal University of Pernambuco (UFPE), Recife, Pernambuco, Brazil

Carlos Esteban Delgado Noriega Federal University of Pernambuco (UFPE), Recife, Pernambuco, Brazil;
Brazilian Research Network on Global Climate Change (Rede CLIMA), São José dos Campos, São Paulo, Brazil

Eduardo Tavares Paes Federal Rural University of Amazônia (UFRA), Belém, Pará, Brazil

Glauber David Almeida Palheta Federal Rural University of Amazônia (UFRA), Belém, Pará, Brazil

Jéssica dos Santos Lima Pantoja Federal Rural University of Amazônia (UFRA), Belém, Pará, Brazil

Syumara Queiroz Federal University of Pernambuco (UFPE), Recife, Pernambuco, Brazil

Michelle Simões Reboita Federal University of Itajubá (UNIFEI), Itajubá, Minas Gerais, Brazil

Ana Carolina Melo Rodrigues-Inoue Federal Universidade of Pará (UFPA), Belém, Pará, Brazil

Claudeilton S. Santana Federal University of Pernambuco (UFPE), Recife, Pernambuco, Brazil

Denise Fabiana de Moraes Costa Schwamborn Federal University of Pernambuco (UFPE), Recife, Pernambuco, Brazil

Ralf Schwamborn Federal University of Pernambuco (UFPE), Recife, Pernambuco, Brazil

Marcelo Oliveira Soares Federal University of Ceará (UFC), Fortaleza, Ceará, Brazil;
Leibniz Center for Tropical Marine Research (ZMT), Bremen, Germany

Pedro Walfir Martins e Souza-Neto University of São Paulo (USP), São Paulo, São Paulo, Brazil

Helenice Vital Federal University of Rio Grande do Norte (UFRN), Natal, Rio Grande do Norte, Brazil

Chapter 1
Tropical Marine Environments of Brazil and Impacts of Climate Change

José Maria Landim Dominguez, Moacyr Cunha de Araújo Filho, Ralf Schwamborn, Ruy Kenji Papa de Kikuchi, and Helenice Vital

Abstract Ongoing climate change has the potential to severely affect the tropical marine environments of Brazil. This chapter provides a brief overview of these environments with a major focus on deltas, mangrove forests, reefs, continental shelves, the zoo- and ichthyoplankton communities of pelagic ecosystems and major processes operating in the Tropical Atlantic. The impacts of ongoing climate change are also summarized. Additionally, the book organization is presented, including a brief description of each chapter.

Keywords Mangroves · Deltas · Coral reefs · Continental shelf · Pelagic ecosystems · Tropical Atlantic · Climate variability

1.1 Introduction

Climate change will affect the physical, biological, and biogeochemical characteristics of coastal zones and oceans, impacting their ecological structure and the different services provided to humankind. These changes have the potential to cause serious socioeconomic impacts at the local (coastal), regional (platform and shallow seas), and global (ocean) scales. The responses of marine environments to climate change will also depend on the natural variability of these systems and other changes introduced by humans as a result of the use of marine resources, making coastal and shelf areas more vulnerable to natural hazards.

J. M. L. Dominguez (✉) · R. K. P. de Kikuchi
Federal University of Bahia (UFBA), Salvador, Bahia, Brazil
e-mail: landim@ufba.br

M. C. de Araújo Filho · R. Schwamborn
Federal University of Pernambuco (UFPE), Recife, Pernambuco, Brazil

H. Vital
Federal University of Rio Grande do Norte (UFRN), Natal, Rio Grande do Norte, Brazil

J. M. Landim Dominguez et al. (eds.), *Tropical Marine Environments of Brazil*,
The Latin American Studies Book Series, https://doi.org/10.1007/978-3-031-21329-8_1

The tropical region of Brazil presents a unique opportunity to assess how the spatial and temporal heterogeneity of tropical marine environments influences the response patterns of these environments and their resilience to climate change that will affect the region in this century. This region contains the main reef constructions of the southwestern Atlantic Ocean, the Brazilian deltas, the second largest area of mangroves in the world in a single country, a continental shelf that varies from the narrowest to the widest in Brazil, the main islands and seamounts, extreme variations in sediment and nutrient flows, and an undeniable importance in interhemispheric heat and mass transfer (Fig. 1.1).

This book summarizes the major advances in knowledge of the marine environments of tropical Brazil as a result of the execution of the National Institute for Science and Technology in Tropical Marine Environments (inctAmbTropic) project. The National Institutes of Science and Technology Program (INCTs) was started in 2008 and financed by the National Fund for Scientific and Technological Development (FNDCT) in partnership with state research support foundations. This program enabled the configuration of interregional networks of collaboration with national coverage and academic, scientific, and technological performance compatible with the best international programs.

The inctAmbTropic conducts research at three spatial scales: (i) coastal zone—characterized by great physical and biological heterogeneity and the interface of interactions between natural and anthropogenic forcings, (ii) continental shelf—an area also of great heterogeneity and still poorly understood and (iii) ocean—a component of the Earth system influenced by mass transport and its interactions with the atmosphere.

The well-being of human communities intrinsically depends on the availability of services that coastal and marine ecosystems provide. This is particularly important for the northern and northeastern regions of Tropical Brazil, which has, in some of its coastal municipalities, population densities that are among the highest in the country (Fig. 1.1). Understanding how different marine environments will react to climate change in the coming decades is therefore of strategic importance for the region.

1.2 The Tropical Marine Environments of Brazil

The main tropical marine environments of Brazil include (i) the wave-dominated deltas, (ii) the main coral reefs of the western South Atlantic, (iii) one of the most important mangrove forests on the planet, and (iv) a mostly narrow continental shelf bathed by oligotrophic waters, starved of sediments and therefore dominated by carbonate sedimentation. Additionally, this region plays an important role in biogeochemical cycles, CO_2 flux, and circulation in the Tropical Atlantic Ocean.

From a physiographic point of view, this area is divided into 3 sectors (Dominguez 2009): (i) the deltaic coast of eastern Brazil, (ii) the sediment-starved coast of northeastern Brazil, and (iii) the tidal embayment from the Amazon (Fig. 1.1).

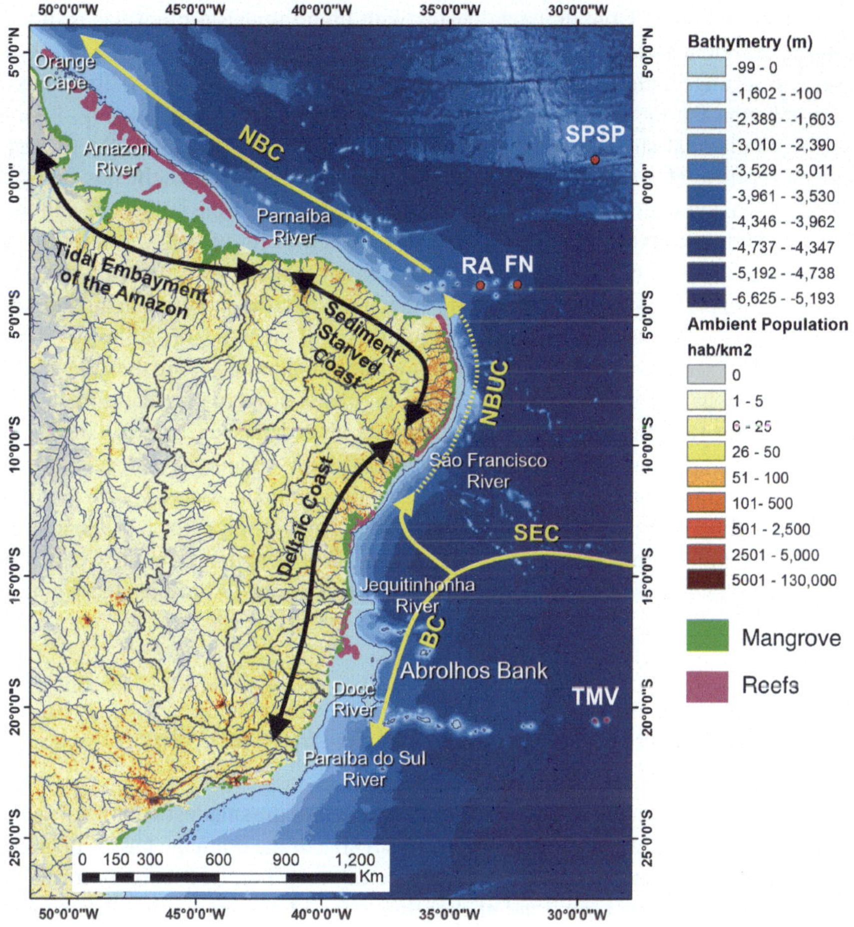

Fig. 1.1 Major tropical marine environments of Brazil. Ambient population refers to an average, or ambient, population that integrates diurnal movements and collective travel habits into a single measure (Dobson et al. 2000). *Source* Landscan Global Population Dataset. FNFernando de Noronha Archipelago; AR: Rocas Atoll; SPSP: Saint Peter and Saint Paul Archipelago; TMV: Trindade Martin Vaz islands; BC: Brazil Current; SEC: South Equatorial Current; and NBUC: North Brazil Undercurrent

1.2.1 The Deltaic Coast of Eastern Brazil

In this section of the coast, the major escarpment typical of rifted passive continental margins (Gilchrist and Summerfield 1994; Seidl et al. 1996; Matmon et al. 2002) retreated back from the coastal zone almost 500 km. All major rivers emptying into this section of the coast have their headwaters in this escarpment, except for the Paraíba do Sul and the São Francisco rivers.

A combination of large drainage basins with high intrabasin relief has resulted in large sediment yields in the major rivers emptying in this section of the coast of Brazil, resulting in classical examples of wave-dominated deltas (Parnaíba, Doce, Jequitinhonha-Pardo, and São Francisco) examined in detail in Chap. 4 (Fig. 1.1).

The continental shelf along this stretch varies from very large (south) to very narrow (north). The Abrolhos Bank, the widest portion of the shelf, has had its origin associated with volcanic activity that occurred between the Paleocene and the Eocene (Mohriak 2006). The narrowness of the shelf north of the Abrolhos Bank results from the presence of the São Francisco craton, a geotectonic unit of Archean-Paleoproterozoic age (Heilbron et al. 2017) that intercepts the shoreline in this stretch of the coastal zone. Overall, along this sector, the shelf break begins at the 50–60 m isobath. Shelf sedimentation away from major river mouths and along the outer shelf is carbonatic, having as a major constituent fragments of incrusting coralline algae and rhodoliths (Dominguez et al. 2013; Bastos et al. 2015) (see Chap. 6).

The most important coral reefs of the western South Atlantic have developed on the Abrolhos Bank in waters shallower than 20 m in a sediment-starved portion of the shelf located between the Jequitinhonha and Doce River deltas (Fig. 1.1). Chapter 5 presents a detailed characterization of the coral reefs present in this sector. Chapter 6 details the characteristics of two shelf areas of this sector (Abrolhos and Central Bahia).

1.2.2 The Sediment-Starved Coast of Northeastern Brazil

This section of the Brazilian coastline currently receives the smallest volume of sediment from the hinterland as a result of the small size of the drainage basins in association with low intrabasinal relief and low precipitation (Fig. 1.1). This coast is thus characterized by a long-term trend of shoreline retreat (Dominguez and Bittencourt 1996), displaying cemented upper shoreface sands ("beachrocks"), reef build-ups, and active sea cliffs carved into early-middle Miocene tablelands. The continental shelf is mostly narrow (< 20 km), a result of the combined effects of this region being the last to separate from Africa during the Mesozoic continental break-up (Rand and Mabesoone 1982), a dominance of transcurrent movements during separation and very limited sediment supply.

Sedimentation is dominantly carbonatic as a result of the very low terrigenous sediment influx. Additionally, because of the reduced sediment influx, most incised valleys carved into the shelf during the Quaternary lowstands are unfilled and have a morphological expression in the bathymetry. Two shelf sectors of this compartment are characterized in detail in Chap. 6.

This sector also contains most of the oceanic islands of Brazil, the Fernando de Noronha Archipelago (FN), the Rocas Atoll (RA), and Saint Peter and Saint Paul Archipelago (SPSP) (Fig. 1.1).

The shelf area is characterized by extreme oligotrophy, whose zoo- and ichthyoplankton communities are described in detail in Chap. 7. On the other hand, the presence of oceanic islands alters the oligotrophy of the typical tropical structure, which includes a strong, permanent thermocline. This alteration involves upwelling by local processes such as surface current divergence, winds, and interactions between the currents and submarine relief (Travassos et al. 1999; Araujo and Cintra 2009), providing nutrients to the photic zone (Neumann-Leitão et al. 1999). As a result, an increase in productivity occurs at numerous trophic levels, ensuring the maintenance and increase of the diversity of planktonic communities (Boltovskoy 1981; Melo et al. 2012) (see Chap. 7).

1.2.3 The Tidal Embayment of the Amazon

This sector extends approximately from the Parnaiba Delta to the Orange Cape, and it is characterized by a broad re-entrant in the coastal zone, which extends for more than 1000 km of shoreline. The combined flows of the Amazon and Tocantins/Pará Rivers bring to the coastal zone the largest sediment load in all of South America. This sector is also characterized by the highest tides in Brazil (Cartwright et al. 1991; Salles et al. 2000), with tidal ranges varying from 3 to 6 m. A combination of large sediment supply and tidal range has favored the development of extensive progradation of the shoreline under tidal dominated conditions, with shoreline progradation of up to 30 km (Souza-Filho et al. 2006) (see Chap. 3).

Two deltas occur in this sector: the tide-dominated delta of the Parnaíba (see Chap. 4) and the Amazon, which was the subject of a recent synthesis by Nittrouer et al. (2021) and therefore was not included in this book. On the shelf, sedimentation is mostly siliciclastic, although carbonate sedimentation dominates on the outer shelf. More recently, a major reef complex has been described for the outer shelf of this region (Moura et al. 2016).

In the coastal zone, the largest continuous mangrove belt of the world occurs occupying a total area of 7500 km^2 (see Chap. 3).

The large continental runoff with macrotidal mangroves originates from estuarine plumes on scales of dozens to hundreds of kilometres, which makes this region an extremely productive ecosystem. This region is among the most important fishing grounds in Brazil and may play an important role in regulating these North Brazilian shelf ecosystems. The zooplankton and ichthyoplankton communities along two transects in this compartment are presented in Chap. 7.

1.2.4 The Tropical Atlantic Ocean

The tropical Atlantic is characterized by permanent oligotrophic conditions outside the areas affected by river discharge (Da Cunha and Buitenhuis 2013) due to the relatively strong static stability with a well-marked thermocline, which is seasonally modulated by the meridional displacement of the ITCZ, controlling precipitation and trade wind regimes (Araujo et al. 2011; Assunção et al. 2020). The permanent thermocline restricts vertical mixing and nutrients up to photic layers and constrains biological productivity (Araujo et al. 2019).

Additionally, the tropical Atlantic is a complex region with two main current systems in play, the equatorial system and the western boundary system. The equatorward western boundary North Brazil Current and Undercurrent (NBC–NBUC) contributes to the upper Atlantic Meridional Overturning Circulation (AMOC) (Fig. 1.1). The currents, winds, precipitation, biogeochemical cycles, and CO_2 flux are seasonally regulated by the Intertropical Convergence Zone.

The tropical Atlantic is also the second-largest oceanic source of CO_2 to the atmosphere after the tropical Pacific (Lefèvre et al. 2010; Ibánhez et al. 2015; Araujo et al. 2019). Despite the net CO_2 outgassing registered in this region, zones of net atmospheric CO_2 uptake exist, which are mainly linked to the seasonal cycle of sea surface temperature (SST) and its associated thermodynamic effects on the partial pressure of CO_2 (pCO_2) (Ibánhez et al. 2015).

Under low discharge conditions, the spread of the Amazon plume acts as an atmospheric CO_2 sink of global relevance (Lefèvre et al. 2010; Ibánhez et al. 2015). In contrast, the northeastern Brazil region acts primarily as a CO_2 source to the atmosphere, with the highest CO_2 fluxes associated with periods of high SST in the area (Araujo et al. 2019). These aspects are discussed in detail in Chap. 8.

1.3 Climate Change Impacts

1.3.1 Deltas

In addition to the reduction in rainfall in the catchments, another major impact of climate change on deltas is sea-level rise and its implications for delta plain inundation and erosional shoreline retreat (Ibáñez et al. 2014). These aspects have been aggravated in recent decades by the construction of dams and artificial dikes that have reduced the ability of deltas to grow vertically (aggradation) and to feed the coastline with sediments (Vörösmarty et al. 2003; Syvitski et al. 2009; Day et al. 2016).

In the particular case of the Brazilian deltas, a few aspects must be considered when analyzing their vulnerability to climate change, since most of the delta plain is composed of beach dune deposits. The top of these beach-dune deposits is significantly higher than the mean sea level due to wave run-up (berm height is a function of the wave height at breaking) and Aeolian deposition (actually, most delta plain beach ridges were formed as dune ridges), which further contributes to increasing the average elevation of the terrain. Additionally, because they were built in a Glacial Isostatic Adjustment (GIA) far field region, the more internal beach deposits are approximately 3–4 m higher than those located in the outermost region of the delta plain (see Chap. 4). These factors contribute to increasing the resilience of these deltas to a rise in sea level.

Notwithstanding, the pristine tide-dominated delta of the Parnaíba has a vulnerability to sea-level rise much higher than the other wave-dominated deltas because most of this delta plain is occupied by mangrove swamps. The survivability of Parnaíba to current sea-level rise will depend on the capacity of these mangrove swamps to build up in response to a rising sea level.

For all Brazilian deltas, the results of climate models point to a reduction in precipitation in the watersheds, implying a reduction in river flows and, consequently, in the contribution of sediments to the mouth (Arias et al. 2021).

1.3.2 Mangroves

As detailed in Chap. 3, the response of the mangrove coasts to ongoing and future sea-level rise will depend on the environmental setting of the mangroves.

On open coasts, a rise in sea level will cause an increase in upstream penetration of the salt wedge and landward migration of mangroves along riverine and supratidal flats that are progressively converted to intertidal flats. However, erosive processes at the seaward front can result in mangrove loss along the shoreline (Allison et al. 2000; Anthony et al. 2010; Santos et al. 2016).

If sea level is rising over an open coast sheltered by barrier islands and densely colonized by mangroves and bounded landward by inactive cliffs, the muddy tidal flat will experience an elevation in water level and a sedimentary aggradation process. The landward retreat of the shoreline due to the rising sea level will result in barrier sand deposition over muddy flats and mangroves. This will cause coastal erosion, increased salinity, hydroperiod frequency, and inundation depth in mangrove forests (Souza-Filho and Paradella 2003; Souza-Filho et al. 2006).

Along deltaic coasts, the rise in sea level will result in increased salt wedge penetration and landward retreat of beach-dune ridges, occasionally burying tidal muddy flats colonized by mangroves.

Finally, tropical tidal flats fringing high gradient land regions or human-made obstacles such as seawalls and other shoreline protection structures will be the most threatened mangroves from sea-level rise. They will likely be drowned "in place" due to a lack of low-lying areas over which they can migrate (Lacerda et al. 2021).

1.3.3 Coral Reefs

Tropical southwestern Atlantic reefs are affected by a plethora of human stressors, including large-scale (global climate change-related processes, such as long-term warming, marine heatwaves, ocean acidification, and sea-level rise) or local but widespread impacts, such as mismanaged touristic and industrial activities, higher nutrient inputs, increased macro- and micro-plastic pollution, environmental disasters (e.g., oil spills and mining dam collapses), overfishing of reef species, and invasive species.

As detailed in Chap. 5, eastern-northeastern Brazil are reef areas that concentrate a higher diversity of habitats (reef forms) and major reef organisms, such as coral and fish communities. The Abrolhos Bank hosts one of the most important shallow water reef complexes of the western South Atlantic Ocean.

Shallow water coral richness is also higher in these regions, where *Mussismilia* species and milleporids are important constituents of the reef-building fauna. Global climate change shifts limit the control and survival of reef-building and dwelling organisms. Projections of future habitat suitability show that the most restricted species (*Mussismilia harttii* and *Mussismilia braziliensis*) can migrate north, a trend that can be enhanced by the increase in aridity and possibly an increase in temperature and salinity in the shallow shelf areas. On the other hand, turbidity might play an opposite role to these prognostications. *Mussismilia hispida*, on the other hand, might benefit from the tropicalization of the temperate realm and migrate south.

1.3.4 Continental Shelf

Important abiotic changes associated with climate change are expected to affect shelf areas. These include changes in atmospheric circulation that will lead to an increase in storm frequency; changes in precipitation affecting terrestrial-derived sediment, nutrients and pollutants; changes in salinity and temperature of the coastal ocean; and changes in ocean chemistry (pH). These changes will directly affect the shelf benthic communities (Holt et al. 2010). Although shelf seas comprise only 7% of the global ocean, they provide extremely important services for human society. The dominance of hard seabed substrates on the northeastern Brazil Shelf (NEBS) significantly increases the biodiversity of these regions. Changes in ocean chemistry can severely affect calcium carbonate-secreting organisms, such as corals and coralline algae, and impact their capacity to create reefs in some locations (Cornwall et al. 2019). In this respect, coralline algae, which are abundant on the NEBS, are considered one of the most crucial foundation taxa in the photic zone (Cornwall et al. 2019). Unfortunately, the NEBS region still lacks robust scientific data to allow a more in-depth evaluation of the impacts of climate change on its seabed and habitats. At the same time, there is a strong pressure for the use of the NEBS seabed, particularly from the energy and marine mineral industries, which creates new research opportunities to carry out mapping and characterization of the seabed.

1.3.5 Zoo- and Ichthyoplankton Communities

The results presented in Chap. 7 show considerable seasonal variability in the zoo- and ichthyoplankton communities, with large peaks in abundance and biomass, indicating that these planktonic systems are not stable at all but rather highly dynamic, with very strong responses to seasonal variations in climate and hydrography. Therefore, these systems are prone to show strong and exacerbated responses to climate variations in the near future, functioning as "amplifiers" of climate signals. Such responses may include drastic changes in the productivity of the ecosystem, but they may also manifest as unprecedented shifts in the timing of the peaks and blooms, leading to a deleterious disarrangement in the food webs, the "trophic mismatch" (Thackeray 2012).

The expected ecosystem responses are highly complex and totally different depending on the study area. In tropical oceanic ecosystems, the expected warming and deepening of the upper mixed layer (Roch et al. 2021) will most likely lead to increased stratification in layers above the permanent thermocline and thus to a reduction in primary (Gittings et al. 2018) and secondary productivity in the coming decades, with deleterious consequences for carbon sequestration, oceanic fish stocks (e.g., tuna and mackerels), seabirds, and other upper trophic levels.

For tropical pelagic ecosystems on the continental shelf, the situation is completely different, since they usually do not have a permanent thermocline. Tides and wind-driven turbulence usually break up any strong thermal stratification, except for areas with estuarine plumes. For the nearshore shelf, coastal and estuarine processes are very important, especially continental runoff. Extreme events, such as very strong rainfall, have drastic consequences for these nearshore ecosystems.

1.4 Book Organization

This book is arranged into 8 chapters.

This chapter provides a general overview of the subject.

Chapter 2 provides definitions of what climate variability and change are and an overview of observed and projected changes in climate in tropical South America. Temperature, rainfall, and drought projections are assessed from an ensemble of global and regional model projections under global warming scenarios of 1.5 and 4.0 °C for various regions of South America.

Chapter 3 presents a characterization of the mangrove swamps along the coast of Brazil and their subdivision based on geological, morphological, oceanographic, and climatic characteristics. It also discusses their spatiotemporal stability and changes in area from 1985 to 2020, sea-level changes and mangrove sedimentation evolution during the late Quaternary, and the impact of future sea-level rise.

Chapter 4 presents a synthesis of the wave-dominated deltas of Brazil, which are subject to different climatic zones, tidal regimes and wave climates, and different degrees of regulation and human occupation, ranging from pristine to human-influenced deltas. Vulnerability to climate change is also examined, particularly concerning the rise in sea level.

Chapter 5 synthesizes the main characteristics of the distribution and variability of reef and coralline ecosystems of the tropical southwestern Atlantic. These reefs are known as marginal reefs because they thrive in environmental conditions (high turbidity) far from those considered to be the optimal conditions for framework builders (calcareous skeleton-secreting organisms, such as corals). Additionally, three endemic coral species (*M. hispida*, *M. harttii*, and *M. braziliensis*) are used as proxies of the reef ecosystem to evaluate the trend of environmental suitability across the Brazilian Tropical Marine region in the RCP8.5 scenario.

Chapter 6 reviews the major characteristics of the continental shelves of northeastern Brazil, emphasizing seafloor morphology, its associated benthic ecosystems and the role of the eustatic variations in their late Quaternary evolution. Major human uses are also discussed.

Chapter 7 investigates zoo- and ichthyoplankton communities in seven Brazilian tropical marine environments that differ considerably in their abiotic and biological settings, ranging from the extremely wide continental shelf lined by mangroves off northern Brazil to the narrow oligotrophic shelf areas located in northeastern Brazil and three oceanic areas related to unique Brazilian island systems (Rocas Atoll, Fernando de Noronha, and St. Peter and St. Paul's Archipelagos).

Chapter 8 summarizes the main characteristics of the circulation, biogeochemical cycles, and CO_2 flux variability in the tropical Atlantic Ocean (TA). This is a very complex region where ocean–atmosphere interactions, oceanic currents, and phenomena such as wave propagations and mesoscale activities occur. The region is considered oligotrophic due to relatively strong static stability, with a well-marked thermocline. Additionally, the TA is the second-largest oceanic source of CO_2 to the atmosphere subject to equatorial upwelling, seasonal variations, and large river discharges, which drive the exchange of CO_2 between the sea and air.

1.5 Final Remarks

This book synthesizes the current knowledge about the major marine environments of tropical Brazil and how they might be impacted by ongoing climate change. Most results presented herein were derived from research conducted under the umbrella of the National Institute for Science and Technology in Tropical Marine Environments (inctAmbTropic) during the last 10 years. We hope this broad overview on this subject will be widely used by students, researchers, and managers interested in the subject and a source of inspiration for further studies on the topics discussed.

Acknowledgements This work was supported by the inctAmbTropic—National Institute on Science and Technology for Tropical Marine Environments, CNPq/FABESB (565054/2010-4, 8936/2011, 465634/2014-1 and inc004/2019).

References

Allison MA, Lee MT, Ogston AS et al (2000) Origin of Amazon mudbanks along the northeastern coast of South America. Mar Geol 163:241–256. https://doi.org/10.1016/0025-3227(95)00020-Y

Anthony EJ, Gardel A, Gratiot N et al (2010) The Amazon-influenced muddy coast of South America: a review of mud-bank-shoreline interactions. Earth-Sci Rev 103(3–4):99–121. https://doi.org/10.1016/j.earscirev.2010.09.008

Araujo M, Cintra M (2009) Modelagem matemática da circulação oceânica na região equatorial do Arquipélago de São Pedro e São Paulo. In: Viana DL, Hazin FHV, Souza, MAC (eds) Arquipélago de São Pedro e São Paulo: 10 anos de Estação Científica. SECIRM, Brasília, pp 107–114

Araujo M, Limongi C, Servain J et al (2011) Salinity-induced mixed and barrier layers in the southwestern tropical Atlantic Ocean off the northeast of Brazil. Ocean Sci 7:63–73. https://doi.org/10.5194/os-7-63-2011

Araujo M, Noriega C, Medeiros C et al (2019) On the variability in the CO_2 system and water productivity in the western tropical Atlantic off North and Northeast Brazil. J Mar Syst 189:62–77. https://doi.org/10.1016/j.jmarsys.2018.09.008

Arias PA, Sathyendranath SN, Bellouin E et al (2021) Technical Summary. In: Zhou MS, Delmotte V, Zhai P et al (eds) Climate change 2021: the physical science basis. Contribution of Working Group I to the Sixth Assessment Report of the Intergovernmental Panel on Climate Change. Cambridge University Press. https://www.ipcc.ch/report/ar6/wg1/downloads/report/IPCC_AR6_WGI_TS.pdf

Assunção RV, Silva AC, Roy A et al (2020) 3D characterisation of the thermohaline structure in the southwestern tropical Atlantic derived from functional data analysis of in situ profiles. Prog Oceanogr 187. https://doi.org/10.1016/j.pocean.2020.102399

Bastos AC, Quaresma VS, Marangoni MB et al (2015) Shelf morphology as an indicator of sedimentary regimes: a synthesis from a mixed siliciclastic-carbonate shelf on the eastern Brazilian margin. J S Am Earth Sci 63:125–136. https://doi.org/10.1016/j.jsames.2015.07.003

Boltovskoy D (ed) (1981) Atlas del zooplancton del Atlantico Sudoocidental y métodos de trabajos con el zooplancton marino. INIDEP, Mar del Plata

Cartwright DE, Ray RD, Sanchez BV (1991) Oceanic tide maps and spherical harmonic coefficients from Geosat altimetry. NASA Tech Memo 104544

Cornwall CE, Diaz-Pulido G, Comeau S (2019) Impacts of ocean warming on coralline algal calcification: meta-analysis, knowledge gaps, and key recommendations for future research. Front Mar Sci. https://doi.org/10.3389/fmars.2019.00186

Da Cunha LC, Buitenhuis ET (2013) Riverine influence on the tropical Atlantic Ocean biogeochemistry. Biogeosciences 10:6357–6373. https://doi.org/10.5194/bg-10-6357-2013

Day JW, Agboola J, Chen Z et al (2016) Approaches to defining deltaic sustainability in the 21st century. Estuarine Coast Shelf Sci 183:275–291. https://doi.org/10.1016/j.ecss.2016.06.018

Dobson J, Bright E, Coleman P et al (2000) LandScan: a global population database for estimating populations at risk. Photogramm Eng Remote Sens 66:849–857

Dominguez JML (2009) Chapter 2—The coastal zone of Brazil. In: Dillenburg S, Hesp P (eds) Geology and geomorphology of Holocene coastal barriers of Brazil. Lecture notes in earth sciences. Springer, Berlin, pp 17–51. https://doi.org/10.1007/978-3-540-44771-9_2

Dominguez JML, Bittencourt ACSP (1996) Regional assessment of long term trends of coastal erosion in Northeastern Brazil. An Acad Bras Ciênc 68:355–371

Dominguez JML, Silva RP, Nunes AS et al (2013) The narrow, shallow, low-accommodation shelf of central Brazil: sedimentology, evolution and human uses. Geomorphology 203:46–59. https://doi.org/10.1016/j.geomorph.2013.07.004

Gilchrist AR, Summerfield MA (1994) Tectonic models of passive margin evolution and their implications for theories of long-term landscape evolution. In: Kirkby MJ (ed) Process models and theoretical geomorphology. Wiley, Chichester, pp 55–84

Gittings JA, Raitsos DE, Krokos G et al (2018) Impacts of warming on phytoplankton abundance and phenology in a typical tropical marine ecosystem. Sci Rep 8. https://doi.org/10.1038/s41598-018-20560-5

Heilbron M, Cordani UG, Alkmim FF (eds) (2017) São Francisco Craton, Eastern Brazil. Regional geology reviews: São Francisco Craton, Eastern Brazil: tectonic genealogy of a miniature continent. Springer. https://doi.org/10.1007/978-3-319-01715-0

Holt J, Wakelin S, Lowe J et al (2010) The potential impacts of climate change on the hydrography of the northwest European continental shelf. Prog Oceanogr 86:361–379. https://doi.org/10.1016/j.pocean.2010.05.003

Ibáñez C, Day JW, Reyes E (2014) The response of deltas to sea-level rise: natural mechanisms and management options to adapt to high-end scenarios. Ecol Eng 65:122–130. https://doi.org/10.1016/j.ecoleng.2013.08.002

Ibánhez JSP, Diverrès D, Araujo M et al (2015) Seasonal and interannual variability of sea-air CO_2 fluxes in the tropical Atlantic affected by the Amazon River plume. Global Biogeochem Cycles 29:1640–1655. https://doi.org/10.1002/2015GB005110

Lacerda LD, Ward RD, Godoy MDP et al (2021) 20-Years cumulative impact from shrimp farming on mangroves of Northeast Brazil. Front Forest Glob Change. https://doi.org/10.3389/ffgc.2021.653096

Lefèvre N, Diverrès D, Gallois F (2010) Origin of CO_2 undersaturation in the western tropical Atlantic. Tellus Ser B Chem Phys Meteorol 62:595–607. https://doi.org/10.1111/j.1600-0889.2010.00475.x

Matmon A, Bierman P, Enzel Y (2002) Pattern and tempo of great escarpment erosion. Geology 30:1135–1138. https://doi.org/10.1130/0091-7613(2002)030%3c1135:PATOGE%3e2.0.CO;2

Melo PAMC, Diaz XFG, Macedo SJD et al (2012) Diurnal and spatial variation of the mesozooplankton community in the Saint Peter and Saint Paul Archipelago, Equatorial Atlantic. Mar Biodivers Rec 5:1–14. https://doi.org/10.1017/S1755267212001054

Mohriak WU (2006) Interpretação geológica e geofísica da Bacia do Espírito Santo e da região de Abrolhos: petrografia, datação radiométrica e visualização sísmica das rochas vulcânicas. Bol Geociênc Petrobrás 14:133–142

Moura RL, Amado-Filho GM, Moraes FC et al (2016) An extensive reef system at the Amazon River mouth. SciAdv 2. https://doi.org/10.1126/sciadv.1501252

Neumann-Leitão S, Gusmão LMO, Silva TA et al (1999) Mesozooplankton biomass and diversity in coastal and oceanic waters off north-eastern Brazil. Arch Fish Mar Res 47:153–165

Nittrouer CA, DeMaster DJ, Kuehl SA et al (2021) Amazon sediment transport and accumulation along the continuum of mixed fluvial and marine processes. Annu Rev Mar Sci 13:501–536. https://doi.org/10.1146/annurev-marine-010816-060457

Rand HM, Mabesoone JM (1982) Northeastern Brazil and the final separation of South America and Africa. Palaeogeogr Palaeoclimatol Palaeoecol 38:163–183. https://doi.org/10.1016/0031-0182(82)90002-5

Roch M, Brandt P, Schmidtko S et al (2021) Southeastern tropical Atlantic changing from subtropical to tropical conditions. Front Mar Sci 26. https://doi.org/10.3389/fmars.2021.748383

Salles FJP, Bentes FCM, Santos JA (2000) Catálogo de Estações Maregráficas. Fundação de Estudos do Mar, Rio de Janeiro, Brazil

Santos VF, Short AD, Mendes AC (2016) Beaches of the Amazon Coast: Amapá and West Pará. In: Short AD, Klein AHF (eds) Brazilian beach systems. Coastal Research Library, Springer, Switzerland, pp 67–94. https://doi.org/10.1007/978-3-319-30394-9_3

Seidl MA, Weissel JK, Pratson LF (1996) The kinematics and pattern of escarpment retreat across the rifted continental margin of SE Australia. Basin Res 12:301–316. https://doi.org/10.1046/j.1365-2117.1996.00266.x

Souza-Filho PWM, Paradella WR (2003) Use of synthetic aperture radar for recognition of Coastal Geomorphological Features, land-use assessment and shoreline changes in Bragança coast, Pará, Northern Brazil. An Acad Bras Cienc 75(3): 341-356. https://doi.org/10.1590/S0001-37652003000300007

Souza-Filho PWM, Farias Martins EdS, Costa FR (2006) Using mangroves as a geological indicator of coastal changes in the Bragança macrotidal flat, Brazilian Amazon: a remote sensing data approach. Ocean Coast Manag 49(7–8):462–475. https://doi.org/10.1016/j.ocecoaman.2006.04.005

Syvitski JPM, Kettner AJ, Overeem I et al (2009) Sinking deltas due to human activities. Nat Geosci 2:681–686. https://doi.org/10.1038/ngeo629

Thackeray SJ (2012) Mismatch revisited: what is trophic mismatching from the perspective of the plankton? J Plankton Res 34:1001–1010. https://doi.org/10.1093/plankt/fbs066

Travassos P, Hazin FHV, Zagaglia JR et al (1999) Thermohaline structure around seamounts and islands off north-eastern Brazil. Arch Fish Mar Res 47:211–222

Vörösmarty CJ, Meybeck M, Fekete B et al (2003) Anthropogenic sediment retention: major global impact from registered river impoundments. Glob Planet Change 39:169–190. https://doi.org/10.1016/S0921-8181(03)00023-7

Chapter 2
Climate Variability and Change in Tropical South America

Jose Antônio Marengo, Tercio Ambrizzi, Michelle Simões Reboita, Marcos Heil Costa, Claudine Dereczynski, Lincoln Muniz Alves, and Ana Paula Cunha

Abstract This chapter provides definitions of what climate variability and change are and an overview of observed and projected changes in climate in tropical South America. We present a review of the findings of the Intergovernmental Panel on Climate Change (IPCC) from the First Report launched in 1990 until the Sixth Report AR6 released in 2021. This review includes the evolution of models, projections and emission scenarios used by the IPCC since the 1990s. We also include a review of observed long term hydroclimate variability for some key regions in tropical South America that include biomes such as Pantanal, Amazon, the semiarid lands of Northeast Brazil and the Paraná-Plata river basin. For these regions, the focus is on extremes, such as droughts and floods. Temperature, rainfall and drought projections are assessed from the ensemble of global and regional model projections under global warming scenarios of 1.5 and 4.0 °C for various regions of South America. Finally, we also discuss some societal impacts of climate variability and change as well as knowledge gaps that need to be filled with new studies.

Keywords Tropical South America · El Niño · Climate variability · Climate change · Drought · Rainfall

---

J. A. Marengo (✉) · A. P. Cunha
National Center for Monitoring and Early Warning of Natural Disasters (CEMADEN), São José dos Campos, São Paulo, Brazil
e-mail: jose.marengo@cemaden.gov.br

T. Ambrizzi
University of São Paulo (USP), São Paulo, São Paulo, Brazil

M. S. Reboita
Federal University of Itajubá (UNIFEI), Itajubá, Minas Gerais, Brazil

M. H. Costa
Federal University of Viçosa (UFV), Viçosa, Minas Gerais, Brazil

C. Dereczynski
Federal University of Rio de Janeiro (UFRJ), Rio de Janeiro, Brazil

L. M. Alves
National Institute for Space Research (INPE), São Jose dos Campos, São Paulo, Brazil

J. M. Landim Dominguez et al. (eds.), *Tropical Marine Environments of Brazil*, The Latin American Studies Book Series, https://doi.org/10.1007/978-3-031-21329-8_2

2.1 Introduction

2.1.1 Components of the Climate System

The climate system corresponds to the interaction among their components: atmosphere, hydrosphere, cryosphere, lithosphere, and biosphere (IPCC 2007, 2013, 2021). Changes in one component directly or indirectly affect the other components and are affected by them. Human-induced climate change already affects several components of the climate system.

The atmosphere is a thin layer that surrounds the Earth and is composed of gases and aerosols that become less dense as the distance from the Earth's surface increases. In this sense, 90% of the atmospheric constituents are within 15 km of Earth's surface, which corresponds to only 1/400 of the radius of Earth (Trenberth 2020). Carbon dioxide (CO_2), ozone (O_3), methane (CH_4) and water vapor (H_2O) are examples of gases with variable concentrations (also referred to as trace gases). Although at small concentrations, these gases are essential for maintaining life, as they contribute to the so-called greenhouse effect. The greenhouse effect is a natural process that has existed since Earth's formation. Without this natural effect, the global mean temperature of the Earth would be -18 °C, which is almost twice as low as its current mean temperature (~15 °C). When solar energy reaches the atmosphere, the atmosphere is not able to absorb a large part of the radiation. The shortwave radiation provided by the Sun is absorbed by the Earth's surface and radiated back as longwave radiation (infrared wavelength). This infrared radiation is absorbed by greenhouse gases such as CO_2 and by clouds (remember that clouds are composed of H_2O). Some of the energy absorbed by the atmospheric constituents will also be radiated back, primarily at infrared wavelengths, in all directions. In this process, the radiation emitted downward from the atmosphere adds to the warming of Earth's surface by solar radiation. This enhanced warming is due to the greenhouse effect. The problem facing the intensification of the greenhouse effect is that anthropogenic actions have contributed to increasing the concentration of the greenhouse gases, therefore increasing the greenhouse effect and the mean temperature of the planet.

The atmosphere interacts with other components of the climate system by means of feedback mechanisms among the climate system components. The feedback mechanisms are also radiative forcings. The term forcing refers to factors that drive or cause changes in the climate system and, as a result, cause climate change. As defined by the IPCC (2013), radiative forcing is a measure of the net change in the energy balance in response to a perturbation. There are three ways to change the radiative balance of the Earth: (a) changing the incoming solar radiation, (b) changing the quantity of solar radiation that is reflected back (albedo) and (c) modifying the amount of longwave radiation that the Earth radiates back to space (changes in the greenhouse gas concentrations) (IPCC 2007). Climate responds to these changes through feedback mechanisms that can either amplify (positive feedback) or decrease (negative feedback) the effects of a change in the climate.

2.1.2 Natural Climate Variability and Change

The Earth's climate is determined by the balance between incoming energy from the Sun and outgoing energy emitted by the Earth/atmosphere system. Balance means that what enters in a system leaves this system in the same magnitude. Therefore, changes in the incoming or outgoing energy lead to changes in the climate. However, what are the drivers of these energy changes in addition to the Sun? Before we explain this subject, it is important to define the meaning of climate change and climate variability. According to the glossary of the IPCC (2021):

Climate change: change in the state of the climate that can be identified (e.g., by using statistical tests) by changes in the mean and/or the variability of its properties and that persists for an extended period, typically decades or longer. Climate change may be due to natural internal processes or external forcings, such as modulations of the solar cycles, volcanic eruptions and persistent anthropogenic changes in the composition of the atmosphere or in land use.

Climate variability: change in the state of the climate that can be identified (e.g., by using statistical tests) by changes in the mean and/or the variability of its properties and that persists for an extended period, typically decades or longer. Climate change may be due to natural internal processes or external forcings, such as modulations of the solar cycles, volcanic eruptions and persistent anthropogenic changes in the composition of the atmosphere or in land use.

Changes in climate have occurred even in the absence of humans (IPCC 2021). Therefore, there are natural drivers contributing to these changes that can be internal (components of the climate system) or external (solar luminosity variations, variations in Earth's orbit around the Sun, and volcanic eruptions) to the climate system. The internal drivers change the climate and are affected by the climate (feedback mechanism); on the other hand, the external drivers can only affect the climate and cannot be affected by it (Hartmann 2015). Internal drivers that can lead to climate change are related to modifications in the thermohaline circulation, ice melting and water vapor increase in the atmosphere. However, they are also greatly responsible for climate variability on different time scales (weakly, intraseasonal, seasonal, interannual, and decadal). This variability is associated with teleconnection patterns.

Teleconnection is a term used to refer to local anomalies in the atmosphere, which, in general, are caused by a heat source in the ocean (Trenberth et al. 1998) and that affects the climate of remote places. Thus, teleconnections also refer to local anomalies in the ocean that disturb the climate system (IPCC 2021). We can also think about teleconnection as a perturbation in the climate system caused by its own components. For South America, a description of the main teleconnection patterns that cause climate variability over South America can be found in Reboita et al. (2021a). In particular, considering an interannual time scale, the most studied and widely known teleconnection pattern is the phenomenon of the El Niño-Southern Oscillation (ENSO). ENSO is an ocean–atmosphere coupled phenomenon that develops in the east and central portions of the tropical Pacific Ocean (Wang et al. 2017; McPhaden et al. 2020). The positive sea surface temperature (SST) anomalies indicate the ENSO warm phase, while the negative anomalies indicate the cold phase (La Niña). During an El Niño event, there is a sea-level gradient with lower values in the eastern Pacific and higher values in the western Pacific, which weakens the trade winds (opposite conditions are observed during a La Niña event). These changes in atmospheric circulation cause anomalous patterns in temperature and precipitation around the globe. Over South America, El Niño episodes are responsible for increased precipitation over southeastern South America and less rainfall over portions of Amazonia and northeast Brazil (Marengo et al. 2017, 2018; McPhaden et al. 2020; Reboita et al. 2021a). The positive SST anomalies during El Niño events can also contribute to boosting global temperatures, increasing global warming in specific years such as 2016 (McPhaden et al. 2020). In climate change scenarios, changes in ENSO are still uncertain, although some models, considering high emission scenarios, indicate that extreme El Niño and La Niña events may double in frequency in the future (Cai et al. 2014, 2015, 2020).

Other important systems may be affected by teleconnection patterns in the Atlantic Ocean sector, such as the North Atlantic Oscillation (NAO), the Southern Annular Mode (SAM) and the Tropical Atlantic SST gradient, which involve variations of opposite signs in the sea-level pressure and SST in both hemispheres (Foltz et al. 2019; Zhang et al. 2021), leading to variations in the position and intensity of the Intertropical Convergence Zone (ITCZ). All these systems and their relationship with teleconnection patterns are well described in Reboita et al. (2021a).

2.1.3 Anthropogenic Climate Change

The Earth's natural greenhouse effect has been modified by human activities. Humans have increased the concentration of greenhouse gases in the atmosphere, leading to the warming of the Earth's surface, as a feedback process in the climate system (IPCC 2021). The sources of greenhouse gases can be natural and anthropogenic. H_2O is the most abundant greenhouse gas in the atmosphere. Its concentration increases as the Earth's atmosphere warms, leading to cloud formation and precipitation. However, the greatest "villain" for increasing the greenhouse effect is CO_2, which has had its concentration greatly increased due to human activities. In 1850, the CO_2 concentration was 280 parts per million (ppm); in August 2021, it reached 416 ppm, thus indicating an increase of approximately 48% during this time interval (https://climate.nasa.gov/vital-signs/carbon-dioxide/). When greenhouse gases are injected into the atmosphere, they have long residence times, i.e., the amounts released into the atmosphere today will remain in the atmosphere for up to two centuries depending on the gas.

Land use changes are also responsible for anthropogenic climate change. When the natural vegetation is changed by agriculture and urbanization, the local albedo is modified as well as the water and energy surface budgets. In the case of urbanization, the large urban centers are also responsible for heat islands that contribute to increasing the air temperature. Agriculture and cattle ranching also contribute to climate change through the release of nitrous oxide (N_2O) available in fertilizers. Another serious issue is deforestation. Forests store large amounts of carbon since trees and other plants absorb carbon dioxide in photosynthesis. Carbon dioxide is converted into carbon and stored in all parts of the plants and the soil. However, the stored carbon is released into the atmosphere in the form of carbon dioxide when the forests are burned or cleared. Rainforests also play an important role in cooling the local climate. As their canopy helps to trap moisture, it leads to slow evaporation, providing a natural air-conditioning effect (Henson 2011). If the forest is removed or burned over large areas, hotter and drier conditions are expected. Studies for South America using climate models, such as Llopart et al. (2018) and Marengo et al. (2021a and references quoted in), have indicated that changing the Amazon Forest to grassland would result in an increase in temperature and in the occurrence of a contrasting spatial pattern on precipitation over Amazonia and changes in the atmospheric circulation for all South America. Increased deforestation may also affect the hydrological cycle in the region, and recent studies by Gatti et al. (2021) have shown that while the Amazon region functions as a sink of CO_2, the southeastern part of the region, along the so-called deforestation arc, behaves as a source of CO_2, with temperature increases, rainfall and atmospheric moisture reduction during the last three decades.

Anthropogenic warming has resulted in an expansion of the dry climate areas and a decrease in polar climates. The poleward shift of the Hadley cell could be associated with this impact (Reboita et al. 2019). On the other hand, dry climate regions are more vulnerable to desertification. This causes a loss of biodiversity and reduces agricultural productivity, such as in the semiarid region of Northeast Brazil (Vendruscolo et al. 2021). Marengo et al. (2020 and references quoted in) show that with regional warming above 4 °C, semiarid regions become arid, and the risk of Caatinga vegetation being replaced by arid vegetation is high, which affects the populations in rural areas. On longer time scales, this aridification of the Northeast would lead to land degradation, resulting in a desertification process.

Climate change is also playing an increasing role in determining wildfire regimes (Shukla et al. 2019). In addition to the CO_2 from fires, bacteria in newly exposed soil may release more than twice the usual amount of another greenhouse gas, nitrous oxide, for at least two years. Brazilian biomes (Pantanal, Cerrado, Amazon Forest) suffered much damage from fires in 2019 and 2020 (Henson 2011). A comprehensive review of the increasing fire outbreaks in Brazilian biomes, their contributing causes, overall environmental impacts and consequences for human well-being is provided by Pivello et al. (2021).

2.1.4 Main Concepts in Climate Change and Modeling

Figure 2.1 shows the air temperature frequency distribution for two scenarios: the preindustrial period and present/future time. The difference in mean global temperatures between these two scenarios is represented by a shift to the right of the frequency distribution. At the same time, changes in both tails of the distribution can be seen, which indicates an increase in the frequency of extreme weather events. Thus, in statistical terms, climate extremes can be envisioned as a given probability distribution of a specific event, for example, droughts. Extreme events can be classified as weather extremes or climate extremes. The former refers to systems with short durations, such as intense rain on a specific day; the latter refers to events with longer durations, such as cold waves, heat waves, and droughts.

From observations, we know that our climate is changing, with one example being the frequency, duration and intensity of extreme weather and climate events (IPCC 2021). However, these extremes can also affect the climate itself. Reichstein et al. (2013) highlighted that extremes produce a direct biogeochemical signal in the atmosphere with local concentrations, for instance, of pollution. In other cases, a climate extreme alters the turnover rate of terrestrial carbon pools, leading to prolonged release of CO_2 into the atmosphere, such as the injection of CO_2 into the atmosphere due to vegetation mortality during drought episodes.

To project changes in future climate, numerical climate models are used. Hence, climate models simulate the physical processes taking place in the atmosphere, providing us with data concerning atmospheric circulation, temperature, precipitation, etc. If a model can successfully simulate the main features of atmospheric circulation, the mean state of the climate (average temperature and precipitation) and the frequency of extreme events, it can give us the confidence to apply them to future projections. More details on climate models, scenarios, projections, and uncertainties are shown in Sect. 2.2.

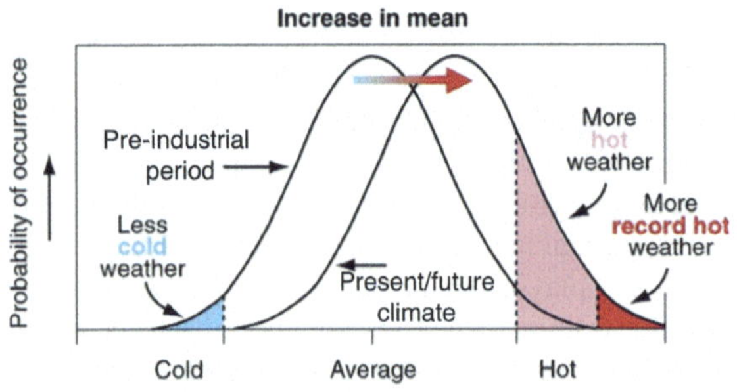

Fig. 2.1 Schematic representation of the probability density function of daily temperature in the preindustrial period and present/future climate

2.2 Climate Modeling: A Primer

2.2.1 *Climate Models: Concepts and Evolution*

Since the 1960s, climate models have evolved a long way until the current state-of-the-art in Earth System Models (ESMs). The first atmospheric models, called atmospheric general circulation models (AGCMs), consisted of physical equations governing atmosphere dynamics, radiative transfer, and other parametrizations. These models were numerically solved using computers and were forced by prescribed boundary conditions such as solar radiation, sea surface temperature, sea ice, and land surface cover. In parallel, oceanic general circulation models (OGCMs) solved similar equations for the ocean and were forced by other boundary conditions, such as air temperature and wind. Climate calculations with a combined Atmosphere–Ocean Model were first performed by Nobelist Syukuro Manabe at the Geophysical Fluids Dynamic Laboratory GFDL (Manabe and Bryan 1969), giving birth to the atmosphere–ocean global climate models AOGCMS.

During the 1990s, global climate models were generally AOGCMs, with horizontal resolutions typically 300–700 km and 12–24 vertical levels in the atmosphere. These models were used in the First and Second Assessment Reports of the IPCC (IPCC 1990, 1995). The use of individual model ensembles (to account for uncertainties in the initial conditions) was rare. In the early 2000s, the first climate system models (CSMs) integrated the four main climate system components. Atmosphere, ocean, land cover, and sea ice were dynamically simulated, varying in time and feed-backing with the other components. In CSMs, the prescribed boundary conditions are solar radiation, aerosols (volcanic and anthropogenic), and atmospheric composition. Typical resolutions at the time were 200 km (horizontal) and 30 levels in the vertical direction for the atmospheric model, and asymmetric resolutions for the oceans, depending on whether it simulates an oceanic current (15–20 km) or not (150–200 km). Land surface and sea ice were usually run at the same resolution as the atmospheric model. These models need to be initialized for at least 30 years until the shallow ocean temperature reaches equilibrium. The use of individual model ensembles became increasingly common (5–100 ensembles). CSMs were the standard in the Third and Fourth IPCC Assessment Reports (IPCC 2001, 2007) and were still used in the Sixth IPCC Assessment Report AR6 (IPCC 2021).

In the late 2000s, the first ESMs were built. In this type of model, the climate system is just a component of the Earth system. In addition to the climate system components, the ESMs explicitly simulate the carbon, water and nitrogen cycles, atmospheric chemistry, and aerosols, directly simulating sea-level rise. CO_2 and aerosol emissions are inputs to the model, and since the full carbon cycle is simulated, the model computes CO_2 concentrations. To do that, these models must be initialized by > 1000 years so that the soil and ocean carbon can reach equilibrium. Not all these characteristics are present in all ESMs, but to be considered an ESM, the model must simulate the full carbon cycle. ESMs are the current standard for the Fifth IPCC AR5 and Sixth IPCC AR6 Assessment Reports (IPCC 2013, 2021), although only a few models of the Coupled Model Intercomparison Project Versions 5 and 6 (CMIP-5 and -6, respectively) ensembles are fully integrated ESMs. These models have much finer resolutions, sometimes < 100 km for the atmospheric and oceanic models.

Following the evolution of climate models, the attribution of causes of climate change has also been better established:

- The IPCC First Assessment Report (FAR IPCC) (IPCC 1990) concluded that while both theory and models suggested that anthropogenic warming was already well underway, its signal could not yet be detected in observational data against the 'noise' of natural variability.
- The IPCC Second Assessment Report (SAR IPCC) stated that the balance of evidence suggests a discernible human influence on global climate. The SAR stated that 'the balance of evidence suggests a discernible human influence on global climate' (IPCC 1995).
- The IPCC Third Assessment Report (TAR IPCC) concluded that there is new and stronger evidence that most of the warming observed over the last 50 years is attributable to human activities (IPCC 2001).
- The IPCC Fourth Assessment Report (IPCC AR4) further strengthened previous statements, concluding that most of the observed increase in global average temperatures since the mid-twentieth century is very likely due to the observed increase in anthropogenic greenhouse gas concentrations (IPCC 2007).
- The IPCC Fifth Assessment Report (IPCC AR5) assessed that a human contribution had been detected to changes in warming the atmosphere and ocean; changes in the global water cycle; reductions in snow and ice; global mean sea-level rise; and changes in some climate extremes. AR5 concluded that it is extremely likely that human influence has been the dominant cause of the observed warming since the mid-twentieth century (IPCC 2013).
- The IPCC Sixth Assessment Report (IPCC AR6) establishes that it is unequivocal that human influence has warmed the atmosphere, ocean, and land. Widespread and rapid changes in the atmosphere, ocean, cryosphere and biosphere have occurred (IPCC 2021).

If we are interested in a more precise representation of the local climate, regional climate models (RCMs) are recommended. Their advantage is that they have a higher horizontal resolution (< 50 km) compared to GCMs, being able to better represent aspects of topography, for example. However, RCMs need both initial and spatial boundary conditions, which are provided by GCMs or by reanalysis. A review of the fundamental aspects of GCMs is available in Ambrizzi et al. (2019). With the purpose of establishing a common framework to facilitate the application and comparison of the results obtained with regional climate dynamic and statistical downscaling and a common protocol for downscaling experiments in different regions of the world, the World Climate Research Program (WCRP) established the Coordinated Regional Climate Downscaling Experiment (CORDEX) in the late 2000s (Giorgi and Gutowski 2015; Giorgi et al. 2021).

2.2.2 Evolution of Climate Emissions Scenarios for IPCC

The IPCC has used a common set of scenarios across the scientific community to provide better comparisons between various studies and to make it easier to communicate model results. In the FAR IPCC (IPCC 1990), idealized emission scenarios were assumed, such as greenhouse gas (GHG) emissions growing at a fixed rate (e.g., 1% a year). Starting with the SAR IPCC (IPCC 1995), a set of six alternative emission scenarios, IS92a-f, were used. These scenarios embodied a wide array of assumptions reflecting how future greenhouse gas emissions might evolve in the absence of climate policies beyond those already adopted (Leggett et al. 1992).

Socioeconomic and emission scenarios are used in climate research to provide plausible descriptions of how variables such as socioeconomic change, technological change, energy, land use, and emissions of greenhouse gases and air pollutants may evolve in the future. They are used as input for climate model runs and as a basis for assessing possible climate impacts, mitigation options, and associated costs. The scenarios from the Special Report on Emission Scenarios (SRES) (Nakicenovic et al. 2000) were used in the TAR IPCC (IPCC 2001) and IPCC AR4 (IPCC 2007) reports. Scenarios were chosen based on our future motivations and emphasis. The A family of scenarios assumes we would focus on economic motivations, while the B family assumes more environmental motivations. Completing the set, family 1 assumed a globalized economy, while family 2 emphasized local communities. The four main scenarios (A1, A2, B1, and B2) considered different futures for global population, global per capita income, per capita energy consumption, and CO_2 emissions per unit of energy produced, with outcomes varying from endless comfort and efficiency (A1) to a return to nature and community (B2), with a sustainable and equitable world in between (B1).

The modeling process using these emissions scenarios follows the sequence: socioeconomic forcings—population, gross domestic product (GDP), and technology—determine GHG and aerosol emissions, which then determine GHG and aerosol concentrations, then radiative forcing, then climate (temperature, precipitation, snow cover, sea ice, sea level, etc.), geophysical impacts (river discharge, fires), and impacts on humans and other species (diseases, heat stress, species distribution, etc.). Although this is a logical way of thinking, the uncertainties in the modeling processes increase from the beginning to the end of the modeling sequence, maximizing uncertainty at the end of the chain.

The IPCC AR5 (IPCC 2013) scenarios modified this modeling sequence, minimizing the uncertainty at the center of the modeling chain, i.e., at the radiative forcing. Scenarios were aggregated at four radiative forcing levels by 2100, above preindustrial levels—2.6, 4.5, 6.0, and 8.5 W/m^2. Representative concentration pathways (RCPs) were modeled to account for the socioeconomic-technologic emissions and concentration pathways that lead to the selected radiative forcings (van Vuuren et al. 2011).

The IPCC AR6 (IPCC 2021) projected global scenarios based on the shared socioeconomic pathways (SSPs) concept. This new set of scenarios synthesizes knowledge across the physical sciences, impact, adaptation, and mitigation research. The core set of SSP scenarios assumes five main pathways—sustainability (SSP1), regional rivalry (SSP3), inequality (SSP4), development based on fossil fuels (SSP5), and a middle-of-the-road pathway (SSP2). Each pathway is assigned to one or more radiative forcings by 2100 above preindustrial levels: SSP1-1.9, SSP1-2.6, SSP2-4.5, SSP3-7.0, and SSP5-8.5, covering a broad range of emission pathways, including new low-emissions pathways. In IPCC AR6 (IPCC 2021), emissions vary between scenarios depending on socioeconomic assumptions, levels of climate change mitigation, and air pollution controls (for aerosols and nonmethane ozone precursors).

2.2.3 Uncertainties in Model Projections and Model Limitations

The Intergovernmental Panel on Climate Change (IPCC) has used an ensemble of different climate models to address climate system uncertainties and to avoid individual model biases since the TAR IPCC report (IPCC 2001). To standardize comparisons between the different models and their differences in grid type, resolution, and output variables, the climate modeling community developed increasingly sophisticated CMIPs.

Both CMIP3—used in the IPCC AR4 (IPCC 2007)—and CMIP5—used in the AR5 (IPCC 2013) included experiments testing the ability of models to reproduce the twentieth century global surface temperature trends both with and without anthropogenic forcings (GHGs and aerosols). The CMIP6 models used in the IPCC AR6 (IPCC 2021) include new and better representations of physical, chemical, and biological processes and higher model resolution than climate models considered in previous IPCC assessment reports, in addition to improvements in the historical radiative forcings. These enhancements improved the simulation of the last century's mean state of most large-scale indicators of climate change and many other aspects across the climate system (IPCC 2021).

In ESMs, the magnitude of feedback between climate change and the carbon cycle becomes larger but also more uncertain in high CO_2 emissions scenarios (very high confidence). However, climate model projections show that the uncertainties in atmospheric CO_2 concentrations by 2100 are dominated by the differences between shared socioeconomic pathways, i.e., by humankind's own choices. Additional ecosystem responses to warming do not yet fully included in climate models, such as CO_2 and CH_4 fluxes from wetlands, permafrost thaw, and wildfires which would further increase concentrations of these gases in the atmosphere (IPCC 2021).

In summary, although scenarios are the major source of uncertainty in climate projections, model uncertainty widens the range of possible future climates. These uncertainties include inaccurate representation of interannual and interdecadal modes of climate system variability and misrepresentation of parametrizations and feedbacks, mainly those related to clouds. Another example is the impossibility of modeling the occurrence of random climate-relevant events, such as volcanic eruptions.

2.3 Observed and Projected Climate Scenarios in Tropical South America

2.3.1 Observed Changes: A Summary

In the upcoming subsections, a review of climate trends is presented for some regions of Brazil. These regions correspond to Brazilian biomes (Fig. 2.2): the Amazon biome in the Amazon River basin, the Caatinga biome in the semiarid lands of Northeast Brazil, the Pantanal biome, and the Parana-La Plata River basin that covers parts of the Cerrado, Atlantic Forest and Pampas biomes.

To study rainfall variability, we consider the standardized precipitation index (SPI). The SPI is a drought index proposed by Mckee et al. (1993) to quantify the probability of occurrence of a precipitation deficit at a specific monthly time scale. To calculate the SPI, precipitation data are fitted to a gamma probability distribution function, and then the inverse normal distribution function is used to rescale the probability values, resulting in SPI values with a mean of zero and a standard deviation of one. As the SPI is a normalized index, it allows the comparison of the index between different locations and climates, which is important for drought monitoring in a large country such as Brazil (Cunha et al. 2019). The time series of SPI-12 months for the regions considered below are shown in Fig. 2.2.

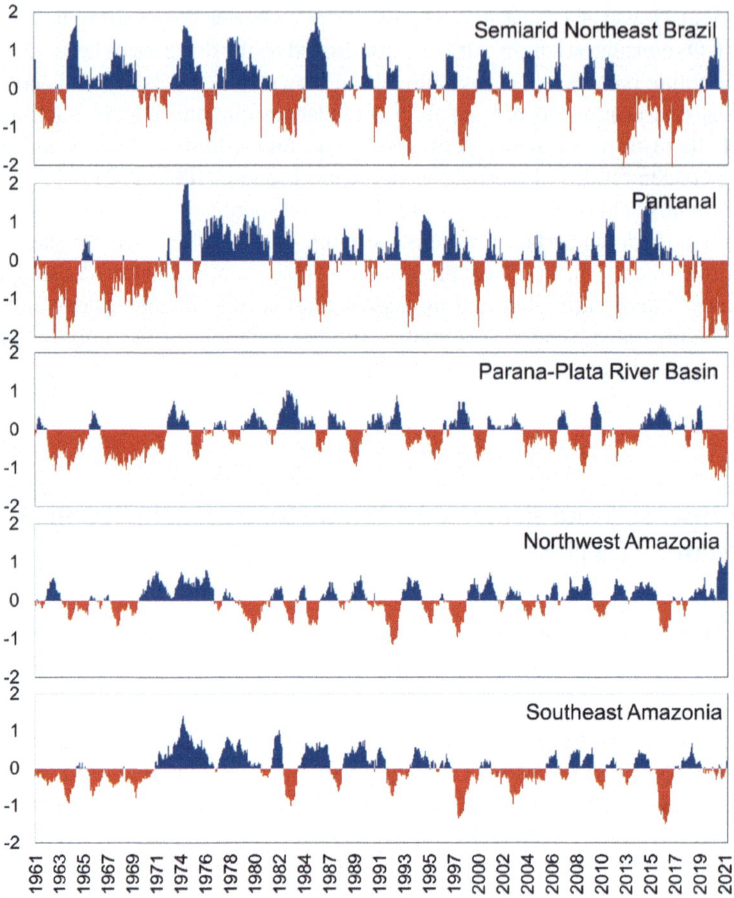

Fig. 2.2 Time series of SPI-12 months for the 5 indicated regions from 1961 to 2021. The regions correspond to the Brazilian biomes. *Source* IBGE-Servico Florestal do Brasil: https://snif.florestal.gov.br/es/los-biomas-y-sus-bosques/608-florestas-nos-biomas-brasileiros

2.3.2 Changes in Rainfall and Hydrology in the Amazon Region

Historical trends in Amazonian precipitation vary considerably among studies, depending on the dataset, time series period and length, season, and the region evaluated. Most modern rainfall records start in the 1960s, hampering the quantification of trends in the Amazonian region. Studies analyzing rainfall trends in the Amazon for the past four decades show a north–south opposite trend, including increasing rainfall in the northwestern Amazon and a decrease in the southeastern Amazon, particularly in the last decade (Fig. 2.2d and e). These trends may be a consequence of the intensification of the hydrological cycle in the region (Gloor et al. 2013; Barichivich et al. 2018; Science Panel for the Amazon 2021). The droughts in 1983, 1998, 2010 and 2016 affected the entire Amazon, and the drought in 2005 affected mostly the southern Amazon.

This intensification means increased climate variability, reflected by the increase in recent extreme hydroclimatic events due to stronger northeast trade winds that transport moisture into the Amazon. Recent work by Espinoza et al. (2019) shows that while the southern Amazon exhibits negative trends in total rainfall and extremes, the opposite is found in the northern Amazon, particularly during the wet season. Wang et al. (2018) combined both satellite and in situ observations and revealed changes in tropical Amazonian precipitation over the northern Amazon. Due to increasing rainfall in the northern Amazon, the overall precipitation trend on a basin scale showed a 2.8 mm/year increase for the 1981–2017 period (Paca et al. 2020).

As shown by Schöngart and Junk (2020), water level data for the Rio Negro at Manaus, close to its confluence with the Solimões (Amazonas) River, started being recorded in September 1902. The mean amplitude between the annual maximum (floods) and minimum (droughts) water levels is 10.22 m (1903–2015). Barichivich et al. (2018) indicated a significant increase in the daily mean water level of approximately 1 m over this 113-yr period. Furthermore, the authors observed a fivefold increase in severe flood events resulting in the occurrence of severe flood hazards over the last two decades in the central Amazon (2009, 2012–2015, 2017, 2019 and 2021) and droughts in 2005, 2010, and 2016.

Substantial warming of the tropical Atlantic since the 1990s has played a central role in the region's hydrology, increasing atmospheric water vapor imported by trade winds into the northern Amazon basin and increasing precipitation, especially during the dry-to-wet and wet seasons. The simultaneous cooling of the equatorial Pacific during this period strengthens the Walker circulation and deep convection over the Amazon (Marengo et al. 2021a).

2.3.3 Rainfall and Hydrological Variability in the Pantanal Region

Bergier et al. (2018) used a seasonal rainfall time series from 1926 to 2016 for the Pantanal and found a positive trend in the number of rainy days for all seasons. Lázaro et al. (2020), using a 42-year historical series, found that the number of days without precipitation has greatly increased in the northern Pantanal, as well as the loss of water mass in the landscape over the last 10 years, specifically during the dry season. Overall, currently, the northern Pantanal has 13% more days without rain than in the 1960s (Lázaro et al. 2020). The rainfall during the summers of 2019 and 2020 was well below normal (Fig. 2.2c) and lower than that during the 1970s. For 2020, rainfall was reduced until November. It rained just 160 mm in January, and in March and November, it rained half of the expected value. Since August 2019, rainfall was below normal, and in October 2019, it rained 50 mm (half of the 100 mm average).

This suggests a late onset of the rainy season of the hydrological year 2019–2020 (Marengo et al. 2021b).

Marengo et al. (2021b) showed that the river levels at Ladário represent the hydrological regime of the Upper Paraguay River Basin, which enables the characterization of a given period as drought or flood in the Pantanal. The annual mean level of the Paraguay River at Ladário is 273 cm (1900–2020), ranging from 145 cm in November to 405 cm in June. In terms of the daily absolute maximum, the five events with levels above 600 cm were registered in April 1988 (664 cm) and May 1905 (662 cm). In April 1988, the river level rose to 664 cm, flooding small communities along the river's shores. The year 1970–1971 recorded the largest number of days with levels equal to or below 100 cm during the observation period. On the dry side, the 5 years with the lowest minimum levels were reported in September 1964 (−61 cm), September 1971 (−57 cm), October 1967 (−53 cm), September 1969 (−53 cm), and October 1910 (−48 cm). Negative values indicate observations below the zero level of the river gauge. The lowest values were measured from 1962 to 1973; all 12 years had levels of 100 cm and below. The most recent minimum level value was −32 cm in October 2020. This is the lowest level in 49 years since the previous lowest minimum in 1971. This is consistent with the SPI values from Fig. 2.2c showing negative SPI values in those years.

2.3.4 Rainfall Variability in Northeast Brazil

Northeast Brazil (NEB) is under the influence of Atlantic trade winds that converge along the Intertropical Convergence Zone (ITCZ). The meridional migration of the ITCZ in the semiarid region of NEB determines the rainfall peak season from February to May. Years with drought were observed during El Niño in 1983, 1998 and from 2012–2018 as well as in other years characterized by warm surface waters in the Tropical North Atlantic (Fig. 2.2a). The 2012–2018 drought was associated with a warmer tropical North Atlantic and aggravated by an El Niño event in 2016 (Marengo et al. 2020). Influences from the tropical Pacific Ocean by means of El Niño and from a warmer tropical North Atlantic that moves the ITCZ anomalously to the north are the main causes of rainfall deficiency and drought in the region (Brasil Neto et al. 2021).

The drought that started in 2012 left 1717 municipalities of NEB (96% of total) in a state of emergency, which included rural food insecurity (Marengo et al. 2017, 2020; Brito et al. 2018; Alvala et al. 2019; Cunha et al. 2019; Vieira et al. 2020). During the 2012–2018 drought, the volume of water in the reservoirs on the São Francisco River (Cunha et al. 2019), an important Brazilian river that crosses the region, was reduced to minimal levels. This was coupled with an increase in demand for irrigation water and evaporation from the reservoirs. As a result of the sharp reduction in the São Francisco River flow since 2012, it became necessary to modify the operation of the reservoirs—which were designed in the 1970s—to maintain a baseflow capable of sustaining the various water uses, mainly the water supply to several cities along the river and large irrigation projects (Marengo et al. 2021a).

2.3.5 Rainfall and Hydrological Variability in the Paraná-La Plata River Basin

Since the 1960s, seven droughts (1977, 1984, 1990, 1992, 2001, 2012 and 2014) have reduced reservoir storage for São Paulo state in Brazil (Fig. 2.2c) (Naumann et al. 2021). In some parts of the La Plata Basin (LPB), such as the Upper Paraná River Basin, severe-to-exceptional hydrological drought conditions have been present since 2014. Nevertheless, in the last two years, 2020–2021, this situation has worsened. Indices of precipitation indicate that the precipitation in the Paraná River basin (Fig. 2.2c) has been below average in recent years (Naumann et al. 2021). Several dry and rainy cycles since the early 1900s have been observed, with the most severe drought on record taking place from December 1968 to September 1971, peaking in March 1969. However, it is important to note that, at that time, the water demand throughout the Paraná River basin was much lower than at present (Cunha et al. 2019). Levels of the Paraná River at Corrientes (Naumann et al. 2021) show that the low levels detected in 2020 and 2021 are comparable to those experienced during the two most severe low-level events in recorded history, i.e., 1934 and 1944.

2.3.6 Cyclones Over the South Atlantic Ocean

Different types of cyclones develop over the South Atlantic Ocean: extratropical, subtropical and tropical (Reboita et al. 2021b, c, e). Extratropical cyclogenesis is the most frequent, and tropical cyclogenesis is the rarest, even though the east coast of Brazil is a region with the potential for tropical cyclogenesis most of the year (Andrelina and Reboita 2021). Extratropical cyclones have a higher frequency in the latitude band of 45° S, whereas subtropical and tropical cyclones occur along the east and south coasts of Brazil. As these systems develop close to the coast, they may cause strong winds, heavy rain and floods.

Projections for the end of the century (2080–2099) under the RCP8.5 scenario through an ensemble with the Regional Climate Model (RegCM4) nested in different CMIP5 global climate models (GCMs) and an ensemble with GCMs indicate (Fig. 2.3): (a) extratropical cyclones over the South Atlantic Ocean are projected to decrease in frequency due to the decrease in near-surface baroclinicity (Reboita et al. 2021c; Marrafon et al. 2022); on the other hand, intensity may be equal to or higher than the historical period (1995–2014), and they can cause more intense rainfall and winds. While the total frequency of extratropical cyclones is projected to decrease, the number of explosive cyclones (deepening rate of ~ 24 hPa/24 h) is

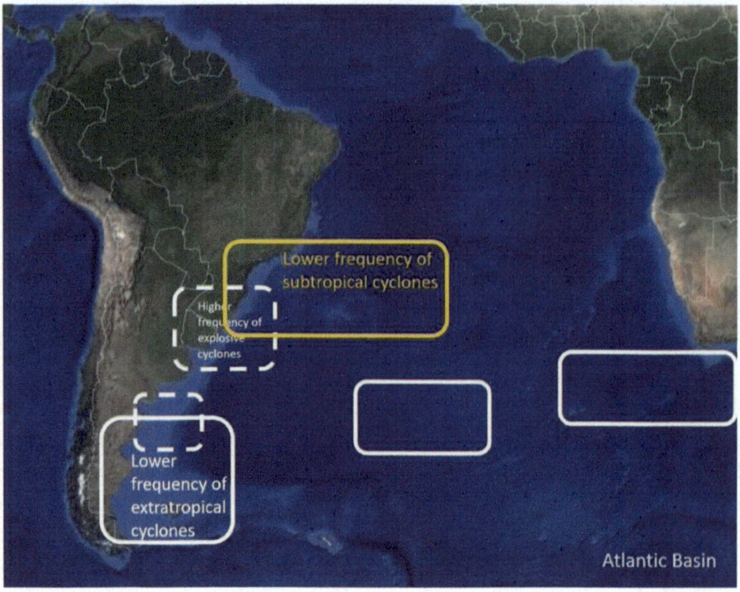

Fig. 2.3 Projected trends in the frequencies of extratropical, explosive and subtropical cyclones at the end of the century (2080–2099) under the RCP8.5 scenario compared to the historical period (1995–2014)

projected to increase (Reboita et al. 2021c); (b) subtropical cyclones are projected to decrease in frequency in part due to the intensification of the South Atlantic subtropical anticyclone (de Jesus et al. 2021; Reboita et al. 2019). On the other hand, intense convection may cause this kind of cyclone to become stronger; and (c) for tropical cyclones, there are no trends in their frequency by the end of the century (Marrafon et al. 2022).

2.3.7 Changes in Mean Climate and Extremes Based on CMIP6 and CORDEX Models Under Various Levels of Warming (from 1.5 to 4 °C)

Recent work by Almazroui et al. (2021) shows the results of an analysis of a large ensemble of models from CMIP6 over South America for future changes in slices 2040–2059 and 2080–2099 relative to the reference period (1995–2014) under four shared socioeconomic pathways (SSP1-2.6, SSP2-4.5, SSP3-7.0 and SSP5-8.5). The CMIP6 models successfully capture the main climate characteristics across South America for the reference period. Future precipitation exhibits a decrease east of the northern Andes in tropical South America and the Amazon and an increase over southeastern South America and the northern Andes, consistent with earlier CMIP

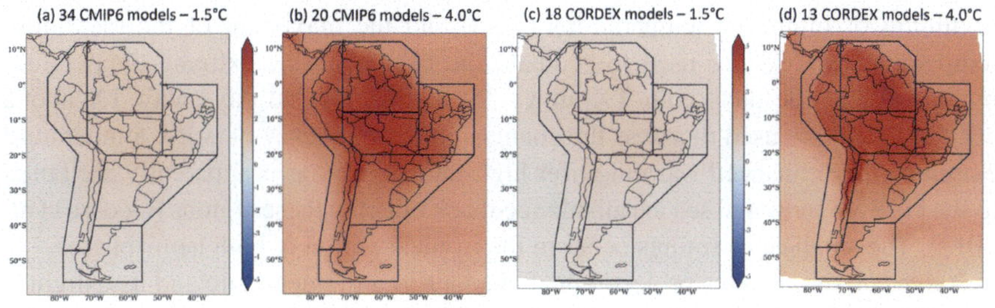

Fig. 2.4 Changes (relative to 1995–2014) in mean annual temperature (*T* in °C) projected by CMIP6 models (under scenario SSP5-8.5) considering 1.5 (**a**) and 4.0 °C (**b**) global warming levels and for CORDEX models (under RCP8.5 scenario) and considering 1.5 (**c**) and 4.0 °C (**d**) global warming levels (GWLs). Boundaries of IPCC AR6 Atlas regions are shown in the upper left corner of the panel. The same regions are used in Figs. 2.5 and 2.6

(3 and 5) projections. In contrast, temperature increases are robust in terms of magnitude even under SSP1-2.6. Future changes mostly progress monotonically from the weakest to the strongest forcing scenario and from the mid-century to late-century projection period. Furthermore, an increasingly heavy-tailed precipitation distribution and a rightward shifted temperature distribution provide strong indications of a more intense hydrological cycle as greenhouse gas emissions increase. The authors found no clear systematic linkage between model spread about the mean in the reference period and the magnitude of simulated subregional climate change in the future period.

In this subsection, we consider the Northern South America (NSA) and South American Monsoon (SAM) AR6 regions to represent the Amazon Basin and the Northeastern South America (NES) AR6 region to represent Northeast Brazil (Iturbide et al. 2021) (Fig. 2.4). Various levels of warming are considered. For tropical South America, we also include Southeastern South America (SES) so that the entire region (tropical South America) is treated as NSA, SAM, NES and SES. In addition, we are working under a new methodology, employing global warming levels (GWLs) (1.5, 2.0, 3.0, and 4.0 °C), instead of decades. Nevertheless, it is important to highlight that such levels of warming can be translated in terms of time (decades), as demonstrated by Seneviratne et al. (2021). For example, the 1.5 °C GWL (relative to the recent past, 1995–2014) will be achieved around the 2050s under the SSP5-8.5 scenario, while the 4.0 °C GWL will be achieved around the 2090s using SSP5-8.5 (see Table 4.2 in Chap. 4—IPCC 2021). Considering the preindustrial period (1850–1900), the 1.5 °C GWL will be reached in 2030 for all scenarios.

2.3.8 *Temperature Projections*

The mean temperature (*T*), minimum temperature (TN), and maximum temperature (TX) are used to assess the change in the average temperature magnitude. The minimum of minimum temperatures (TNn) and the maximum of the maximum temperatures (TXx) are used to evaluate the change in the extreme temperature magnitudes, while days with TX above 35 °C (TX35) and days with TX above 40 °C (TX40) are used to assess the changes in the frequency of warm temperature extremes. The enhanced warming over high-latitude oceans is apparently attributed to the positive snow and sea ice albedo feedback effect in these regions (Goosse et al. 2018). The smallest warmings occur in the tropical region of both hemispheres.

Table 2.1 summarizes projections of changes in air temperature and its extremes for NSA, SAM, NES and SES considering 1.5 and 4.0 °C GWLs under the SSP5-8.5 (RCP8.5) scenario for CMIP6 (CORDEX) models relative to 1995–2014. For the 1.5 °C GWL, the NSA, SAM, NES and SES heat up approximately half of that value (approximately 0.7 °C). However, as the level of warming rises, this difference is reduced. For the 4.0 °C GWL, in the NSA, SAM, NES, and SES, the temperature increases between 2.9 and 4.2 °C in the CMIP6 and CORDEX models. The same behavior is observed for TN, TX, TNn and TXx. Regarding the frequency of warm extremes (TX35 and TX40), the increase from 1.5 to 4.0 GWLs are very high, mainly for SAM, where it will reach almost 92 (107) days with TX above 40 °C, considering CMIP6 (CORDEX) results.

Both the projected magnitude of warming and the frequency of occurrence of hot extremes are higher for SAM than for all South American regions (Fig. 2.4). This feature is also presented in Chou et al. (2014), Reboita et al. (2014), López-Franca et al. (2016), Teichmann et al. (2021) and Coppola et al. (2021). Chou et al. (2014), using the Eta Regional Climate Model forced by two global climate models, HadGEM2-ES and MIROC5, under two RCP scenarios (8.5 and 4.5), show that in the future, the major warming area will be located in the central part of Brazil. In Coppola et al. (2021) and Teichmann et al. (2021), the results from two regional models nested in three GCMs from CMIP5 and RCP2.6 and RCP8.5 show the most warming over the NSA, SAM and NES. López-Franca et al. (2016), using 4 RCMs driven by 3 GCMs and projections for 2079–2098, show greater increases in warm nights and warm days over northern South America. Additionally, Reboita et al. (2014) projections, using RegCM3 nested in ECHAM5 and HadCM3 MCGs under the A1B scenario, indicate general warming throughout all South American regions and seasons, which is more pronounced in the far-future period.

According to the IPCC (2021), it is virtually certain, compared with the recent past (1995–2014) and compared to the preindustrial period (1850–1900), that all of South America will have an increase in the intensity and frequency of hot extremes and a decrease in the intensity and frequency of cold extremes under a 4.0 °C GWL. Additionally, according to this report, it is virtually certain that the mean air temperature will rise across all of South America, with the largest increases occurring in the Amazon Basin (NSA and SAM).

Table 2.1 Changes in the magnitude and frequency of temperature indices for 1.5 and 4.0 °C GWLs using CMIP6 models under the SSP5-8.5 scenario and CORDEX models under the RCP8.5 scenario

Temperature index and unit	AR6 region	1.5 °C		4.0 °C	
		CMIP6	CORDEX	CMIP6	CORDEX
T (°C)	NSA	0.8 (0.2–1.2)	0.7 (0.5–1.1)	3.8 (3.1–4.5)	3.2 (3.2–4.8)
	NES	0.7 (0.3–1.1)	0.7 (0.4–1.0)	3.1 (2.7–3.6)	3.5 (3.0–4.3)
	SAM	0.8 (0.3–1.3)	0.8 (0.5–1.2)	3.9 (3.2–4.6)	4.2 (3.4–5.0)
	SES	0.6 (0.2–0.9)	0.6 (0.3–0.8)	3.0 (2.5–3.4)	2.9 (2.4–3.5)
TN (°C)	NSA	0.7 (0.2–1.1)	0.7 (0.4–1.1)	3.6 (3.0–4.3)	3.4 (2.9–4.0)
	NES	0.6 (0.2–1.0)	0.7 (0.4–1.0)	3.1 (2.5–3.8)	3.2 (2.6–3.7)
	SAM	0.8 (0.3–1.3)	0.8 (0.5–1.1)	4.0 (3.4–4.6)	4.0 (3.2–4.7)
	SES	0.6 (0.2–0.9)	0.6 (0.3–0.8)	3.1 (2.4–3.5)	3.0 (2.5–3.6)
TX (°C)	NSA	0.9 (0.2–1.4)	0.8 (0.5–1.1)	4.2 (3.3–5.2)	3.9 (3.0–5.0)
	NES	0.8 (0.2–1.3)	0.7 (0.4–1.0)	3.8 (3.1–4.6)	3.3 (2.9–3.8)
	SAM	0.9 (0.3–1.6)	0.9 (0.5–1.3)	4.5 (3.4–5.5)	4.4 (3.5–5.2)
	SES	0.7 (0.1–1.1)	0.6 (0.2–0.9)	3.1 (2.2–3.8)	2.9 (2.3–3.5)
TNn (°C)	NSA	0.7 (0.2–1.0)	0.7 (0.4–1.1)	3.5 (2.6–4.4)	3.8 (3.1–4.5)
	NES	0.7 (0.3–0.9)	0.6 (0.4–0.9)	3.2 (2.6–4.1)	3.3 (2.6–3.7)
	SAM	0.8 (0.3–1.3)	0.7 (0.3–1.1)	3.7 (2.9–4.5)	4.1 (3.1–5.1)
	SES	0.6 (0.2–1.2)	0.5 (0.0–0.8)	2.9 (2.4–3.4)	2.6 (1.7–3.2)
TXx (°C)	NSA	1.0 (0.3–1.6)	0.9 (0.5–1.2)	4.5 (3.7–5.2)	4.2 (3.3–5.0)
	NES	0.9 (0.3–1.6)	0.8 (0.5–1.2)	3.9 (3.3–4.5)	3.9 (3.4–4.3)
	SAM	1.3 (0.4–2.4)	1.2 (0.6–1.9)	6.1 (4.3–8.0)	5.5 (4.3–7.0)
	SES	0.9 (0.4–1.4)	0.8 (0.3–1.3)	4.1 (3.2–5.0)	3.9 (3.1–4.5)
TX35 (days)	NSA	16.1 (1.7–28.4)	18.1 (7.1–26.6)	89.7 (33.0–109.1)	98.1 (69.3–122.1)
	NES	8.5 (1.7–24.8)	10.5 (3.8–25.6)	43.3 (33.9–100.3)	57.5 (33.9–87.3)
	SAM	17.8 (4.5–31.4)	20.1 (9.6–37.3)	91.7 (43.3–153.0)	106.7 (80.7–132.4)
	SES	5.7 (1.9–10.7)	6.0 (1.0–9.1)	28.2 (16.6–36.2)	34.1 (19.8–47.5)
TX40 (days)	NSA	4.3 (0.0–16.1)	6.0 (0.0–12.9)	39.2 (1.6–77.7)	52.6 (11.0–104.9)
	NES	1.9 (0.0–9.7)	1.8 (0.1–4.6)	18.4 (0.8–37.1)	21.6 (5.7–43.3)

(continued)

Table 2.1 (continued)

Temperature index and unit	AR6 region	1.5 °C		4.0 °C	
		CMIP6	CORDEX	CMIP6	CORDEX
	SAM	3.9 (0.2–13.1)	8.5 (1.5–17.8)	29.4 (4.1–55.2)	57.0 (28.5–101.2)
	SES	2.3 (0.4–5.2)	1.4 (0.2–2.3)	10.8 (6.0–19.4)	12.6 (5.4–22.1)

Source Interactive Atlas—IPCC AR6 (Iturbide et al. 2021)
Displayed are median values and, in parentheses, the 5–95% ranges over NSA, SAM, NES, and SES relative to the period 1995–2014

2.3.9 Precipitation Projections

The total precipitation (PR) is used to indicate the change in the amount of precipitation, and the maximum 1-day and 5-day precipitation (RX1day and RX5day) are used to show the change in the magnitude of precipitation extremes. Unlike the results for temperature, precipitation and its extremes exhibit a more heterogeneous behavior and, in most regions, with no agreement between the models (see precipitation global maps at https://interactive-atlas.ipcc.ch/). Moreover, for the near future (2021–2040) or for the low GWLs (1.5 °C and 2.0 °C), even considering the most pessimistic scenarios (SSP5-8.5 and RCP8.5), the sign of change is very weak, with no agreement between models. Figure 2.5 presents the change in annual precipitation (PR) for 1.5 and 4.0 °C GWLs relative to 1995–2014 using CMIP6 and CORDEX models under the SSP5-8.5 and RCP8.5 scenarios, respectively. The only robust signs of projected changes in annual precipitation and where there is an agreement between the models (Fig. 2.5b and d) occur in the east of the NSA, with a 12% reduction (median) in PR, as well as in southwestern South America (SWS). Projections of increases in PR are observed over SES and in the extreme west of northwestern South America (NWS). The drying (wetting) conditions over NSA and SAM (SES) are also presented in Chou et al. (2014), Llopart et al. (2014), Reboita et al. (2014), Coppola et al. (2021), Teichmann et al. (2021), and Teodoro et al. (2021).

Chou et al. (2014) showed a reduction in summer precipitation over northern and central South America and an increase in PR over southern South America. Llopart et al. (2014), using the regional climate model RegCM4 driven by the HadGEM, GFDL and MPI GCMs under RCP8.5 and, more recently, Teodoro et al. (2021), have also shown projections of reduced precipitation over the broad Amazon and central Brazil region and increased precipitation over the La Plata basin and central Argentina. According to Llopart et al. (2014), the tendency toward an extension of the dry season over central South America is due to a late onset and an early retreat of the SAM. Reboita et al. (2014) also projected dry conditions in all seasons over northern South America and an increase in precipitation over the SES mainly in spring and summer. More recent work by Coppola et al. (2021) points out that extreme wet and flood prone maxima are projected to increase over the La Plata basin (SES).

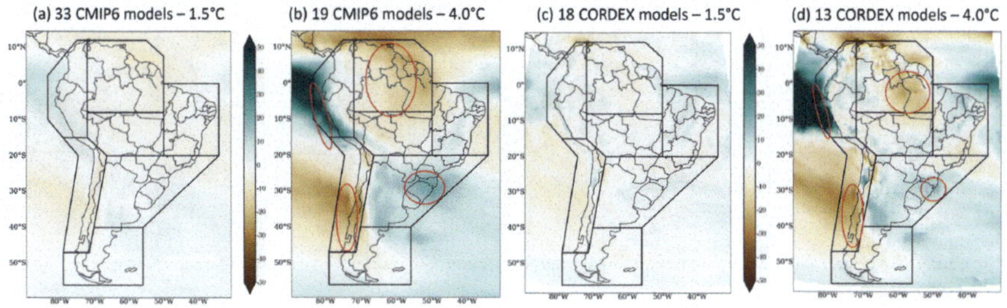

Fig. 2.5 Changes (relative to 1995–2014) in mean annual precipitation (PR, in %) projected by CMIP6 models (under scenario SSP5-8.5) considering 1.5 (**a**) and 4.0 °C (**b**) GWLs and for CORDEX models (under RCP8.5 scenario) considering 1.5 (**c**) and 4.0 °C (**d**) GWLs. GWLs. Red circles and ellipses indicate areas with robust signs of projected changes in annual precipitation and where there is an agreement between the models. See Fig. 2.4 for names of regions delimited in figure

Regarding the behavior of the magnitude of extreme precipitation (RX1day and RX5day), there is an increase over most of South America, including the Amazon and Northeast Brazil. It is important to highlight that extreme precipitation is determined by local exchanges in heat, moisture, and other related quantities (thermodynamic changes) and those associated with atmospheric and oceanic motions (dynamic changes) (Seneviratne et al. 2021). The increase in water vapor leads to robust increases in precipitation extremes everywhere, with a magnitude that varies between 4 and 8% per degree Celsius of surface warming (thermodynamic contribution) (Fischer and Knutti 2016; Sun et al. 2020). However, the dynamic contributions show large differences across models and are more uncertain than thermodynamic contributions (Shepherd 2014; Trenberth et al. 2015; Pfahl et al. 2017).

Increases in extreme precipitation magnitude and frequency over NSA, SAM, NES, and SES are projected for the end of the century using a 4 °C GWL (Li et al. 2021). However, using RCMs, Chou et al. (2014) and Coppola et al. (2021) projected negative changes in precipitation extremes over the NSA. Regarding SAM, NES, and SES, Coppola et al. (2021) project positive changes, while Chou et al. (2014) project negative changes, except for SES.

In summary, IPCC (2021) projections indicate that for mean precipitation there will be a drying (wetting) signal for NSA and NES (SES). IPCC (2021) states that an intensification of heavy precipitation is projected with medium confidence, compared to the recent past (1995–2014) and with the preindustrial period (1850–1900) for 2 and 4 °C GWLs, for NSA, SAM, and NES and particularly for SES.

2.3.10 Drought Projections

Consecutive dry days (CDD) are used to assess changes in meteorological drought. Figure 2.6 presents changes in CDD for 1.5 and 4.0 °C GWLs relative to 1995–2014

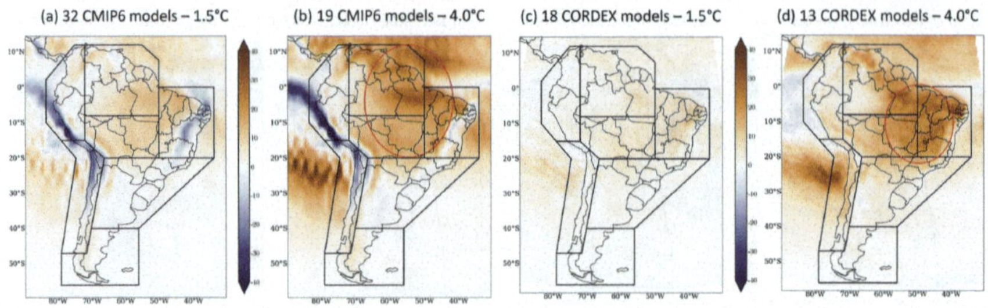

Fig. 2.6 Changes (relative to 1995–2014) in consecutive dry days (CDD, in number of days projected by CMIP6 models (under scenario SSP5-8.5) considering 1.5 (**a**) and 4.0 °C (**b**) GWLs and for CORDEX models (under RCP8.5 scenario) considering 1.5 (**c**) and 4.0 °C (**d**) GWLs. See Fig. 2.4 for the names of the regions delimited in the figure

using CMIP6 and CORDEX models under the SSP5-8.5 and RCP8.5 scenarios, respectively. The increase in CDD over the Amazon basin is remarkable, indicating a drier climate in the future, both in CMIP6 and CORDEX models. Over NES, the models also show an increase in CDD. Over the SES, there is no agreement between the models, but the model results show an overall decrease in precipitation at 1.5 °C GWLs. According to Chou et al. (2014), Coppola et al. (2021), and Reboita et al. (2021d), positive changes in CDD over NSA, SAM and NES are projected in the far-future.

2.4 Impacts on Natural and Human Systems

2.4.1 Impacts on Natural Ecosystems

The drought that started in 2012 highlights the vulnerability of NEB. Arid conditions have been detected during recent years mainly in the central semiarid region, covering almost 2% of NEB (Marengo et al. 2020). The projections of vegetative stress conditions derived from the empirical model for the vegetation health index (VHI) are consistent with projections from vegetation models, where semidesert types typical of arid conditions would replace the current semiarid bushland vegetation ("caatinga") by 2100 (Marengo et al. 2020). Therefore, there is a possibility that under permanent drought conditions with warming above 4 °C, arid conditions would prevail in NEB since 2060, which would lead to land degradation and desertification.

The drought situation in 2019–20 in the Pantanal and the Upper Paraguay River basin has been unusually harsh, with dry and warm conditions favoring the propagation of fires. The increased number of fires affects human activities and biodiversity. These fires and droughts aggravated the situation of vulnerable fauna and flora. Changes in the quality of the rainy season can also affect wetland hydrology, and droughts can seriously affect the living conditions of biological populations (Marengo et al. 2021a).

Drought and flood events in Amazonia have produced impacts such as an increased risk of forest fires, extreme warming, floods and inundations, which can affect the human population and flora and fauna both on land and in lakes and rivers (Davidson et al. 2012; Marengo et al. 2013; Brando et al. 2014; Doughty et al. 2015). While droughts increase the risk of tree mortality, the combination of severe droughts and floods can put additional stress on Amazon forests, especially if the flooding regime of regularly inundated areas is perturbed outside of their natural range (Langerwisch et al. 2013). It also affects riverine carbon balance by outgassing carbon from the Amazon River and the amount exported to the Atlantic Ocean, with nonlinear effects to be expected if deforestation is also considered (Langerwisch et al. 2016).

2.4.2 *Potential Impacts on Society*

Due to the impacts of the 2012–2017 drought in NEB, public policies have been implemented to reduce social and economic vulnerability for small farmers. In the long term, to make the semiarid region less vulnerable to drought, strengthened integrated water resource management and a proactive drought policy are needed to restructure the economy. Integrating drought monitoring and seasonal climate forecasting provides a means of assessing the impacts of climate variability and change, leading to disaster risk reduction through early warning. In a future scenario with a high risk of drought and possible desertification in regions with warming above 4 °C, agricultural activities can be affected by severe water stress and may be devastating for local populations. This condition can make drought irreversible, and prolonged water stress could lead to aridity and land degradation (Marengo et al. 2020).

Changes in the rainfall regime and the flood pulse in the Pantanal are likely to disrupt the processes that maintain these landscapes; furthermore, landscape modification may dramatically alter wetlands (Ivory et al. 2019; Marengo et al. 2021b). Many human activities in the region rely on the ecosystem services provided by the Pantanal, including professional and touristic fishing and contemplative tourism (Bergier et al. 2018). The available land for cattle ranching and farming is dependent on the extent of flooding during each wet season. Due to these flooding events, local ranchers struggle to survive. Because of drought, fires spread and affect the natural biodiversity in the Pantanal region as well as the agribusiness and cattle ranching sectors. The uncontrolled fires occurring in the dry season are of anthropogenic origin. They are directly related to deforestation, cleaning, and reforming pastures. Improper practices and the use of fire as a management practice without control techniques endanger conservation (Aragão et al. 2018; Alho et al. 2019).

It is important to mention that in the Amazon region, the perception of drought and flood by the population may be different when compared to other regions, and low or high river levels are better indicators of drought or floods, respectively, compared to rainfall anomalies. Through their close dependency on water levels, local people are well placed to detect variability in both climate and hydrological regimes and can respond to early warning signals to cope with potential impacts on their activities (Pinho et al. 2015; Marengo et al. 2018). Development in the Amazon region has pushed the agricultural frontier, resulting in widespread land cover change. As agriculture in the region has low productivity and is unsustainable, the loss of biodiversity and continued deforestation will lead to high risks of irreversible changes in Amazon forests (Nobre et al. 2016).

2.4.3 Governance Actions to Mitigate Climate Change: Examples for Northeast Brazil (NEB) and Amazon

In this section, we focus on two regions of South America that have high social and biodiversity vulnerability in tropical South America, NEB and Amazon, respectively. In NEB, current-drought emergency relief measures include water distribution by trucks ("carros pipa"), plus cash transfer and state-sponsored microinsurance programs for smallholders (Magalhães 2016). Actions are heavily concentrated on water distribution, initially to rural but also to urban and coastal communities that depend upon water supply originating in the semiarid regions. In addition, recent anti-poverty programs, such as education, health and extreme poverty alleviation, have significantly improved the life conditions in the region (Alvala et al. 2019; Marengo et al. 2020). Despite significant improvement in quality-of-life indicators in the past 15 years, levels of vulnerability remain high, especially in rural households that are more dependent on agriculture (Engle and Lemos 2010). In a recent work, Marengo et al. (2021c) presented a list of initiatives that can be considered adaptation options to cope with drought in NEB in the long term. Some are technical options, while others can be considered more political. However, others are related to monitoring and early warning of drought and seasonal climate forecasts from national and regional agencies. The target of all these actions is the protection of vulnerable populations and small-scale farming in the semiarid region of NEB.

The challenges of governance of climate change in the Amazon region are related to the fact that even though the region will be impacted, drivers of deforestation are linked to economic activities such as logging, cattle ranching, soy harvesting, and mining, as well as to public investments in infrastructure such as roads and hydropower plants (Science Panel for the Amazon 2021), all of which have played a significant role in the economic growth of the region and of Brazil.

Amazonian municipalities of Brazil are involved in global climate governance. The actions are linked to the national policy to control deforestation. Other actions can be characterized by the initiatives taken by municipalities to become involved in international negotiation processes focusing on the United Nations Framework Convention on Climate Change (UNFCCC) and the Reducing Emissions from Deforestation and Forest Degradation (REDD+) mechanisms. These initiatives have strengthened subnational, national, regional and transnational networks and have created commitments to mitigate and find ways to adapt to global climate change. This is relevant to the goals of deforestation reduction proposed by Brazil during the Paris Agreement in 2015 (IPCC 2021).

2.5 Gaps, Limitations and Future Lines of Work

With the likelihood of more frequent droughts, there is a need for a better perception that adaptive capacity is still low, as shown by the consequences of recent droughts in tropical South America. Science has assembled enough knowledge to underline the global and regional importance of the Amazon, Pantanal and Parana-Prata regions, which can support policy-making to keep these sensitive ecosystems functioning. This major challenge requires substantial resources and strategic cross-national planning and a unique blend of expertise and capacities established in the regions' countries and from international collaboration. While science can still advance further in this area, we have also assembled enough knowledge to underline the global and regional importance of an intact Amazon region to support policy-making and to keep this sensitive ecosystem functioning. Measures and strategies for drought preparedness could be strengthened by regional, national and international financing mechanisms to provide for disaster risk reduction and disaster risk management in the long term.

A multiannual drought has been affecting the Plata Basin since mid-2019. The lack of rainfall, mainly in the upper part of the basin, has led to a considerable decrease in the flow of the Paraguay and Paraná rivers. Due to its prolonged duration and severity, this drought has already produced many impacts on several different socioeconomic sectors and has also severely affected ecosystems. These include water supply disruptions, forest fires, reduced agricultural yields, decreased river transport on the Paraguay and Paraná rivers, and a considerable reduction in hydroelectric energy production (Naumann et al. 2021). The impacts of this drought need to be assessed in the context of cross bordering collaboration.

On the governance side, governments and institutions are critical determinants of adaptive capacity and resilience in NEB, from the municipal to the federal levels. Other actions on the scientific side include preventive activities, such as risk monitoring and early warning systems for weather and climate extremes that can trigger natural disasters that would impact human and natural systems. These actions include seasonal climate forecasts for the region, spanning from model development to risk communication to the public and decision-makers. Also, it is necessary to consolidate disaster risk management in Brazil, which requires exploring synergies between all the institutions involved. It is necessary to create mechanisms for the integration and articulation of technical and scientific knowledge of the various dimensions of risk by ensuring a linkage between policies related to disaster risk reduction, climate change adaptation, and the UN Sustainable Development Goal 13 (Climate Action).

Acknowledgements This research was funded by the National Institute of Science and Technology for Climate Change Phase 2 under CNPq, grant number 465501/2014-1; Fundação de Amparo à Pesquisa do Estado de São Paulo (FAPESP) grant numbers 2014/50848-9 and 2017/09659-6; the National Coordination for High Level Education and Training (CAPES), grant number 88887.136402/2017-00; and CNPq grant 301397/2019-8 for José A. Marengo and Tercio Ambrizzi. It was partially funded by the Project Rede Pantanal from the Ministry of Science, Technology and Innovations of Brazil (FINEP grant 01.20.0201.00). Tercio Ambrizzi also had partial support from CNPq 301397/2019-8. Michelle Reboita was funded by CNPq grants 306488/2020-5 and 420262/2018-0.

References

Alho CJR, Mamede SB, Benites M et al (2019) Threats to the biodiversity of the Brazilian Pantanal due to land use and occupation. Ambiente & Sociedade. https://doi.org/10.1590/1809-4422asoc201701891vu2019L3AO

Almazroui M, Ashfaq M, Islam MN et al (2021) Assessment of CMIP6 performance and projected temperature and precipitation changes over South America. Earth Syst Environ 5:155–183. https://doi.org/10.1007/s41748-021-00233-6

Alvala R, Cunha AP, Brito SS et al (2019) Drought monitoring in the Brazilian Semiarid region. An Acad Bras Ciênc. https://doi.org/10.1590/0001-3765201720170209

Ambrizzi T, Reboita MS, da Rocha RP et al (2019) The state of the art and fundamental aspects of regional climate modeling in South America. Ann N Y Acad Sci 1436:96–120. https://doi.org/10.1111/nyas.13932

Andrelina B, Reboita MS (2021) Climatologia do Índice do Potencial de Gênese de Ciclones Tropicais nos Oceanos Adjacentes à América do Sul. An Inst Geociências. https://doi.org/10.11137/1982-3908_2021_44_39515

Aragão LEOC, Anderson LO, Fonseca MG et al (2018) 21st century drought-related fires counteract the decline of Amazon deforestation carbon emissions. Nat Commun. https://doi.org/10.1038/s41467-017-02771-y

Barichivich J, Gloor E, Peylin P et al (2018) Recent intensification of Amazon flooding extremes driven by strengthened Walker circulation. Sci Adv. https://doi.org/10.1126/sciadv.aat8785

Bergier I, Assine ML, McGlue MM et al (2018) Amazon rainforest modulation of water security in the Pantanal wetland. Sci Total 619–620:1116–1125. https://doi.org/10.1016/j.scitotenv.2017.11.163

Brando PM, Balch JK, Nepstad DC et al (2014) Abrupt increases in Amazonian tree mortality due to drought-fire interactions. PNAS 111(17):6347–6352. https://doi.org/10.1073/pnas.1305499111

Brasil Neto RM, Santos CAG, Silva JFCB et al (2021) Evaluation of the TRMM product for monitoring drought over Paraíba State, northeastern Brazil: a trend analysis. Sci Rep. https://doi.org/10.1038/s41598-020-80026-5

Brito SSB, Cunha APMA, Cunningham CC et al (2018) Frequency, duration and severity of drought in the Semiarid Northeast Brazil region. Int J Climatol 38(2):517–529. https://doi.org/10.1002/joc.5225

Cai W, Borlace S, Lengaigne M et al (2014) Increasing frequency of extreme El Niño events due to greenhouse warming. Nat Clim Change 4:111–116. https://doi.org/10.1038/nclimate2100

Cai W, Wang G, Santoso A et al (2015) Increased frequency of extreme La Niña events under greenhouse warming. Nat Clim Change 5:132–137. https://doi.org/10.1038/nclimate2492

Cai W, McPhaden M, Grimm AM et al (2020) Climate impacts of the El Niño-Southern Oscillation on South America. Nat Rev Earth Environ 1:215–231. https://doi.org/10.1038/s43017-020-0040-3

Chou SC, Lyra A, Mourão C et al (2014) Evaluation of the Eta simulations nested in three global climate models. Am J Clim Change 3:438–454. https://doi.org/10.4236/ajcc.2014.35039

Coppola E, Raffaele F, Giorgi F et al (2021) Climate hazard indices projections based on CORDEX-CORE, CMIP5 and CMIP6 ensemble. Clim Dyn 57:1293–1383. https://doi.org/10.1007/s00382-021-05640-z

Cunha APMA, Zeri M, Deusdará Leal K et al (2019) Extreme drought events over Brazil from 2011 to 2019. Atmosphere. https://doi.org/10.3390/atmos10110642

Davidson EA, de Araújo AC, Artaxo P et al (2012) The Amazon basin in transition. Nature 481:321–328. https://doi.org/10.1038/nature10717

Doughty CE, Metcalfe DB, Girardin CAJ et al (2015) Drought impact on forest carbon dynamics and fluxes in Amazonia. Nature 519:78–82. https://doi.org/10.1038/nature14213

Engle NL, Lemos MC (2010) Unpacking governance: building adaptive capacity to climate change of river basins in Brazil. Global Environ Change 20:4–13. https://doi.org/10.1016/j.gloenvcha.2009.07.001

Espinoza JC, Ronchail J, Marengo JA et al (2019) Contrasting North-South changes in Amazon wet-day and dry day frequency and related atmospheric features (1981–2017). Clim Dyn 52:5413–5430. https://doi.org/10.1007/s00382-018-4462-2

Fischer EM, Knutti R (2016) Observed heavy precipitation increase confirms theory and early models. Nat Clim Change 6:986–991. https://doi.org/10.1038/nclimate3110

Foltz GR, Brandt P, Richter I et al (2019) The tropical Atlantic observing system. Front Mar Sci. https://doi.org/10.3389/fmars.2019.00206

Gatti LV, Basso LS, Miller JB et al (2021) Amazonia as a carbon source linked to deforestation and climate change. Nature 595:388–393. https://doi.org/10.1038/s41586-021-03629-6

Giorgi F, Gutowski WJ (2015) Regional dynamical downscaling and the CORDEX initiative. Annu Rev Environ Resour 40:467–490. https://doi.org/10.1146/annurev-environ-102014-021217

Giorgi F, Coppola E, Jacob D et al (2021) The CORDEX-CORE EXP-I initiative: description and highlight results from the initial analysis. Bull Am Meteorol Soc. https://doi.org/10.1175/BAMS-D-21-0119.1

Gloor M, Brienen RJW, Galbraith D et al (2013) Intensification of the Amazon hydrological cycle over the last two decades. Geophys Res Lett 40:1729–1733. https://doi.org/10.1002/grl.50377

Goosse H, Kay JE, Armour KC et al (2018) Quantifying climate feedbacks in polar regions. Nat Commun. https://doi.org/10.1038/s41467-018-04173-0

Hartmann L (2015) Global physical climatology, 2nd edn. Elsevier, Amsterdam

Henson R (2011) The rough guide to climate change. Dorling Kindersley, UK

IPCC (1990) Climate change 1990: first assessment report. Cambridge University Press, Cambridge

IPCC (1995) Climate change 1995: the science of climate change. Contribution of Working Group I to the Second Assessment Report of the Intergovernmental Panel on Climate Change. Cambridge University Press, Cambridge

IPCC (2001) Climate change 2001: the scientific basis. Contribution of Working Group I to the Third Assessment Report of the Intergovernmental Panel on Climate Change. Cambridge University Press, Cambridge

IPCC (2007) Climate change 2007: the physical science basis. Contribution of Working Group I to the Fourth Assessment Report of the Intergovernmental Panel on Climate Change. Cambridge University Press, Cambridge

IPCC (2013) Climate change 2013: the physical science basis. Contribution of Working Group I to the Fifth Assessment Report of the Intergovernmental Panel on Climate Change. Cambridge University Press, Cambridge

IPCC (2021) Climate change 2021: the physical science basis. Contribution of Working Group I to the Sixth Assessment Report of the Intergovernmental Panel on Climate Change. Cambridge University Press, Cambridge

Iturbide M, Fernández J, Gutiérrez JM et al (2021) Repository supporting the implementation of FAIR principles in the IPCC-WG1 Atlas. Zenodo. https://doi.org/10.5281/zenodo.3691645

Ivory SJ, McGlue MM, Spera S et al (2019) Vegetation, rainfall, and pulsing hydrology in the Pantanal, the world's largest tropical wetland. Environ Res Lett. https://doi.org/10.1088/1748-9326/ab4ffe

de Jesus EM, da Rocha RP, Crespo NM et al (2021) Future climate trends of subtropical cyclones in the South Atlantic basin in an ensemble of global and regional projections. Clim Dyn. https://doi.org/10.1007/s00382-021-05958-8

Langerwisch F, Rost S, Gerten D et al (2013) Potential effects of climate change on inundation patterns in the Amazon Basin. Hydrol Syst Sci 17:2247–2262. https://doi.org/10.5194/hess-17-2247-2013

Langerwisch F, Walz A, Rammig A et al (2016) Deforestation in Amazonia impacts riverine carbon dynamics. Earth Syst Dyn 7:953–968. https://doi.org/10.5194/esd-7-953-2016

Lázaro WL, Oliveira-Júnior ES, Silva CJ et al (2020) Climate change reflected in one of the largest wetlands in the world: an overview of the Northern Pantanal water regime. Acta Limnol Bras. https://doi.org/10.1590/S2179-975X7619

Leggett J, Pepper WJ, Swart RJ et al (1992) Emissions scenarios for the IPCC: an update. In: Houghton JT, Callander BA, Varney SK (eds) WMO/UNEP Intergovernmental Panel on Climate Change 1992—the supplementary report to the IPCC scientific assessment, pp 75–95

Li C, Zwiers F, Zhang X et al (2021) Changes in annual extremes of daily temperature and precipitation in CMIP6 models. J Clim 34:3441–3460. https://doi.org/10.1175/JCLI-D-19-1013.1

Llopart M, Coppola E, Giorgi F et al (2014) Climate change impact on precipitation for the Amazon and La Plata basins. Clim Change 125:111–125. https://doi.org/10.1007/s10584-014-1140-1

Llopart M, Reboita M, Coppola et al (2018) Land use change over the Amazon forest and its impact on the local climate. Water. https://doi.org/10.3390/w10020149

López-Franca N, Zaninelli P, Carril A et al (2016) Changes in temperature extremes for 21st century scenarios over South America derived from a multi-model ensemble of regional climate models. Clim Res 68:151–167. https://doi.org/10.3354/cr01393

Magalhães AR (2016) Life and drought in Brazil. In: De Nys E, Engle NL, Magalhães AR (eds) Drought in Brazil—proactive management and policy. CRC Press, Boca Raton, pp 1–18. https://doi.org/10.1201/9781315367415

Manabe S, Bryan K (1969) Climate calculations with a combined ocean-atmosphere model. J Atmos Sci 26:786–789. https://doi.org/10.1175/1520-0469(1969)026%3c0786:CCWACO%3e2.0.CO;2

Marengo JA, Torres RR, Alves LM (2017) Drought in Northeast Brazil—past, present, and future. Theor Appl Climatol 129:1189–1200. https://doi.org/10.1007/s00704-016-1840-8

Marengo JA, Alves LM, Soares WR et al (2013) Two contrasting severe seasonal extremes in tropical South America in 2012: flood in Amazonia and drought in Northeast Brazil. J Clim 26:9137–9154. https://doi.org/10.1175/JCLI-D-12-00642.1

Marengo JA, Souza CM, Thonicke K et al (2018) Changes in climate and land use over the Amazon region: current and future variability and trends. Front Earth Sci. https://doi.org/10.3389/feart.2018.00228

Marengo JA, Cunha APMA, Nobre CA et al (2020) Assessing drought in the drylands of northeast Brazil under regional warming exceeding 4 °C. Nat Hazards 103:2589–2611. https://doi.org/10.1007/s11069-020-04097-3

Marengo JA, Camarinha PI, Alves LM et al (2021a) Extreme rainfall and hydro-geo-meteorological disaster risk in 1.5, 2.0, and 4.0°C global warming scenarios: an analysis for Brazil. Front Clim. https://doi.org/10.3389/fclim.2021.610433

Marengo JA, Cunha AP, Cuartas LA et al (2021b) Extreme drought in the Brazilian Pantanal in 2019–2020: characterization, causes, and impacts. Front Water. https://doi.org/10.3389/frwa.2021.639204

Marengo JA, Galdos MV, Challinor A et al (2021c) Drought in Northeast Brazil: a review of agricultural and policy adaptation options for food security. Clim Resilience Sustain. https://doi.org/10.1002/cli2.17

Marrafon VH, Reboita MS, da Rocha RP et al (2022) Classificação dos tipos de ciclones sobre o Oceano Atlântico Sul em projeções com o RegCM4 E MCGs. Rev Bras Climatol. https://doi.org/10.55761/abclima.v30i18.14603

McKee TB, Doesken NJ, Kleist J (1993) The relationship of drought frequency and duration to time scales. In: Proceedings of the 8th conference on applied climatology, Anaheim, CA, 17–22 Jan 1993

McPhaden MJ, Santoso A, Cai W (2020) Introduction to El Niño Southern oscillation in a changing climate. In: McPhaden MJ, Santoso A, Cai (eds) El Niño southern oscillation in a changing climate. Wiley, New York, pp 3–20

Nakicenovic N, Alcamo J, Davis G et al (2000) Special report on emissions scenarios: a special report of Working Group III of the Intergovernmental Panel on Climate Change. Cambridge University Press, Cambridge

Naumann G, Podesta G, Marengo J et al (2021) The 2019–2021 extreme drought episode in La Plata Basin. European Union, Luxembourg. https://doi.org/10.2760/773

Nobre CA, Sampaio G, Borma LS et al (2016) Land-use and climate change risks in the Amazon and the need of a novel sustainable development paradigm. PNAS 113:10759–10768. https://doi.org/10.1073/pnas.1605516113

Paca VHM, Espinoza-Dávalos GE, Moreira DM et al (2020) Variability of trends in precipitation across the Amazon River basin determined from the CHIRPS precipitation product and from station records. Water. https://doi.org/10.3390/w12051244

Pfahl S, O'Gorman PA, Fischer EM (2017) Understanding the regional pattern of projected future changes in extreme precipitation. Nat Clim Change 7:423–427. https://doi.org/10.1038/nclimate3287

Pinho PF, Marengo JA, Smith MS (2015) Complex socio-ecological dynamics driven by extreme events in the Amazon. Reg Environ Change 15:643–655. https://doi.org/10.1007/s10113-014-0659-z

Pivello VR, Vieira I, Christianini AV et al (2021) Understanding Brazil's catastrophic fires: causes, consequences and policy needed to prevent future tragedies. Perspect Ecol Conserv 19:233–255. https://doi.org/10.1016/j.pecon.2021.06.005

Reboita MS, da Rocha RP, Dias CG et al (2014) Climate projections for South America: RegCM3 driven by HadCM3 and ECHAM5. Adv Meteorol. https://doi.org/10.1155/2014/376738

Reboita MS, Ambrizzi T, Silva BA et al (2019) The South Atlantic subtropical anticyclone: present and future climate. Front Earth Sci. https://doi.org/10.3389/feart.2019.00008

Reboita MS, Ambrizzi T, Crespo NM et al (2021a) Impacts of teleconnection patterns on South America climate. Ann N Y Acad Sci 1504:116–153. https://doi.org/10.1111/nyas.14592

Reboita MS, Crespo NM, Dutra LMM et al (2021b) Iba: the first pure tropical cyclogenesis over the Western South Atlantic Ocean. J Geophys Res: Atmos. https://doi.org/10.1029/2020JD033431

Reboita MS, Crespo NM, Torres JA et al (2021c) Future changes in winter explosive cyclones over the Southern Hemisphere domains from the CORDEX-CORE ensemble. Clim Dyn 57:3303–3322. https://doi.org/10.1007/s00382-021-05867-w

Reboita MS, Kuki CAC, Marrafon VH et al (2021d) South America climate change revealed through climate indices projected by GCMs and Eta-RCM ensembles. Clim Dyn. https://doi.org/10.1007/s00382-021-05867-w

Reboita MS, Reale M, da Rocha RP et al (2021e) Future changes in the wintertime cyclonic activity over the CORDEX-CORE southern hemisphere domains in a multi-model approach. Clim Dyn 57:1533–1549. https://doi.org/10.1007/s00382-020-05317-z

Reichstein M, Bahn M, Ciais P et al (2013) Climate extremes and the carbon cycle. Nature 500:287–295. https://doi.org/10.1038/nature12350

Schöngart J, Junk WJ (2020) Clima e hidrologia nas várzeas da Amazônia Central. In: Junk WJ, Piedade MTF, Wittmann F et al (eds) Várzeas Amazônicas Desafios para um Manejo Sustentável, pp 44–65

Science Panel for the Amazon 2021. Executive Summary of the Amazon Assessment Report 202. United Nations Sustainable Development Solutions Network, New York, USA

Seneviratne SI, Zhang X, Adnan A et al (2021) Weather and climate extreme events in a changing climate. In: Masson-Delmotte BZ, Zhai VP, Pirani A (eds) Climate change 2021: the physical science basis. Contribution of Working Group I to the Sixth Assessment Report of the Intergovernmental Panel on Climate Change. Cambridge University Press. https://www.ipcc.ch/report/sixth-assessment-report-working-group-i/. Accessed 14 Dec 2022

Shepherd TG (2014) Atmospheric circulation as a source of uncertainty in climate change projections. Nat Geosci 7:703–708. https://doi.org/10.1038/ngeo2253

Shukla PR, Skea J, Calvo Buendia E et al (2019) Climate change and land: an IPCC special report on climate change, desertification, land degradation, sustainable land management, food security, and greenhouse gas fluxes in terrestrial ecosystems. https://www.ipcc.ch/srccl/. Accessed 14 Dec 2022

Sun Q, Zwiers F, Zhang X et al (2020) A comparison of intra-annual and long-term trend scaling of extreme precipitation with temperature in a large-ensemble regional climate simulation. J Clim 33:9233–9245. https://doi.org/10.1175/JCLI-D-19-0920.1

Teichmann C, Jacob D, Remedio AR et al (2021) Assessing mean climate change signals in the global CORDEX-CORE ensemble. Clim Dyn 57:1269–1292. https://doi.org/10.1007/s00382-020-05494-x

Teodoro TA, Reboita MS, Llopart M et al (2021) Climate change impacts on the South American Monsoon System and its surface-atmosphere processes through RegCM4 CORDEX-CORE projections. Earth Syst Environ 5:825–847. https://doi.org/10.1007/s41748-021-00265-y

Trenberth KE (2020) ENSO in the global climate system. In: McPhaden MJ, Santoso A, Cai (eds) El Niño southern oscillation in a changing climate. Wiley, New York, pp 21–37

Trenberth KE, Fasullo JT, Shepherd TG (2015) Attribution of climate extreme events. Nat Clim Change 5:725–730. https://doi.org/10.1038/nclimate2657

Trenberth KE, Branstator GW, Karoly D et al (1998) Progress during TOGA in understanding and modeling global teleconnections associated with tropical sea surface temperatures. J Geophys Res—Oceans 103:14291–14324. https://doi.org/10.1029/97JC01444

Vendruscolo J, Perez Marin AM, Santos Felix E et al (2021) Monitoring desertification in semiarid Brazil: using the Desertification Degree Index (DDI). Land Degrad Dev 32:684–698. https://doi.org/10.1002/ldr.3740

Van Vuuren DP, Stehfest E, Den Elzen MGJ, et al (2011) RCP2.6: Exploring the possibility to keep global mean temperature change below 2°C. Clim Change. https://doi.org/10.1007/s10584-011-0152-3

Vieira RMSP, Sestini MF, Tomasella J et al (2020) Characterizing spatio-temporal patterns of social vulnerability to droughts, degradation and desertification in the Brazilian northeast. Environ Sustain Indic. https://doi.org/10.1016/j.indic.2019.100016

Wang C, Deser C, Yu JY et al (2017) El Niño and Southern Oscillation (ENSO): a review. In: Glynn PW, Manzello DP, Enochs IC (eds) Coral reefs of the eastern tropical Pacific. Springer, Amsterdam, pp 85–106

Wang XY, Li X, Zhu J et al (2018) The strengthening of Amazonian precipitation during the wet season driven by tropical sea surface temperature forcing. Environ Res Lett 13:94015. https://doi.org/10.1088/1748-9326/aadbb9

Zhang W, Jiang F, Stuecker MF et al (2021) Spurious North Tropical Atlantic precursors to El Niño. Nat Commun. https://doi.org/10.1038/s41467-021-23411-6

Chapter 3
Mangrove Swamps of Brazil: Current Status and Impact of Sea-Level Changes

Pedro Walfir Martins e Souza-Filho, Cesar Guerreiro Diniz, Pedro Walfir Martins e Souza-Neto, João Paulo Nobre Lopes, Wilson Rocha da Nascimento Júnior, Luiz Cortinhas, Nils Edvin Asp, Marcus Emanuel Barroncas Fernandes, and José Maria Landim Dominguez

Abstract Mangrove swamps are found in intertidal zones along tropical and subtropical regions around the world. The spatial distribution of these coastal swamps is initially controlled by pristine landscape morphology and coastal processes related mainly to tidal ranges, currents, and salinity. This chapter presents the subdivision of the Brazilian mangrove coast based on geological, morphological, oceanographic and climatic characteristics; their spatiotemporal stability and changes in area from 1985 to 2020; sea-level changes and mangrove sedimentation evolution during the late Quaternary; and the impact of future sea-level rise. Brazilian mangroves represent the second largest area worldwide in a single country, and even in the face of a climate change scenario and a trend of relative sea-level rise over the last century, their forest area increased by 2.5%. The paleoenvironmental record of sea-level changes during the late Quaternary indicates that mangrove shorelines have a broad capacity to adjust to sea-level changes even under trajectories of sea-level rise and fall. During this century, the tendency of sea-level rise under a scenario of global warming is clear, resulting in at least four models of mangrove sedimentation adjustment based on their response to changes in hydrogeomorphic processes and pristine geological conditions.

P. W. M. e Souza-Filho (✉) · J. P. N. Lopes (✉) · W. R. da Nascimento Júnior (✉)
Vale Institute of Technology, Belém, Pará, Brazil
e-mail: pedropwm@gmail.com

P. W. M. e Souza-Filho
Federal University of Pará (UFPA), Belém, Pará, Brazil

C. G. Diniz · L. Cortinhas
Solved—Solutions in Geoinformation, Belém, Pará, Brazil

P. W. M. e Souza-Neto
University of São Paulo (USP), São Paulo, São Paulo, Brazil

N. E. Asp · M. E. B. Fernandes
Federal University of Pará (UFPA), Bragança, Pará, Brazil

J. M. L. Dominguez
Federal University of Bahia (UFBA), Salvador, Bahia, Brazil

J. M. Landim Dominguez et al. (eds.), *Tropical Marine Environments of Brazil*, The Latin American Studies Book Series, https://doi.org/10.1007/978-3-031-21329-8_3

Keywords Coastal environments · Tidal flat · Quaternary evolution · Climate change

3.1 Introduction

Mangroves comprise mainly trees and shrubs that colonize intertidal flats in tropical and subtropical regions of the world (Perry et al. 2008). They include vegetation uniquely adapted to tidal conditions such as brackish and saltwater, periodic inundation and exposure, waves and wind, strong currents and runoff, and fine sediments (Duke et al. 1998; Tomlinson 2016). In the tidal flat zone, mangroves take the form of distinctly vegetated structures ranging from sparse shrubs to verdant closed canopy forests that occupy shorelines and upstream estuaries. Worldwide, mangroves cover a total area of ~137,700 km^2 distributed in 118 countries and territories (Giri et al. 2011; Bunting et al. 2018).

Many mangrove typologies occupy coastal plains, including deltaic, estuarine, lagoonal and open coast depositional landforms (Worthington et al. 2020). In this context, mangroves are generally limited to tidal flats that always present a narrow elevation change close to within the mean tidal level range (McKee et al. 2012). The extent of mud, mixed sediment and sand intertidal flats are conditioned by the vertical tidal range, referred to as accommodation space, whose sedimentary processes are a function of the stage of sea-level change, or maturity of the geomorphological setting, of these larger depositional systems (Woodroffe et al. 2016). The distribution of mangroves over time is controlled by the interplay of sea-level history and sediment supply, resulting in retrogradation, aggradation and progradation of coastal environments (Woodroffe 2019); vegetative stabilization; hydrology; and hydrochemical processes (Semeniuk 2018).

Along the 10,959 km of the Brazilian shoreline (IBGE 2020), mangroves present a discontinuous distribution from Orange Cape in Amapá State, latitude 4.45° N, to Laguna in Santa Catarina State, latitude 28.65° S, comprising 17 states (Fig. 3.1). The most recent mapping of Brazilian mangrove forests and shrubs quantified an area of 9627 km^2 in 2000 (Giri et al. 2011), 11,144 km^2 in 2008 (Magris and Barreto 2010), 11,072 km^2 in 2010 (Bunting et al. 2018), and 9900 km^2 in 2018 (Diniz et al. 2019). It is important to emphasize that these authors used different methods of digital image processing and mangrove mapping, which preclude accurate intercomparison of these figures. According to Bunting et al. (2018), Brazilian mangroves represent the second largest area worldwide in a single country, equivalent to 8.1% of the global mangrove area.

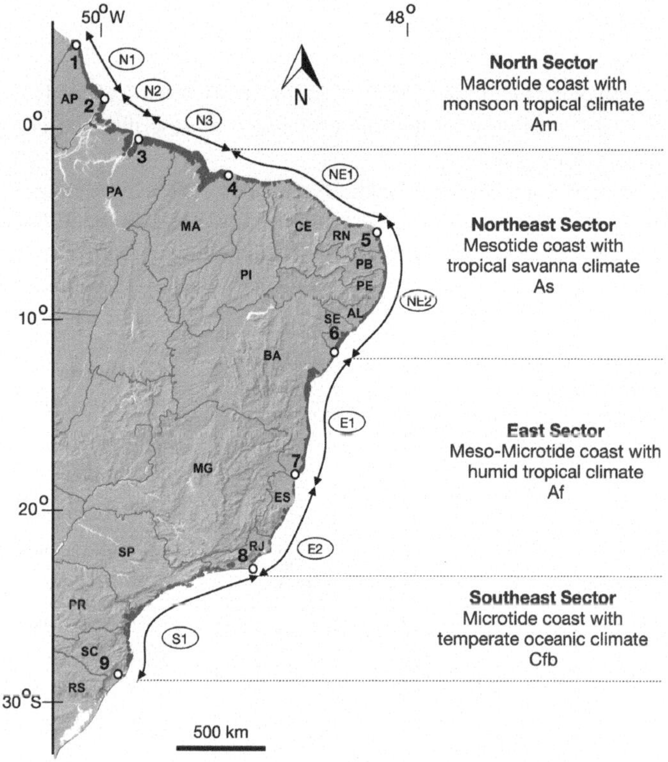

Fig. 3.1 Mangrove distribution and sectorization of the Brazilian coast based on geological, morphological, oceanographic, and climatic characteristics. N1—Subsector North 1; N2—Subsector North 2; N3—Subsector North 3; NE1—Subsector Northeast 1; NE2—Subsector Northeast 2; E1—Subsector East 1; E2—Subsector East 2; and SE1—Subsector Southeast 1. The numbers indicate the geographic boundaries of the coastal subsectors: 1—Orange Cape; 2—Norte Cape; 3—Marajó Bay; 4—Santo Amaro City, Lençois Maranhenses; 5—Calcanhar Cape; 6—Estuary of Real River; 7—Estuary of Mucuri River; 8—Arraial do Cabo; and 9—Santo Antônio Lagoon. AP—Amapá; PA—Pará; MA—Maranhão; PI—Piauí; CE—Ceará; RN—Rio Grande do Norte; PB—Paraíba; PE—Pernambuco; AL—Alagoas; SE—Sergipe; BA—Bahia; ES—Espírito Santo; RJ—Rio de Janeiro; SP—São Paulo; PR—Paraná; SC—Santa Catarina; and RS—Rio Grande do Sul

3.2 Environmental Information

To better understand the influence of geology, climate, and oceanographic conditions on mangrove distribution along the Brazilian coast, regional data and maps are required to investigate spatial variability in mangrove distribution. In this section, we present the sources of the environmental data used in this chapter.

3.2.1 Geology

The third edition of the Geological Map of South America at a scale of 1:5 M (Gomes Tapias et al. 2019) was used as the main reference to describe the geology of the Brazilian coastal zone (Fig. 3.2).

The coastal zone along the North Sector is marked by extensive Cenozoic deposits, while in the Northeast sector, these deposits occupy a narrow band bordering the coastline. In the East sector, Cenozoic and Mesozoic sedimentary rocks and Archaean high-grade metamorphic rocks crop out along the coast. The Southeast sector is

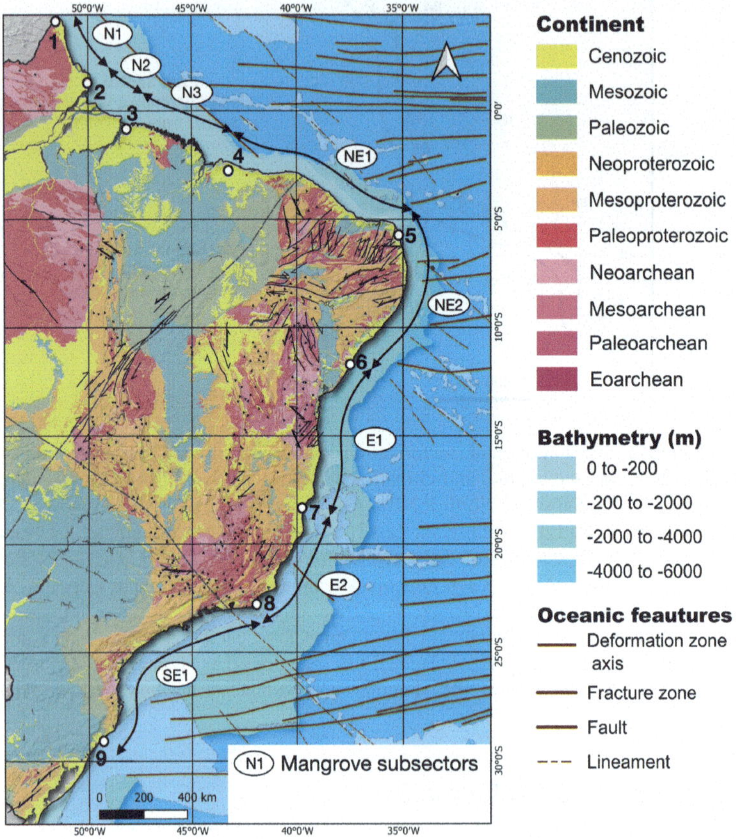

Fig. 3.2 Geological map of Brazil (Gomes Tapias et al. 2019) with indication of mangrove subsectors. N1—Subsector North 1; N2—Subsector North 2; N3—Subsector North 3; NE1—Subsector Northeast 1; NE2—Subsector Northeast 2; E1—Subsector East 1; E2—Subsector East 2; and SE1—Subsector Southeast 1. The numbers indicate the geographic boundaries of the coastal subsectors: 1—Orange Cape; 2—Norte Cape; 3—Marajó Bay; 4—Santo Amaro City, Lençois Maranhenses; 5—Calcanhar Cape; 6—Estuary of Real River; 7—Estuary of Mucuri River; 8—Arraial do Cabo; and 9—Santo Antônio Lagoon

characterized by outcrops of Proterozoic metamorphic rocks bordering the coastal zone.

3.2.2 Climate

Figure 3.3 shows the Köppen climate class distribution for Brazil from Alvares et al. (2013). Air temperature, precipitation, evaporation, and relativity humidity over the continent are shown in Fig. 3.4.

The climate in the North sector is defined as tropical monsoon (Am) (Fig. 3.3), with average annual precipitation higher than 1800 mm, air temperature ranging from 24 to 33 °C, evaporation between 1000 and 1600 mm per year, and air relativity humidity between 80 and 90%. Figure 3.4 shows the spatial distribution of these meteorological variables. The Northeast sector is characterized as tropical savanna (As) (Fig. 3.3), with average annual precipitation ranging from 800 to 1600 mm, air temperature ranging from 24° to 35 °C, evaporation between 1600 and 2800 mm per year, and air relativity humidity between 70 and 85% (Fig. 3.4). In the East sector, the climate is defined as humid tropical without a dry season (Af) (Fig. 3.3), with average annual precipitation ranging from 1000 to 1600 mm, air temperature ranging from 20 to 33 °C, evaporation between 1000 and 1600 mm per year, and air relativity humidity between 75 and 85% (Fig. 3.4). The climate in the Southeast sector is defined as temperate oceanic (Cfb) (Fig. 3.3), with average annual precipitation ranging from 1200 to 2000 mm, air temperature ranging from 18 to 29 °C, evaporation between 800 and 1400 mm per year, and air relativity humidity between 75 and 85% (Fig. 3.4).

3.2.3 Oceanographic Parameters

Figure 3.5a shows the spatial variation in the M2 tidal amplitude from the AVISO Global Tide Model with a horizontal resolution of 0.0625°. Figure 3.5b shows the spatial variation in the significant wave height generated from the ECMWF (ERA5) reanalysis for the global climate and weather for the 1979 to 2021 period (Hersbach et al. 2020).

Figure 3.6 shows the distribution of mean sea surface temperature, salinity, chlorophyll and particulate organic carbon (POC) for the western South Atlantic Ocean. The World Ocean Atlas 2018 (WOA2018) was used for sea surface temperature and salinity at the ocean surface (Locarnini et al. 2019; Zweng et al. 2018). This climatology used in situ measurements to generate a mean field for the period between 1955 and 2017 with a horizontal resolution of 0.25°. POC from 1997 to 2021 (Stramski et al. 2008) and chlorophyll type 2 from 2002 to 2012 (Doerffer and Schiller 2007) were derived from the Global Color project – GLOBCOLOR (https://www.globcolor.info/).

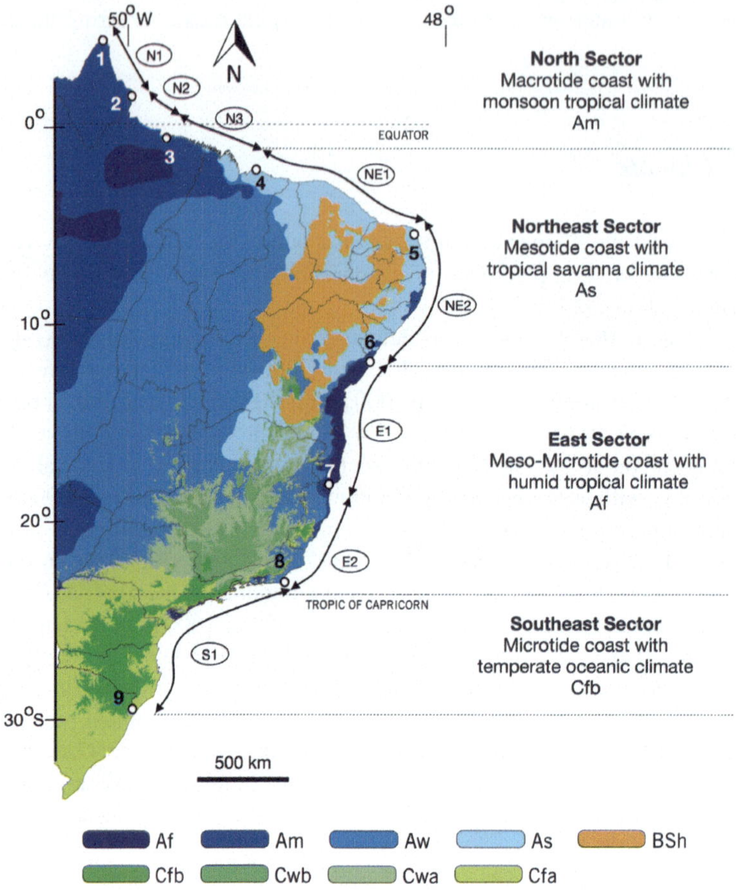

Fig. 3.3 Köppen climate classification map for Brazil (Alvares et al. 2013) with indication of mangrove subsectors. N1—Subsector North 1; N2—Subsector North 2; N3—Subsector North 3; NE1—Subsector Northeast 1; NE2—Subsector Northeast 2; E1—Subsector East 1; E2—Subsector East 2; and SE1—Subsector Southeast 1. The numbers indicate the geographic boundaries of the coastal subsectors: 1- Orange Cape; 2—Norte Cape; 3—Marajó Bay; 4—Santo Amaro City, Lençois Maranhenses; 5—Calcanhar Cape; 6—Estuary of Real River; 7—Estuary of Mucuri River; 8—Arraial do Cabo; and 9—Santo Antônio Lagoon

The North sector is characterized by a macrotidal regime with spring tides ranging from 3 m to 6.6 m and a mean significant wave height reaching up to 0.5 m (Fig. 3.5). The average sea surface temperature is higher than 27 °C, with higher concentrations of chlorophyll *a* (> 1300 mg.m^{-3}), salinity is lower than 35, and POC is higher than 40 mol.m^{-3} (Fig. 3.6).

The Northeast sector has a mesotidal regime with a spring tidal range reaching up to 3.8 m and a mean significant wave height reaching 1.2 m (Fig. 3.5). The sea surface temperature is higher than 27 °C and the concentrations of chlorophyll *a*,

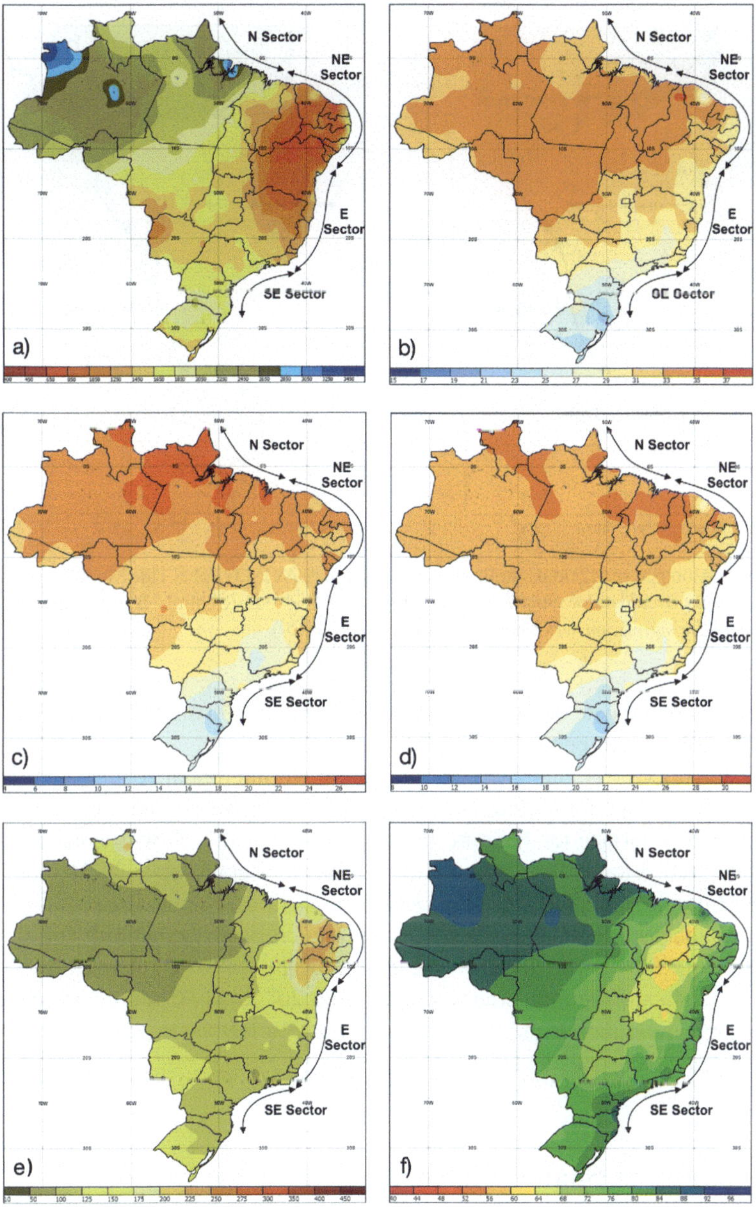

Fig. 3.4 Climatological normals for Brazil (1990–2020). **a** Accumulated average annual rainfall in mm, **b** maximum annual air temperature in °C, **c** minimum annual air temperature in °C, **d** average annual air temperature in °C, **e** accumulated average annual evaporation in mm, and **f** average annual relative humidity in %. Data source: Brazilian Institute of Meteorology—INMET, (https://clima.inmet.gov.br/NormaisClimatologicas/)

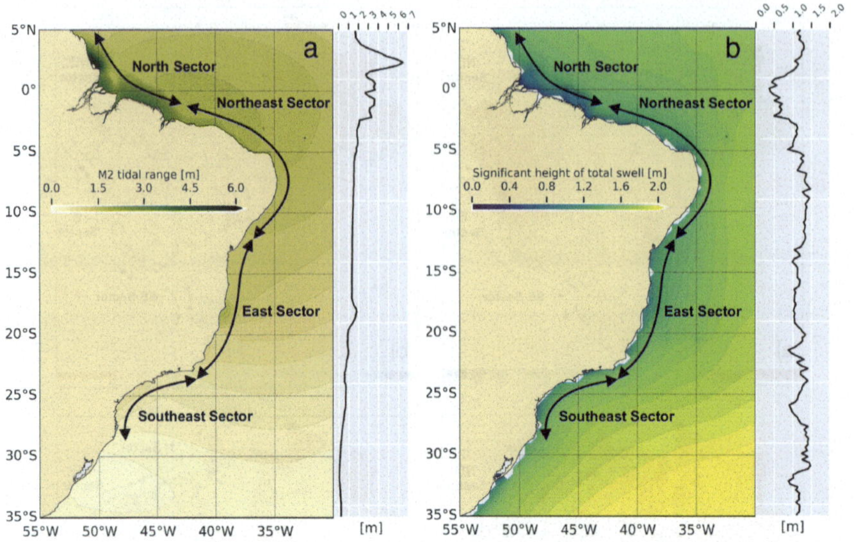

Fig. 3.5 **a** Alongshore M2 tidal amplitude variation *Source* AVISO Global Tide Model https://www.aviso.altimetry.fr/en/data/products/auxiliary-products/global-tide-fes.html. **b** Alongshore spatial variation in the significant wave height (Hersbach et al. 2020)

salinity, and particulate organic carbon range from 8 to 100 mg.m^{-3}, 24 to 37.5, and 20 to 30 mol.m^{-3}, respectively (Fig. 3.6).

The East Sector is characterized by a meso- to microtidal regime with spring tides ranging from 1.6 to 2.9 m and a mean significant wave height reaching 1.5 m (Fig. 3.5). Sea surface temperature ranges from 21° to 27 °C with concentrations of chlorophyll *a* lower than 100 mg.m^{-3}, salinity of approximately 36, and POC ranging from 20 to 50 mol.m^{-3} (Fig. 3.6).

The Southeast sector is characterized by a microtidal regime with spring tides ranging from 0.8 to 1.7 m and a mean significant wave height reaching 1.5 m (Fig. 3.5). Sea surface temperature ranges from 20° to 24 °C with concentrations of chlorophyll *a*, salinity, and POC ranging from 8 to 100 mg.m^{-3}, 35 to 36, and 20 to 30 mol.m^{-3}, respectively (Fig. 3.6).

3.3 Mangrove Spatial Distribution Along the Brazilian Coast

The mangrove cover area used in this chapter was generated from the Mapbiomas Project (https://mapbiomas.org). The mangrove extension was mapped in the Google Earth Engine (GEE) managed pipeline to compute the annual status of Brazilian mangroves from 1985 to 2020. Mangrove forests were identified from the use of

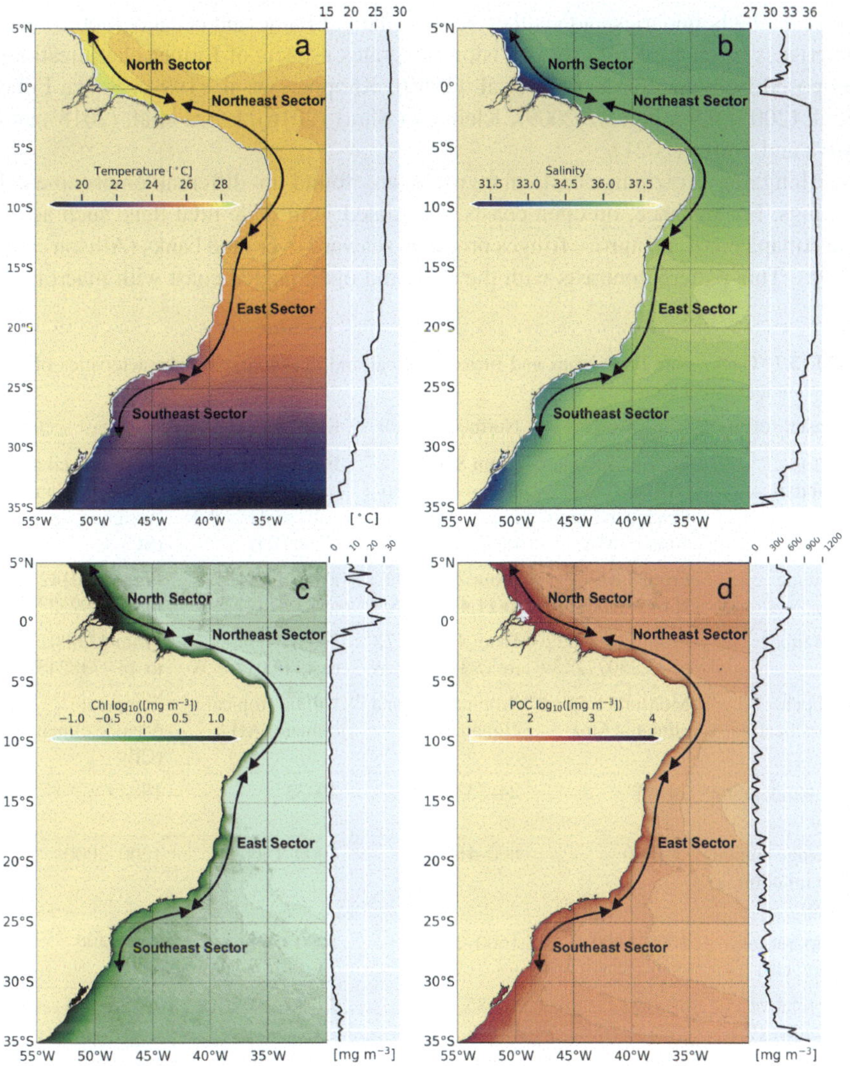

Fig. 3.6 Spatial distribution of: **a** sea surface temperature. **b** Salinity. **c** Chlorophyll *a*. **d** Particulate organic carbon along the Brazilian coast. *Data Source* **a**, **b**) https://www.ncei.noaa.gov/products/world-ocean-atlas; **c**, **d**) https://www.globcolor.info/

a spectral index, the Modular Mangrove Recognition Index (MMRI), which was specifically designed to better discriminate mangrove forests from the surrounding vegetation (Diniz et al. 2019).

The integration of mangrove mapping with geological, oceanographic, and climatic characteristics allowed us to divide the Brazilian coast into five large sectors: North, Northeast, East, Southeast, and South (Figs. 3.1, 3.2, and 3.3). Table 3.1

presents the boundaries and synthesizes the main climatic and oceanographic characteristics of each sector. This division integrates aspects of former classifications proposed by Schaeffer-Novelli et al. (1990), Knoppers et al. (1999), Souza-Filho et al. (2005), Dominguez (2009), Klein and Short (2016), Lessa et al. (2018), and Silva and Torres (2021).

Along the Brazilian coast, mangroves are found in different environmental settings. For instance, on open coasts, associated with wide tidal flats, such as on the Amapá coast, mangrove fringes prograde seaward over mud banks (Allison et al. 1995). This pattern contrasts with the sheltered open jagged coast with macrotidal

Table 3.1 Geographic boundaries and meteorological and oceanographic characteristics of the Brazilian mangrove sectors

Coastal sectors	North sector	Northeast sector	East sector	Southeast sector
Sector boundaries	From Oiapoque (AP) to Cabo Orange to Santo Amaro (MA)	From Santo Amaro (MA) to the estuary of the Real River (SE)	From the estuary of the Real River (SE) to Arraial do Cabo (RJ)	From Arraial do Cabo (RJ) Santo Antônio Lagoon (SC)
Latitude	From 3.847621° N to 2.498599° S	From 2.498599° to 11.44249° S	From 11.44249° to 23.014227° S	From 23.014227° to 28.519253° S
Longitude	From 51.836322° to 43.250772° W	From 43.250772° to 37.340266° W	From 37.340266° to 42.000135° W	From 42.000135° to 48.773824°W
Köppen Classification	Monsoon tropical climate (Am)	Tropical savanna climate (As)	Humid tropical climate (Af)	Temperate oceanic climate (Cfb)
Air temperature (°C)	24–33°	24–35°	20–33°	18–29°
Average annual precipitation (mm)	> 1800	800–1660	1000–1600	1200–2000
Evaporation (mm/year)	1000–1600	1600–2800	1000–1600	800–1400
Air relative humidity (%)	80–90	70–85	75–85	75–85
Tidal range (m)	3.5–6.6	2.2–3.8	1.6–2.9	0.8–1.7
Significant wave height (m)	Up to 0.5	Up to 1.2	Up to 1.5	Up to 1.5
Sea surface temperature (°C)	> 27°	> 27°	21–27°	20–24°
Chlorophyll *a* ($mg.m^{-3}$)	> 1300	8–100	< 100	8–100
Salinity	< 35	24–37.5	36	35–36
Particulate organic carbon ($mol.m^{-3}$)	> 40	20–30	20–50	20–30

estuaries, characteristic of the Pará and Maranhão states, where mangroves developed on the largest tidal mudflats of Brazil, behind dune-beach sandy barriers (Souza-Filho et al. 2009). Estuarine mangroves occur in northeastern and southeastern Brazil along the margins of tidal channels located in coastal embayments (Rodrigues 2014), while mangroves in lagoons and bays have developed behind spits and other sandy barriers, such as Sepetiba Bay in Rio de Janeiro State (Dadalto et al. 2022). Mangroves also occur in delta plains, for example, in the São Francisco delta, behind sandy barriers (Dominguez and Guimarães 2021). The following sections present detailed descriptions of the major mangrove sectors present along the Brazilian coast, emphasizing the long-term sedimentation patterns and their stratigraphic framework and showing the five major types of mangroves occurring in Brazil: open coast, sheltered open coast, estuary, delta, and bay-lagoon. Although these five categories do not encompass all types of occurrences, they represent the most important environments where mangroves are found along the coast of Brazil.

3.3.1 The North Sector

This sector, on the Amazonian coast, is 4048 km long and forms an embayed shoreline and runs from the estuary of the Oiapoque River on the Orange Cape, Amapá State, to Santo Amaro City, in the western boundary of the Lençóis Maranhenses dune field in Maranhão State (Fig. 3.1). This is a subsiding coast (Rossetti 2014; Souza-Filho 2000), and it is marked by the presence of the Amazon River mouth and 24 other estuaries. Although this sector is subjected to rising sea levels, the coastal zone is prograding (Souza-Filho et al. 2009).

Three subsectors were identified (Fig. 3.7):

Subsector N1 extends from the mud banks of Orange Cape to Norte Cape located in the North Channel of the Amazon River (Fig. 3.7). This coastal region is strongly influenced by the Amazon sediment plume, which extends northwestward, creating extensive mud flats prograding in an open coast (Fig. 3.7a) dissected mainly by estuaries draining a wet-muddy-humid hinterland (Allison et al. 1995; Batista et al. 2009). According to Schaeffer-Novelli et al. (1990), this subsector is characterized by homogeneous forests dominated by black mangrove, *Avicennia,* where *Pterocarpus* is sometimes found. The mangrove trees grow on an open coast forming an extensive and continuous mangrove belt 2 km wide, extending up to 30 km upstream in the estuaries. *Rhizophora* occurs in the estuarine portions of rivers close to the shoreline.

Subsector N2 is bounded by the mouths of the Amazon and Tocantins Rivers and Marajó Bay, comprising Marajó Island and a cluster of low-lying islands in the Amazon River mouth. This subsector presents evidence of tectonic activity contemporaneous with sediment deposition during the mid- to late Quaternary (Rossetti et al. 2008). Mangrove vegetation is restricted to muddy sandflats around these islands, forming fringes up to 4 km in width (França and Souza-Filho 2003, 2006; França et al. 2007). Mangroves occur mixed with freshwater swamp vegetation due to the high discharges of the Amazon and Tocantins Rivers (Schaeffer-Novelli et al. 1990).

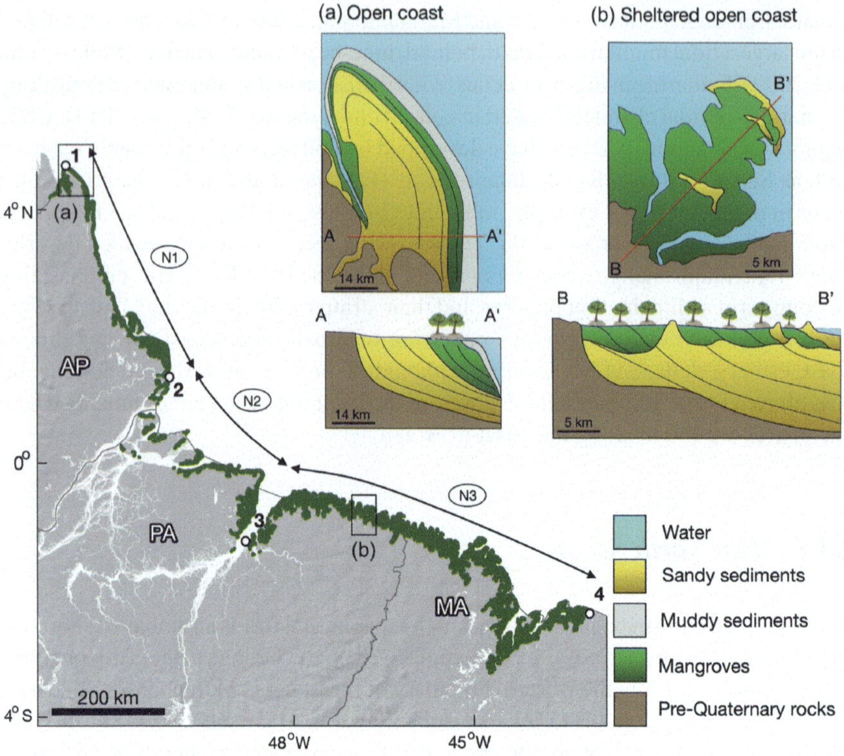

Fig. 3.7 The Brazilian Mangrove North Sector also depicts sedimentary environments and stratigraphic frameworks reflecting long-term sedimentation patterns. **a** open coast mangrove type typical of the Orange Cape mud flats common in Subsector N1. **b** sheltered open coast mangrove type typical of the macrotidal sandy dune-beach barriers of the Bragança coastal plain in Subsector N3. 1—Orange Cape; 2—North Cape, 3—Marajó Bay, and 4—Santo Amaro City

Avicennia germinans occur at higher elevations and lower salinity areas. Low salt tolerance might explain the restriction of *R. racemosa* and *R. harrisonii* to the Marajó Bay region (Menezes et al. 2008).

Subsector N3 comprises the Amazon macrotidal mangrove coast (Souza-Filho 2005) and extends from Marajó Bay to Lençóis Maranhenses (Fig. 3.7). Seventeen macrotidal estuaries occur along this embayed and jagged coast, comprising the largest continuous mangrove belt in the world with an area of approximately 7500 km^2. In this region, the mangrove belt reaches up to 23 km in width (Nascimento et al. 2013; Souza-Filho 2005). This subsector is characterized by wide intertidal mud and sandflats developed under a subsiding macrotidal barrier estuarine system (Souza-Filho et al. 2009), characterizing a sheltered open coast (Fig. 3.7b). *Rhizophora mangle* is the most widely distributed mangrove species, dominating upstream riverbanks and back basins in macrotidal estuaries and downstream riverbanks and open coast fringe forests over muddy tidal flats. *Rhizophora mangle* stands are backed by *Avicennia germinans* and *A. schaueriana,* and *Laguncularia racemosa* stands at higher elevations and less inundated areas under less saline conditions (Menezes et al. 2008), while saline muddy sandflats are colonized by *Spartina* (Schaeffer-Novelli et al. 1990).

3.3.2 The Northeast Sector

The Northeast Sector is 2047 km long and extends from Santo Amaro City, Lençóis Maranhenses, to the estuary of the Real River, Sergipe State (Fig. 3.8). This Brazilian coastal sector has received the smallest volume of sediments in recent geological time as a result of the small size of the drainage basins, low relief and low precipitation (Dominguez 2009). Geomorphologically, the Northeast Sector is marked by coastal cliffs up to ~ 40 m high carved into the Miocene Barreiras Formation (Peulvast et al. 2006; Rossetti et al. 2013). Quaternary accumulations in this sector include active and inactive dune and transgressive aeolian sand sheets, Pleistocene and Holocene beach deposits, beach rocks and estuarine deposits (Bezerra et al. 2003; Barreto et al. 2006). This sector is marked locally by the presence of the deltas of the Parnaíba and São Francisco Rivers on the western and southern boundaries, respectively. Between these two deltas, 11 mesotidal estuaries occur (Lessa et al. 2018).

Two subsectors were identified in this sector (Fig. 3.8):

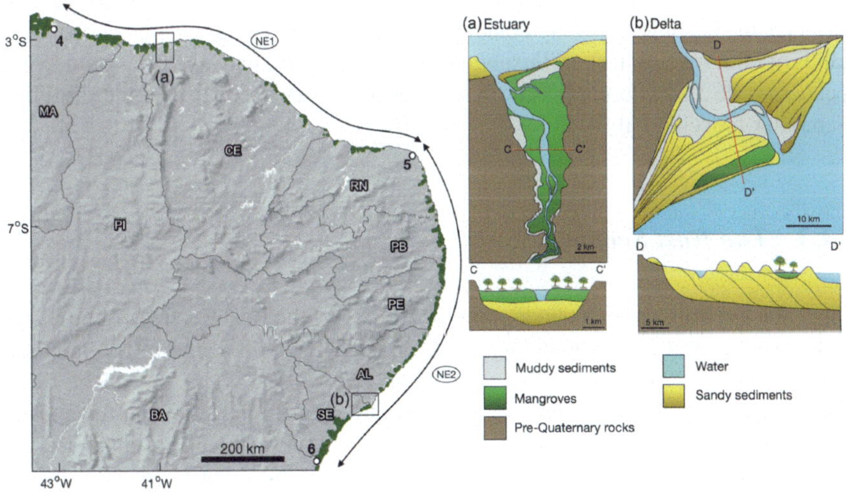

Fig. 3.8 The Brazilian Mangrove Northeast Sector also depicts sedimentary environments and stratigraphic frameworks reflecting long-term sedimentation patterns. **a** estuarine mangrove type model exemplified by the mesotidal estuarine system of the Acaraú River in Ceará State (Subsector NE1). **b** wave-dominated deltaic mangrove type exemplified by the São Francisco Delta (Subsector NE2). 4—Santo Amaro City, 5—Calcanhar Cape, and 6—Estuary of Real River

Subsector NE1 extends from Santo Amaro City to Calcanhar Cape in Rio Grande do Norte State (Fig. 3.8). The west–east-oriented subsector is marked by intense east–west littoral drifts with sand accumulating in narrow sandy barriers at the foot of sea cliffs and at the mouth of mesotidal estuaries located behind transgressive beach-dune fields (Hesp et al. 2009; Vital et al. 2016). Figure 3.8a illustrates the spatial distribution of this kind of mangrove depositional system developed under estuarine mesotidal conditions in the Acaraú River in Ceará State. Mangroves are poorly developed along this coast segment due to the lack of freshwater and prolonged droughts (Schaeffer-Novelli et al. 1990) and reduced sediment input to build a coastal plain. High salt concentrations limit mangrove development that is restricted to estuarine channels, dominated by *Rhizophora mangle* followed by *Avicennia schaueriana* (Costa et al. 2017).

Subsector NE2 extends from the Calcanhar Cape to the Real River estuary (Fig. 3.8). This north–south oriented subsector is also marked by sea cliffs carved into the Miocene Barreiras Formation (Rossetti et al. 2013) and extensive active and inactive transgressive dune fields (Bezerra et al. 2001). These dunes extend up to 3 km inland, reaching heights of up to 80–120 m. Beach rocks are discontinuously exposed along the shoreline. They are oriented nearly parallel to the shoreline, a few kilometers long, with thicknesses from 10 cm to 3 m and widths from 2 to 50 m. Most of them are located in the present intertidal zone, which is characterized by the presence of gently seaward-dipping stratification (Ferreira et al. 2018; Vieira et al. 2017). Due to the high wave energy in this coastal subsector, mangroves are restricted to estuarine channels and behind coastal barriers, such as the delta of the São Francisco River, where mangroves occur behind sand spits (Fig. 3.8b). In tidal flat deposits, basin forests may contain either *Avicennia* or *Laguncularia* or mixed stands of both species. Either *Rhizophora* or *Laguncularia* may appear as pioneering species, while *Spartina* may occur in some accretionary deposits near the shoreline (Schaeffer-Novelli et al. 1990).

3.3.3 The East Sector

This 1804 km long sector extends from the estuary of the Real River (Sergipe State) to the Arraial do Cabo promontory in Rio de Janeiro State (Fig. 3.9). According to Dominguez (2009), along this sector, the major escarpment typical of rifted passive continental margins has retreated hundreds of kilometers from the coastal zone, creating large drainage basins. Furthermore, this coastal sector is characterized by classical examples of wave-dominated deltas, such as Jequitinhonha, Doce and Paraíba do Sul (Dominguez et al. 1987). In between these major deltas, coastal cliffs carved into the Barreiras formation occur locally. This sector is marked by the presence of the Todos os Santos, Camamu and Vitória bays, in addition to the estuaries of smaller rivers (Lessa et al. 2018).

The east sector can be subdivided into two subsectors:

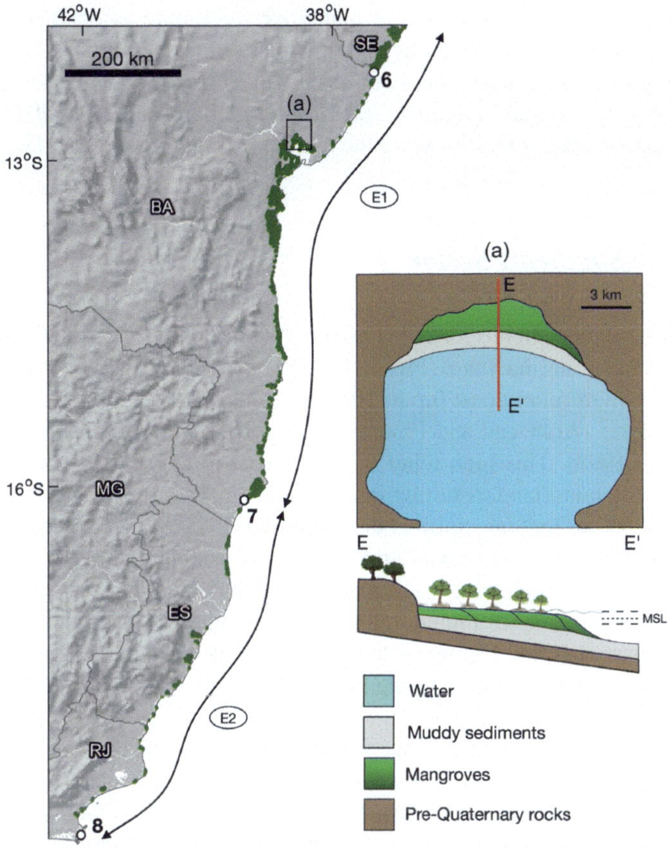

Fig. 3.9 The Brazilian Mangrove East Sector also depicts sedimentary environments and stratigraphic frameworks reflecting long-term sedimentation patterns. **a** bay-lagoon delta mangrove delta type exemplified by the mesotidal Todos os Santos Bay in Bahia State. 6—Estuary of the Real River, 7—Estuary of the Mucuri River and 8—Arraial do Cabo

Subsector E1 is situated between the estuaries of the Real and Mucuri Rivers (Fig. 3.9). Siliciclastic sediments occur along a narrow band bordering the shoreline extending to the 15 isobaths, while the rest of the shelf is characterized by carbonate sedimentation (Dominguez et al. 2013; Halla et al. 2020). Extensive mangroves occur associated with the estuaries of major rivers and strand plains and in the large bays present in this sector, such as Camamu and Todos os Santos Bay (Fig. 3.9a). This ecosystem has a fringe physiognomy with a low bearing that contains *Laguncularia racemosa*, *Rhizophora mangle* and *Avicennia schaueriana* (Freitas et al. 2002).

Subsector E2 extends from the estuary of the Mucuri River (Bahia State) to Arraial do Cabo (Rio de Janeiro State) (Fig. 3.9). This subsector is characterized by the two largest wave-dominated deltas of Brazil, the Doce and Paraíba do Sul deltas, with sea cliffs carved into the Barreiras Formation present in between (Dominguez et al. 1987).

Interestingly, no major occurrences of mangrove forests are associated with these two deltas that do not present significant intrusion of salt water into the channel. The largest mangrove forest of this subsector occurs in Vitória Bay, with the dominance of *Rhizophora mangle* and *Laguncularia racemosa* (Barbirato et al. 2021). In this sector, mangroves occur mainly along the margins of microtidal estuaries.

3.3.4 The Southeast Sector

The Southeast Sector extends for 2443 km from Arraial do Cabo (Rio de Janeiro State) to Laguna (Santa Catarina State) (Fig. 3.10). Geomorphologically, this sector is characterized as a high-relief coast (up to 1000 m), where the Mantiqueira and do Mar mountain ranges (Archaean and Proterozoic rocks) border the coastline (Almeida and Carneiro 1998). This high relief near the coast forces the major drainages to flow toward the hinterland, resulting in an overall small sediment load to the coast (Dominguez 2009). Tectonic uplift at the end of the Cretaceous followed by gravitational collapse produced a series of grabens oriented subparallel to the coast, which were flooded by the sea and created a number of bays and large estuaries, which characterizes this sector (Dominguez 2009). This Southeast Sector is marked by the presence of barrier-island lagoon systems and bays such as Guanabara, Santos and Paranaguá. The tidal flats occur inside bays and lagoons behind beach-dune barrier systems (Fig. 3.10a).

In this sector, the largest mangrove forests occur in Guanabara, Sepetiba, Santos and Paranaguá Bays. The mangrove fringes of these regions are dominated by *Rhizophora mangle,* followed by *Laguncularia racemosa*, *Avicennia germinas* and *A. schaueriana* (Schaeffer-Novelli et al. 1990). The Santo Antônio Lagoon located in the region of Laguna in Santa Catarina State is the southern limit of occurrence of mangroves in the western Atlantic Ocean (Soares et al. 2012).

3.4 Mangrove Changes from 1985 to 2020

Along the Brazilian coast, mangrove forests exhibited a net increase from 9564 km^2 in 1985 to 9800 km^2 in 2020, which represents just 2.5% of the total mangrove area. This total area increased by 6.4% from 1985 to 2002, decreasing afterwards by 3.8% from 2003 to 2020 (Fig. 3.11). Diniz et al. (2019) defined these two periods as accretionary and erosive, respectively. Here, we use this timeline to describe the mangrove dynamics along the coastal subsectors. Figure 3.11 presents the quantification of the mangrove area per sector for 1985, 2002, 2003, and 2020, whereas Fig. 3.12a presents the mangrove area changes per subsector for the same years. The stable mangrove area representing the unchanged area during the analysis period and mangrove area gain (expansion) and mangrove area loss (reduction) for the 1985–2002 and 2003–2020 periods are presented in Figs. 3.12b, c, respectively.

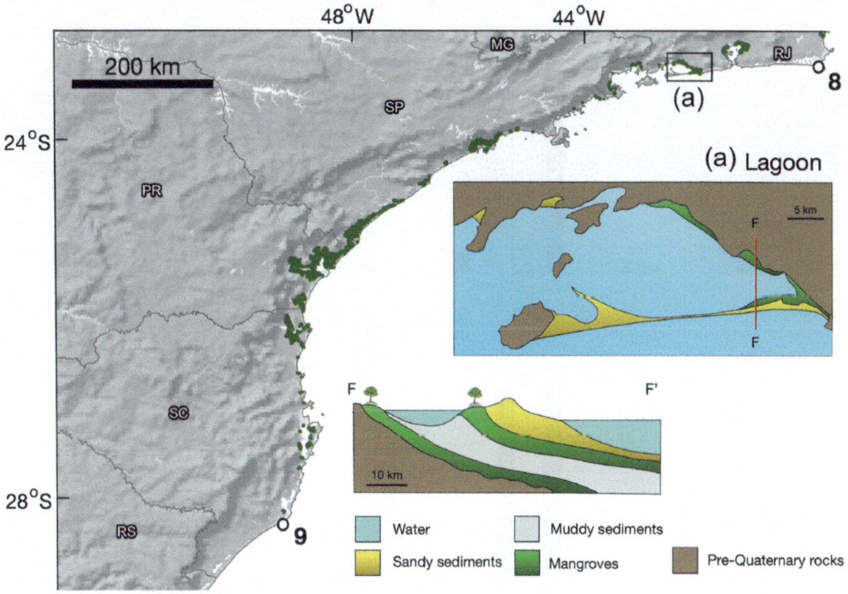

Fig. 3.10 The Brazilian Mangrove Southeast Sector also depicts sedimentary environments and stratigraphic frameworks reflecting long-term sedimentation patterns. **a** bay/lagoon mangrove type model exemplified by the Sepetiba barrier island system in the State of Rio de Janeiro. 8—Estuary of the Mucuri River and 9—Santo Antônio Lagoon in Santa Catarina State

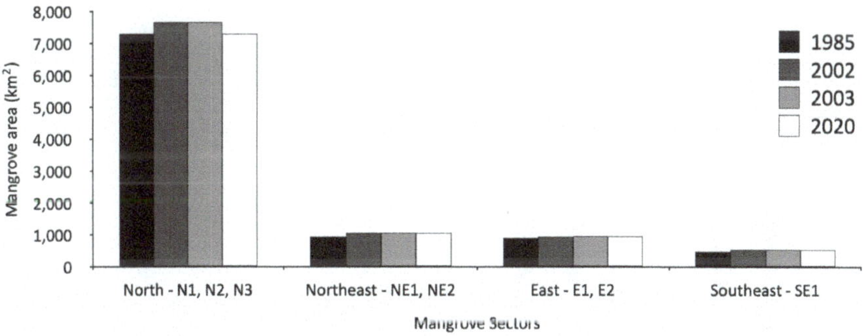

Fig. 3.11 Mangrove area per coastal sector for 1985, 2002, 2003, and 2020

In the North Sector, mangrove forest occupied an area of 7283 km^2 in 2020, which represents 74.3% of the Brazilian mangroves. From 1985 to 2020, the total mangrove area in this sector remained constant.

Subsector N1 comprised 8.4% of the Brazilian mangroves in 2020. From 1985 to 2020, mangrove forest decreased by 3.3%. The rate of change, however, was not uniform. From 1985 to 2002, the mangrove forest increased by 18.2%, and from 2003 to 2020, it decreased by 17.9%. Between 1985 and 2002, 665 km^2 of mangrove forest

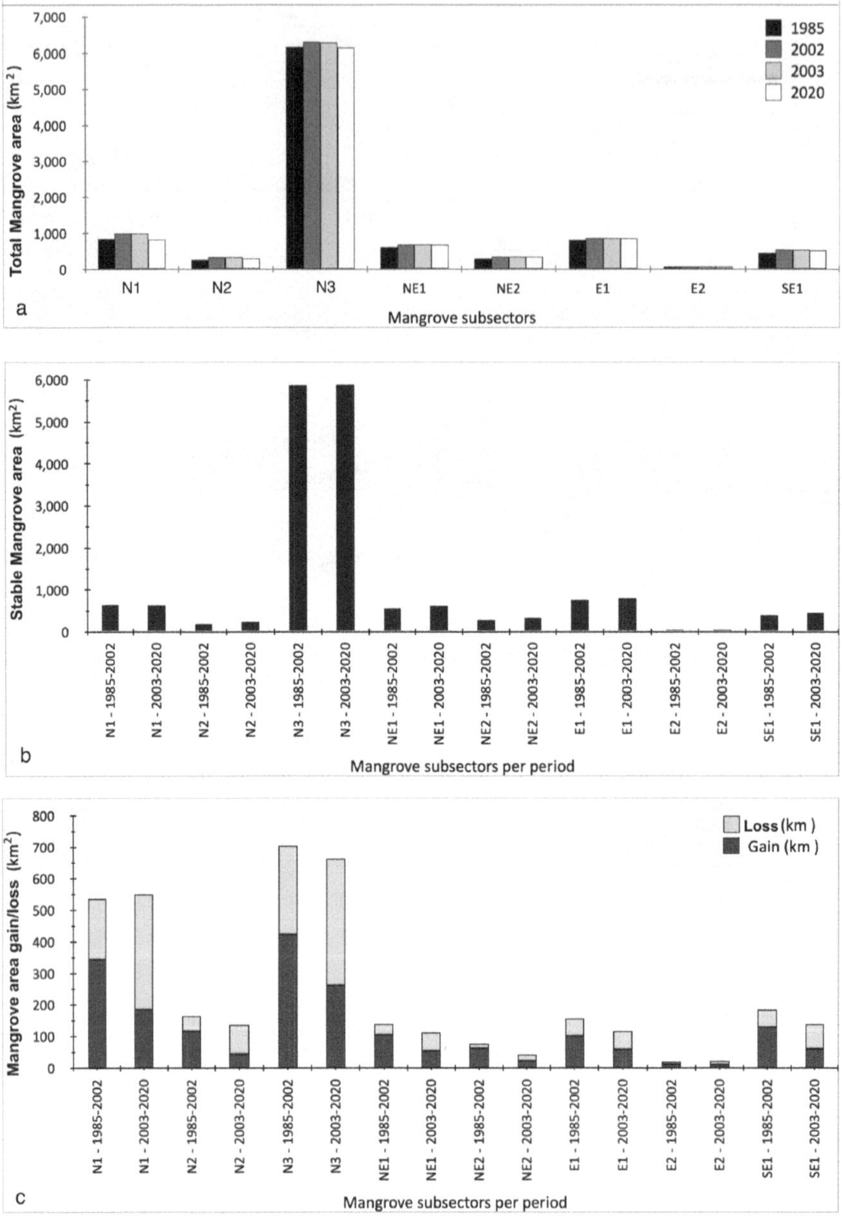

Fig. 3.12 **a** Total mangrove area per subsector for 1985, 2002, 2003, and 2020. **b** Stable mangrove area and **c** mangrove area gain and loss in the 1985–2002 and 2003–2020 periods

remained stable in this subsector with a gain of 345 km^2 and a loss due to coastal erosion of 190 km^2. However, from 2003 to 2020, the stable mangrove forest area remained at approximately 642 km^2 with gains of 185 km^2 and losses of 366 km^2 (Fig. 3.12).

Subsector N2 accounted for 3.1% of the Brazilian mangrove forest area in 2020. In this subsector, the mangrove area enlarged from 267.6 km^2 in 1985 to 300.5 km^2 in 2020, representing a net gain of 12.3%. Stable mangrove areas occupied 220 km^2 and 255 km^2 during the 1985–2002 and 2003–2020 periods, respectively. This subsector presents the same behavior as Subsector N1, with the largest gain occurring during 1985–2002 (117 km^2) and the largest loss occurring during the 2003–2020 period (91 km^2) (Fig. 3.12).

Subsector N3, which comprises the largest mangrove belt in the world (Souza-Filho 2005), had a total area of 6155 km^2 in 2020. This area represents 62.8% of the Brazilian mangrove forests. Along this sector, the mangrove area also decreased from 6160 km^2 in 1985 to 6155 km^2 in 2020, a reduction of 0.1%. The stable mangrove area was 5878 km^2 from 1985 to 2002 and 5892 km^2 from 2003 to 2020, with the largest gain occurring in the first period (422 km^2) and the largest loss in the second period (399 km^2) (Fig. 3.12).

In the Northeast Sector, mangrove areas totaled 1045 km^2 in 2020, which corresponds to 10.7% of the Brazilian mangroves. In general, the mangrove forest area increased by 13.1% from 1985 to 2020 (Fig. 3.11).

Subsector NE1 contained 7% of the total mangrove area of Brazil in 2020. In this subsector, the mangrove area increased continuously from 612.3 km^2 to 681.6 km^2 from 1985 to 2020, exhibiting a pattern completely different from that in the North Sector. From 1985 to 2002, 579.7 km^2 of mangrove area remained stable, with a gain of 105.4 km^2 and a loss of 33 km^2. However, from 2003 to 2020, the stable mangrove forest area increased to 627 km^2, with a gain of 54.3 km^2 and a loss of 58.3 km^2 (Fig. 3.12).

Subsector NE2 comprised 3.7% of the total Brazilian mangrove forests in 2020. Its mangrove area also increased continuously from 311.4 km^2 in 1985 to 363.4 km^2 in 2020, representing a gain of 15.8%. The stable mangrove area was 220 km^2 and 255 km^2 during the 1985–2002 and 2003–2020 periods, respectively. This subsector presented the same behavior as Subsector NE1, with a gain of 63.4 km^2 and a loss of 14.2 km^2 during the 1985–2002 period and similar gains (22.4 km^2) and losses (21 km^2) during the 2003–2020 period (Fig. 3.12).

The East Sector encompassed 946.2 km^2 of mangrove forest in 2020, representing 9.7% of the Brazilian mangroves. The mangrove area increased by 6.3% from 1985 to 2020, although a small decrease of 0.3% occurred from 2003 to 2020 (Fig. 3.11).

Subsector E1 comprised 8.9% of the Brazilian mangrove forests in 2020. In this subsector, the mangrove area increased from 820 km^2 in 1985 to 869.7 km^2 in 2020, representing a net gain of 6.1%. The stable mangrove area occupied 764.2 km^2 and 812 km^2 during the 1985–2002 and 2003–2020 periods, respectively. This subsector exhibited the same behavior as Subsectors NE1 and NE2, with gains (101 km^2) larger than losses (55.8 km^2) during 1985–2002 and equivalent gains (57.8 km^2) and losses (59 km^2) during 2003–2020 (Fig. 3.12).

Subsector E2 had the smallest mangrove area of Brazil, with 76.5 km^2 in 2020, representing only 0.8% of the total mangrove forest of Brazil. Along this sector, the mangrove area increased by 9.3% from 1985 to 2020. The stable mangrove area was similar during the two periods of investigation (63.3 km^2 during 1985–2002 and 66.2 km^2 during 2003–2020), with the largest gain (13.3 km^2) and smaller loss (6.6 km^2) occurring during 1985–2002 and similar gains (10.3 km^2) and losses (11.9 km^2) occurring during the 2003–2020 period (Fig. 3.12).

The Southeast Sector has an area of 525.6 km^2 of mangrove forest, which represents 5.4% of the total Brazilian area. From 1985 to 2020, the mangrove area increased from 467.4 km^2 in 1985 to 525 km^2 in 2020. However, from 2003 to 2020, a small decrease of 3.2% from 543 km^2 to 525.6 km^2 was documented. Between 1985 and 2002, the stable mangrove area totaled 412 km^2, with a gain of 129.2 km^2 and a loss of approximately 55 km^2. However, from 2003 to 2020, the mangrove stable area increased to 465.4 km^2, with gains of 60.2 km^2 and losses of 77.6 km^2 (Fig. 3.12).

Summing up the data presented above indicates that overall, the mangrove area in Brazil remained stable during the 1985–2020 period in all sectors, although it exhibited changes of a small magnitude. An increase of just 2.5% in the total mangrove area in Brazil was documented.

3.5 Impacts of Sea-Level Changes in the Brazilian Mangroves

3.5.1 Sea-Level Changes and Evolution of Mangrove Sedimentation Evolution During the Holocene

A number of papers have discussed the impact of future sea-level changes (Schaeffer-Novelli et al. 2002; Ellison 2015; Duncan et al. 2018) on the sedimentation, morphology, and development of mangrove ecosystems (Cohen et al. 2015; Woodroffe et al. 2016; Woodroffe 2019).

One way to investigate the impact of future sea-level changes in mangrove forests is to assess how tidal flats colonized by mangroves have responded to the rise in sea- level since the Last Glacial Maximum (LGM). Different responses of mangrove shorelines to sea-level change were discussed by Woodroffe (2018), including drowning, backstepping, catch-up, keep-up, progradation and emergence. The past trajectories of the mangrove shorelines depend on the Holocene sea-level history. Along the Atlantic South American coast, it is possible to observe two distinct Holocene sea-level histories: (i) a 3–4-m relative sea-level drop since the Mid-Holocene (Angulo et al. 2006) observed along most of the northeastern to southern Brazilian coast and (ii) a continuous sea-level rise during the Holocene along the northern coast of Brazil and the Caribbean (Khan et al. 2015) (Fig. 3.13). These differences are a direct effect of Greenland ice sheet (GIS) and Antarctic

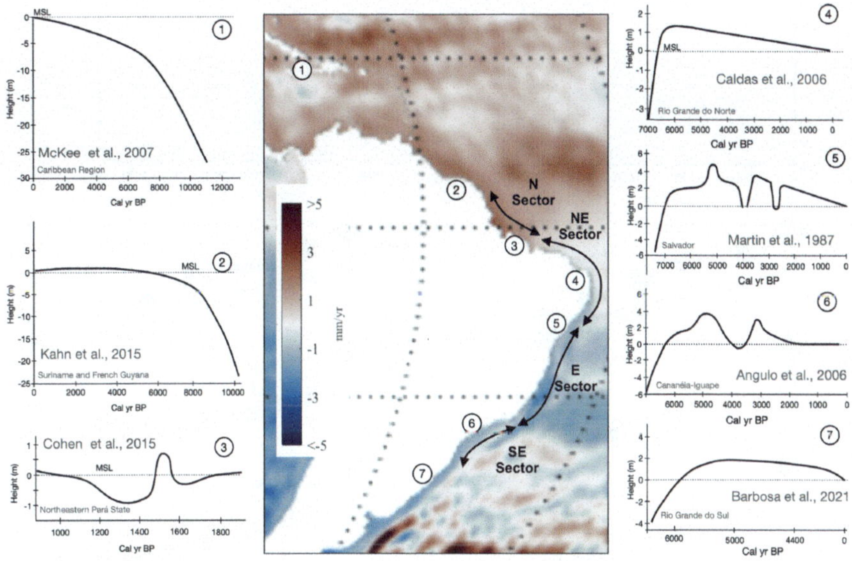

Fig. 3.13 Sea-level behavior from 1993 to 2015 (Elipot 2020) (central panel) and sea-level curves for the last 10,000 years for different regions of South America and the Caribbean. Modified from Martin et al. (1987), Angulo et al. (2006), Caldas et al. (2006), McKee et al. (2007), Cohen et al. (2015), Khan et al. (2015), and Barboza et al. (2021)

ice sheet (AIS) adjustments (Oppenheimer et al. 2019). These same behaviors are mimicked by the modern global mean sea-level (Elipot 2020) (Fig. 3.13).

The mangrove response and corresponding stratigraphy associated with these two sea-level histories observed along the Brazilian coast are illustrated in Fig. 3.14. In the North Sector of the Brazilian coast (Fig. 3.14a), transgressive muddy (Subsector N1) and sandy (Subsector N3) deposits record the rapid rise of sea-level from the LGM to the Mid-Holocene, when the first mangrove deposits accumulated in open muddy banks (Subsector N1) and sheltered tidal flats behind barrier islands (Subsector N3). In Subsector N3, under conditions of an almost stable sea level, successive barriers emerged during the Holocene, allowing for tidal flat progradation resulting in a continuous mangrove belt (Fig. 3.14a) up to 36 km wide (Souza-Filho et al. 2009).

In the Northeast, East and Southeast sectors of the Brazilian coast (Fig. 3.14b), maximum mangrove development occurred around the Mid-Holocene highstand when large bays and estuaries were present in the coastal zone (Dominguez et al. 1987). As a result of the 3–4 m drop in relative sea-level since that time, many of these bays, lagoons and estuaries were infilled, and the shoreline prograded, forming beach-ridge terraces (Andrade et al. 2003; Dominguez et al. 1987). The exposed estuarine and lagoon sediments were replaced by freshwater swamp environments (Fig. 3.14b). From that point on, mangrove development became restricted to sheltered areas behind narrow elongated barrier islands.

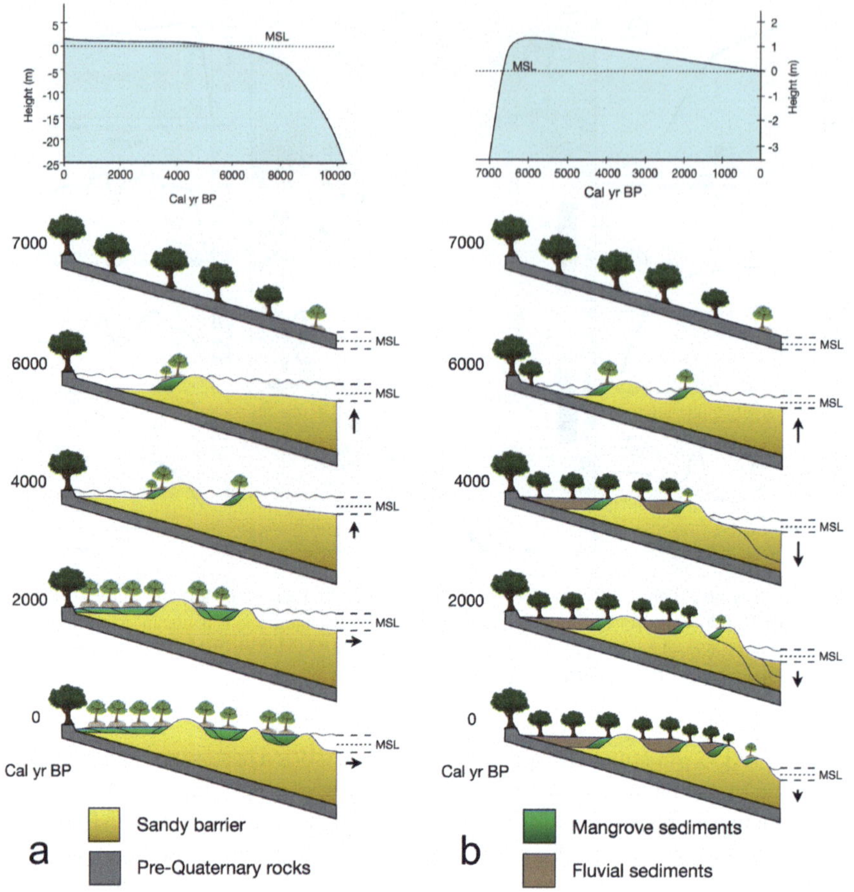

Fig. 3.14 Effects of different sea-level histories on the styles of mangrove sedimentation. **a** North Mangrove Sector and **b** Northeast, East, and Southeast Mangrove sectors

3.5.2 Impact of Future Sea-Level Rise in the Brazilian Mangroves

Mangrove sedimentation during the late Quaternary in response to postglacial sea-level rise brings important lessons as to how these ecosystems will respond to global warming and sea-level rise (SLR). Under the RCP8.5 scenario, the global mean sea-level (GMSL) will rise by 0.84 m (0.61–1.10 m) by 2100 (Oppenheimer et al. 2019). Beyond 2100, GMSL will continue to rise for centuries due to thermal expansion, mass loss of glaciers and ice sheets in the Greenland ice sheet (GIS) and Antarctic ice sheet (AIS) and land water storage changes.

The response of the mangrove coasts to this sea-level rise will depend on the environmental setting of the mangroves. Figure 3.15 presents four models of mangrove

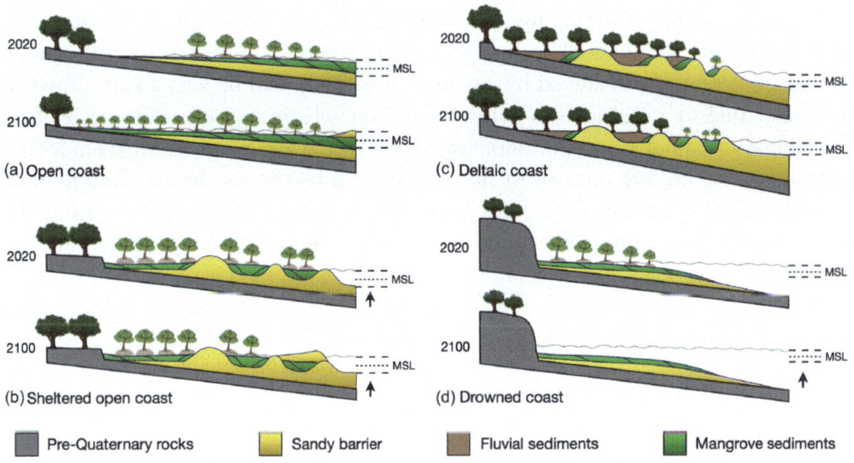

Fig. 3.15 Four response scenarios of mangrove sedimentation to ongoing sea-level rise

response to sea-level rise based on their past response to changes in hydrogeomorphic processes and pristine geological conditions, including changes in relative sea level.

On open coasts, a rise in sea-level will cause an increase in upstream penetration of the salt wedge and landward migration of mangroves along riverine and supratidal flats that are progressively converted to intertidal flats (Fig. 3.15a). Hence, mangrove zones migrate landward by seedling recruitment and vegetative reproduction as new brackish habitat becomes available landward through inundation and concomitant changes in salinity, as observed along the Australian coast (Semeniuk 1994). Erosive processes at the seaward front can result in mangrove loss along the shoreline (Allison et al. 2000; Anthony et al. 2010; Santos et al. 2016).

If sea-level is rising over a sheltered open coast densely colonized by mangroves and bounded landward by inactive cliffs, the muddy tidal flat will experience an elevation in the water level and a sedimentary aggradation process. Under this hydromorphic condition, the inner tidal flat may also expand laterally, prograding toward the estuarine channels and tidal creeks. The landward retreat of the shoreline due to the rising sea-level will result in barrier sand deposition over muddy flats and mangroves (Fig. 3.15b). This will cause coastal erosion, falling of trees, increased salinity, and hydroperiod, frequency, and depth of inundation in mangrove forests (Souza-Filho and Paradella 2003; Souza-Filho et al. 2006).

Along deltaic coasts (Fig. 3.15c), sea-level rise will result in increased salt wedge penetration and landward retreat of beach-dune ridges, occasionally burying tidal muddy flats colonized by mangroves. Mangrove survival will depend on the ability of mangrove species to colonize newly available habitats at a rate that keeps pace with the rate of sea-level rise. Hence, it might be possible to see a progressive landward migration of mangrove trees over deltaic flood plain deposits due to increased salinity.

Tropical tidal flats fringing high gradient land regions or human-made obstacles such as seawalls and other shoreline protection structures (Fig. 3.15d) will contain the mangroves most threatened by sea-level rise. They will be very likely drowned "in place" due to a lack of low-lying areas over which they can migrate. Hence, the mangroves situated in the Southeast Sector of the Brazilian coast, bounded by highland areas, are the ones most threatened by a rise in sea level. Nevertheless, the presence of construction, such as dams, roads, and seawalls, common along the Northeast and East sectors, may increase the vulnerability of mangroves to the rise in sea-level in those sectors (Lacerda et al. 2021).

3.6 Concluding Remarks

Brazil hosts the second largest mangrove area in the world. Mangroves develop around the mean sea-level and as such are potentially vulnerable to the impacts of the ongoing rise in sea level, which will certainly affect their distribution, structure, and function, and also impacting their dynamics, health and connectivity with adjacent systems.

Potential changes in the mangrove ecosystem will directly affect key factors that determine the survivability of mangrove forests. Such factors are related to abiotic processes, such as geomorphology and stratigraphy (Ellison and Stoddart 1991), topography, mineralogy and sedimentation (Woodroffe 1995), nutrient distribution and microclimate (Ellison 2000), and salt and freshwater in adjacent environments (Gilman et al. 2008).

Mangroves occupy almost the entire coastline of Brazil and developed under different environmental conditions. Four large coastal sectors were identified in which mangrove forests occur: North, Northeast, East, and Southeast (Fig. 3.1), each characterized by different geological-geomorphological, oceanographic, and climatic conditions.

Even in the face of a climate change scenario and a trend of sea-level rise over the last century, the Brazilian mangrove forest area has increased from 9564 km^2 in 1985 to 9800 km^2 in 2020, which represents an increase of 2.5%. In the North Sector on the Amazon coast, which is dominated by macrotides with a monsoon climate, the mangrove forest area represents 74.3% of the Brazilian mangroves. In the Northeast Sector, a mesotidal coast with a tropical savanna climate, mangrove areas compose 10.7% of the Brazilian mangroves. The East Sector, a meso- to microtidal coast with a humid tropical climate, encompasses 9.7% of the Brazilian mangroves. The Southeast Sector, a microtidal coast with a temperate oceanic climate, contains only 5.4% of the Brazilian mangroves.

The paleoenvironmental records of sea-level changes during the Holocene indicate that over millennial timescales, mangrove shorelines in the North sector were exposed to a continuous sea-level rise, while in the Northeast, East, and Southeast sectors, the sea-level trajectories were completely different, with a highstand at approximately 6000–7000 Cal year. BP followed by a 3- to 4-m sea-level drop. These records point to a broad capacity of mangroves to adjust to sea-level changes. Four models of mangrove response to the ongoing rise in sea-level are proposed based on the understanding of past mangrove evolution under these varying Holocene sea-level histories. However, given the biogeographic variability of mangroves and environmental conditions along the Brazilian coast, predicting the effect of sea-level rise for the different mangrove sectors described herein is still a challenge. Further research is still needed to distinguish the effects of different phenomena and trends other than sea-level rise, which may affect mangroves on a multidimensional scale.

Acknowledgements We gratefully acknowledge the Conselho Nacional de Desenvolvimento Científico e Tecnológico (CNPq), Coordenação de Aperfeiçoamento de Pessoal de Nível Superior (CAPES), Mapbiomas, and Instituto Tecnológico Vale for the support provided to our research on Brazilian mangrove dynamics and evolution.

References

Allison MA, Nittrouer CA, Faria LEC (1995) Rates and mechanisms of shoreface progradation and retreat downdrift of the Amazon river mouth. Mar Geol 125(3–4):373–392. https://doi.org/10.1016/0025-3227(95)00020-Y

Allison MA, Lee MT, Ogston AS et al (2000) Origin of Amazon mudbanks along the northeastern coast of South America. Mar Geol 163:241–256. https://doi.org/10.1016/0025-3227(95)00020-Y

Almeida FFM, Carneiro CDR (1998) Origem e evolução da Serra do Mar. Rev Bras Geoc 28(2):135–150

Alvares CA, Stape JL, Sentelhas PC et al (2013) Koppen's climate classification map for Brazil. Meteorol Z 22(6):711–728. https://doi.org/10.1127/0941-2948/2013/0507

Andrade ACS, Dominguez JML, Martin L et al (2003) Quaternary evolution of the Caravelas strandplain—Southern Bahia State—Brazil. An Acad Bras Cienc 75:357–382. https://doi.org/10.1590/S0001-37652003000300008

Angulo RJ, Lessa GC, Souza MCd (2006) A critical review of mid- to late-Holocene sea-level fluctuations on the eastern Brazilian coastline. Quat Sci Rev 25(5–6):486–506. https://doi.org/10.1016/j.quascirev.2005.03.008

Anthony EJ, Gardel A, Gratiot N et al (2010) The Amazon-influenced muddy coast of South America: a review of mud-bank-shoreline interactions. Earth-Sci Rev 103(3–4):99–121. https://doi.org/10.1016/j.earscirev.2010.09.008

Barbirato JO, Ferreira NC, Zandonadi DB et al (2021) Structural characterization of areas with different levels of conservation in the mangrove of Vitória Bay, ES. Braz J Anim Environ Res 4(2):2600–2614. https://doi.org/10.34188/bjaerv4n2-085

Barboza EG, Dillenburg SR, do Nascimento Ritter M et al (2021) Holocene sea-level changes in southern brazil based on high-resolution radar stratigraphy. Geosciences 11(8):2600–2614. https://doi.org/10.3390/geosciences11080326

Barreto AMF, Suguio K, Bezerra FHR et al (2006) Geologia e Geomorfologia do Quaternário Costeiro do Estado do Rio Grande do Norte. Rev Inst Geoc USP 4(2):1–12

Batista EM, Souza-Filho PWM, Silveira OFM (2009) Avaliação de áreas deposicionais e erosivas em cabos lamosos da zona costeira Amazônica através da análise multitemporal de imagens de sensores remotos. Rev Bras Geof 27:83–96. https://doi.org/10.1590/S0102-261X2009000500007

Bezerra FHR, Barreto AMF, Suguio K (2003) Holocene sea-level history on the Rio Grande do Norte State coast Brazil. Mar Geo 196(1–2):73–89

Bezerra FHR, Amaro VE, Vita-Finzi C et al (2001) Pliocene-Quaternary fault control of sedimentation and coastal plain morphology in NE Brazil. J South Am Earth Sci 14(1):61–75. https://doi.org/10.1016/S0025-3227(03)00044-6

Bunting P, Rosenqvist A, Lucas MR et al (2018) The global mangrove watch—a new 2010 global baseline of mangrove extent. Remote Sens 10. https://doi.org/10.3390/rs10101669

Caldas LHdO, Stattegger K, Vital H (2006) Holocene sea-level history: evidence from coastal sediments of the northern Rio Grande do Norte coast NE Brazil. Mar Geol 228(1–4):39–53. https://doi.org/10.1016/j.margeo.2005.12.008

Cohen MCL, Alves ICC, França MC et al (2015) Relative sea-level and climatic changes in the Amazon littoral during the last 500years. CATENA 133:441–451. https://doi.org/10.1016/j.catena.2015.06.012

Costa BCF, Amaro VE, Ferreira ATS (2017) Classificação de Espécies de Mangue no Nordeste do Brasil com Base em Imagens Híbridas de Sensoriamento Remoto. Anu Inst Geoc—UFRJ 40(1):135–149. https://doi.org/10.11137/2017_1_135_149

Dadalto TP, Carvalho BC, Guerra JV et al (2022) Holocene morpho-sedimentary evolution of Marambaia Barrier Island (SE Brazil). Quat Res 105:182–200. https://doi.org/10.1017/qua.2021.43

Diniz C, Cortinhas L, Nerino G et al (2019) Brazilian mangrove status: three decades of satellite data analysis. Remote Sens 11. https://doi.org/10.3390/rs11070808

Doerffer R, Schiller H (2007) The MERIS Case 2 water algorithm. Int J Remote Sens 28(3–4):517–535. https://doi.org/10.1080/01431160600821127

Dominguez JML (2009) The Coastal Zone of Brazil. In: Dillenburg SF, Hesp PA (eds) Geology and Geomorphology of Holocene Coastal Barriers of Brazil. Springer, New York, pp 17–51. https://doi.org/10.1007/978-3-540-44771-9_2

Dominguez JML, Guimarães JK (2021) Effects of Holocene climate changes and anthropogenic river regulation in the development of a wave-dominated delta: The São Francisco River (eastern Brazil). Mar Geol 435. https://doi.org/10.1016/j.margeo.2021.106456

Dominguez JML, Martin L, Bittencourt ACSP (1987) Sea-level history and quaternary evolution of river-mouth-associated beach-ridge plains along the eastern/southeastern Brazilian coast: a summary. In: Nummedal D, Pilkey OH, Howard JD (eds) Sea-level fluctuation and coastal evolution. SEPM, Tulsa, Oklahoma, U.S.A., pp 115–127. https://doi.org/10.2110/pec.87.41.0115

Dominguez JML, Silva RP, Nunes AS et al (2013) The narrow, shallow, low-accommodation shelf of central Brazil: sedimentology, evolution and human uses. Geomorphology 203(1):46–59. https://doi.org/10.1016/j.geomorph.2013.07.004

Duke N, Ball M, Ellison J (1998) Factors influencing biodiversity and distributional gradients in mangroves. Glob Ecol Biogeogr Let 7(1):27–47. https://doi.org/10.2307/2997695

Duncan C, Owen HJF, Thompson JR et al (2018) Satellite remote sensing to monitor mangrove forest resilience and resistance to sea level rise. Methods Ecol Evol 9(8):1837–1852. https://doi.org/10.1111/2041-210X.12923

Elipot S (2020) Measuring global mean sea level changes with surface drifting buoys. Geophys Res Lett 47(21):e2020GL091078. https://doi.org/10.1029/2020GL091078

Ellison JC (2000) How South Pacific Mangroves may respond to predicted climate change and sea-level rise. In: Gillespie A, Burns WCG (ed) Climate change in the South Pacific: impacts and responses in Australia, New Zealand, and Small Island States, vol 2, advances in global change research. Springer, Dordrecht, pp 289–300. https://doi.org/10.1007/0-306-47981-8_16

Ellison JC (2015) Vulnerability assessment of mangroves to climate change and sea-level rise impacts. Wetl Ecol Manag 23(2):115–137. https://doi.org/10.1007/s11273-014-9397-8

Ellison JC, Stoddart DR (1991) Mangrove ecosystem collapse during predicted sea-level rise: Holocene analogues and implications. J Coast Res 7:151–165

Ferreira AV Jr, Paes BrCE, Vieira MM et al (2018) Diagenesis of Holocene Beachrock in Northeastern Brazil: petrology, isotopic evidence and age. Quat Environ Geosci. https://doi.org/10.5380/abequa.v9i2.53011

França CF, Souza-Filho PWM (2003) Análise das mudanças morfológicas costeiras de médio período na margem leste da Ilha de Marajó (PA) em Imagem Landsat. Rev Bras Geoc 32 (Suplemento 2):127–136

França CF, Souza-Filho PWM (2006) Compartimentação morfológica da margem leste da Ilha de Marajó: zona costeira dos municípios de Soure e Salvaterra—Estado do Pará. Rev Bras Geomorf 7(1):33–42

França CF, Souza-Filho PWM, El-Robrini M (2007) Análise faciológica e estratigráfica da planície costeira de Soure (margem leste da ilha de Marajó-PA), no trecho compreendido entre o canal do Cajuúna e o estuário Paracauari. Acta Amazonica 37(2):261–268. https://doi.org/10.20502/rbg.v7i1.58

Freitas H, Guedes MLS, Smith DH et al (2002) Characterization of the mangrove plant community and associated sediment of Todos os Santos Bay, Bahia Brazil. Aquat Ecosyst Health Manag 5(2):217–229. https://doi.org/10.1080/14634980290031901

Gilman EL, Ellison J, Duke NC et al (2008) Threats to mangroves from climate change and adaptation options: a review. Aquat Bot 89(2):237–250. https://doi.org/10.1007/s10661-006-9212-y

Giri C, Ochieng E, Tieszen LL et al (2011) Status and distribution of mangrove forests of the world using earth observation satellite data. Glob Ecol Biogeogr 20(1):154–159. https://doi.org/10.1111/j.1466-8238.2010.00584.x

Gomes Tapias J, Schobbenhaus C, Montez Ramírez NE (2019) Geological map of South America. In. CGMW, Servicio Geológico Colombiano, CPRM, Rio de Janeiro. https://rigeo.cprm.gov.br/handle/doc/21606

Halla MMS, Dominguez JML, Corrêa-Gomes LC (2020) Structural controls on the morphology of an extremely narrow, low-accommodation, passive margin shelf (Eastern Brazil). Geo-Mar Lett. https://doi.org/10.1007/s00367-019-00605-y

Hersbach H, Bell B, Berrisford P et al (2020) The ERA5 global reanalysis. Quarterly J Royal Meteorol Soc 146(730):1999–2049. https://doi.org/10.1002/qj.3803

Hesp PA, Maia LP, Claudino-Sales V (2009) The Holocene Barriers of Maranhão, Piauí and Ceará States, Northeastern Brazil. In: Dillenburg SF, Hesp PA (eds) Geology and geomorphology of holocene coastal barriers of Brazil. Springer, New York, pp 325–345. https://doi.org/10.1007/978-3-540-44771-10

IBGE (2020) Anuário estatístico do Brasil. Instituto Brasileiro de Geografia e Estatística, Rio de Janeiro—IBGE 80:1-1—8-50

Khan NS, Ashe E, Shaw TA et al (2015) Holocene relative sea-level changes from near-, intermediate-, and far-field locations. Curr Clim Change Rep 1(4):247–262. https://doi.org/10.1007/s40641-015-0029-z

Klein AHF, Short AD (2016) Brazilian beach systems: introduction. In: Short AD, Klein AHF (eds) Brazilian beach systems. Springer, Switzerland, pp 1–35. https://doi.org/10.1007/978-3-319-30394-9_1

Knoppers B, Ekau W, Figueiredo AG (1999) The coast and shelf of east and northeast Brazil and material transport. Geo-Mar Lett 19(3):171–178. https://doi.org/10.1007/s003670050106

Lacerda LD, Ward RD, Godoy MDP et al (2021) 20-Years cumulative impact from shrimp farming on mangroves of Northeast Brazil. Front For Glob Change. https://doi.org/10.3389/ffgc.2021.653096

Lessa GC, Santos FM, Souza Filho PW et al (2018) Brazilian estuaries: a geomorphologic and oceanographic perspective. In: Lana PdC, Bernardino AF (eds) Brazilian estuaries: a benthic perspective. Springer International Publishing, Cham, pp 1–37. https://doi.org/10.1007/978-3-319-77779-5_1

Locarnini RA, Mishonov AV, Baranova OK et al (2019) World Ocean Atlas 2018. In: Mishonov A (ed) vol NOAA Atlas NESDIS 81. NOAA, Silver Spring, MD, p 52. http://www.nodc.noaa.gov/OC5/indprod.html

Magris RA, Barreto R (2010) Mapping and assessment of protection of mangrove habitats in Brazil. Pan-Am J Aquat Sci 5(4):546–556

Martin L, Suguio K, Flexor J-M et al (1987) Quaternary evolution of the central part of the Brazilian coast. The role of relative sea-level variation and of shoreline drift. In: UNESCO (ed) Quaternary coastal geology of West Africa and South America. vol UNESCO Report in Marine Science. UNESCO, Paris, pp 97–145

McKee KL, Cahoon DR, Feller IC (2007) Caribbean mangroves adjust to rising sea level through biotic controls on change in soil elevation. Glob Ecol Biogeogr 16(5):545–556. https://doi.org/10.1111/j.1466-8238.2007.00317.x

McKee K, Rogers K, Saintilan N (2012) Response of Salt Marsh and Mangrove Wetlands to Changes in Atmospheric CO_2, Climate, and Sea Level. In: Middleton BA (ed) Global change and the function and distribution of Wetlands. Springer Netherlands, Dordrecht, pp 63–96. https://doi.org/10.1007/978-94-007-4494-3_2

Menezes MPMd, Berger U, Mehlig U (2008) Mangrove vegetation in Amazonia: a review of studies from the coast of Pará and Maranhão States, north Brazil. Acta Amazonica 38:403–420. https://doi.org/10.1590/S0044-59672008000300004

Nascimento WR Jr, Souza-Filho PWM, Proisy C et al (2013) Mapping changes in the largest continuous Amazonian mangrove belt using object-based classification of multisensor satellite imagery. Estuar Coast Shelf Sci 117:83–93. https://doi.org/10.1016/j.ecss.2012.10.005

Oppenheimer M, Glavovic BC, Hinkel J et al (2019) Sea level rise and implications for low-lying islands, coasts and communities. In: IPCC special report on the ocean and cryosphere in a changing climate. In: Pörtner HO, Roberts DC, Masson-Delmotte V et al (eds) Cambridge University Press, Cambridge UK-NY, USA, pp 321–445. https://doi.org/10.1017/9781009157964.006

Perry CT, Berkeley A, Smithers SG (2008) Microfacies characteristics of a tropical, mangrove-fringed shoreline, Cleveland Bay, Queensland, Australia: sedimentary and taphonomic controls on mangrove facies development. J Sediment Res 78(2):77–97. https://doi.org/10.2110/jsr.2008.015

Peulvast J-P, Sales VC, Bezerra FHR et al (2006) Landforms and neotectonics in the equatorial passive margin of Brazil. Geodin Acta 19(1):51–71. https://doi.org/10.3166/ga.19.51-71

Rodrigues SWP (2014) Detecção de mudança e sedimentação no estuário do rio Coreaú. In: Geoscience Institute. vol CDD 22. ed. 621.3678098131. Universidade Federal do Pará, Belém-PA, Brazil p 107

Rossetti DF (2014) The role of tectonics in the late Quaternary evolution of Brazil's Amazonian landscape. Earth-Sci Rev 139:362–389. https://doi.org/10.1016/j.earscirev.2014.08.009

Rossetti DF, Bezerra FHR, Dominguez JML (2013) Late Oligocene-Miocene transgressions along the equatorial and eastern margins of Brazil. Earth-Sci Rev 123:87–112. https://doi.org/10.1002/jqs.1132

Rossetti DF, Góes AM, Valeriano MM et al (2008) Quaternary tectonics in a passive margin: Marajó Island, northern Brazil. J Quat Sci 23(2):121–135. https://doi.org/10.1002/jqs.1132

Santos VF, Short AD, Mendes AC (2016) Beaches of the Amazon coast: Amapá and West Pará. In: Short AD, Klein AHF (eds) Brazilian beach systems. Coastal Research Library, Springer, Switzerland, pp 67–94. https://doi.org/10.1007/978-3-319-30394-9_3

Schaeffer-Novelli Y, Cintron-Molero G, Soares MLG (2002) Mangroves as indicators of sea level change in the muddy coasts of the world. In: Healy T, WaJ-AH Y (ed) Muddy coasts of the world: processes, deposits and function. Springer, New York, pp 245–262. https://doi.org/10.1016/S1568-2692(02)80083-3

Schaeffer-Novelli Y, Cintrón-Molero G, Adaime R et al (1990) Variability of mangrove ecosystems along the Brazilian coast. Estuaries Coast 13(2):204–218. https://doi.org/10.2307/1351590

Semeniuk V (1994) Predicting the effect of sea-level rise on mangroves in Northwestern Australia. J Coast Res 10(4):1050–1076

Semeniuk V (2018) Tidal flats. In: Finkl CW, Makowski C (eds) Encyclopedia of coastal science. Springer International Publishing, Cham, pp 1–20. https://doi.org/10.1007/978-3-319-48657-4_317-2

Silva JB, Torres MFA (2021) Assinatura Energética dos Manguezais no Domínio Costeiro Brasileiro. Rev Bras Geogr Fís 14(4):2286–2303. https://doi.org/10.26848/rbgf.v14.4.p2286-2303

Soares MLG, Estrada GCD, Fernandez V et al (2012) Southern limit of the Western South Atlantic mangroves: assessment of the potential effects of global warming from a biogeographical perspective. Estuar Coast Shelf Sci 101:44–53. https://doi.org/10.1016/j.ecss.2012.02.018

Souza-Filho PWM (2000) Tectonic control on the coastal zone geomorphology of the northeastern Pará State. Rev Bras Geoc 30(3):523–526

Souza-Filho PWM (2005) Costa de manguezais de macromaré da Amazônia: cenários morfológicos, mapeamento e quantificação de áreas usando dados de senores remotos. Rev Bras Geof 23(4):427–435. https://doi.org/10.1590/S0102-261X2005000400006

Souza-Filho PWM, Paradella WR (2003) Use of synthetic aperture radar for recognition of Coastal Geomorphological Features, land-use assessment and shoreline changes in Bragança coast, Pará Northern Brazil. Acad Bras Cienc 75(3):341–356. https://doi.org/10.1590/S0001-37652003000300007

Souza-Filho PWM, Farias Martins EdS, Costa FR (2006) Using mangroves as a geological indicator of coastal changes in the Bragança macrotidal flat, Brazilian Amazon: a remote sensing data approach. Ocean Coast Manag 49(7–8):462–475. https://doi.org/10.1016/j.ocecoaman.2006.04.005

Souza-Filho PWM, Paradella WR, Silveira OFM (2005) Synthetic aperture radar for recognition of coastal features in the wet tropics: apllications in the Brazilian Amazon coast. Bol Mus Para Emílio Goeldi. Ser Ciênc Nat 1(1):201–207

Souza-Filho PWM, Lessa GC, Cohen MCL et al (2009) The subsiding macrotidal barrier estuarine system of the Eastern Amazon coast, Northern Brazil. In: Dillenburg SF, Hesp PA (eds) Geology and geomorphology of holocene coastal barriers of Brazil. Springer, New York, pp 347–375. https://doi.org/10.1007/978-3-540-44771-9

Stramski D, Reynolds RA, Babin M et al (2008) Relationships between the surface concentration of particulate organic carbon and optical properties in the eastern South Pacific and eastern Atlantic Oceans. Biogeosciences 5(1):171–201. https://doi.org/10.5194/bg-5-171-2008

Tomlinson PB (2016) The botany of mangroves. Cambridge University Press, Cambridge. https://doi.org/10.1017/CBO9781139946575

Vieira MM, Sial AN, De Ros LF et al (2017) Origin of holocene beachrock cements in northeastern Brazil: evidence from carbon and oxygen isotopes. J South Am Earth Sci 79:401–408. https://doi.org/10.1016/j.jsames.2017.09.002

Vital H, Silveira IM, Tabosa WF et al (2016) Beaches of Rio Grande do Norte. In: Short AD, Klein AHF (eds) Brazilian Beach Systems. Coastal Research Library, vol Coastal Research Library. Springer, Switzerland, pp 200–229. https://doi.org/10.1007/978-3-319-30394-9_8

Woodroffe CD (1995) Response of tide-dominated mangrove shorelines in Northern Australia to anticipated sea-level rise. Earth Surf Process Landf 20(1):65–85. https://doi.org/10.1002/esp.3290200107

Woodroffe CD (2018) Mangrove response to sea level rise: palaeoecological insights from macrotidal systems in northern Australia. Mar Freshw Res 69(6):917–932. https://doi.org/10.1071/MF17252

Woodroffe CD (2019) Chapter 2—the morphology and development of coastal wetlands in the tropics. In: Perillo GME, Wolanski E, Cahoon DR et al (eds) Coastal Wetlands, 2nd edn. Elsevier, pp 79–103. https://doi.org/10.1016/B978-0-444-63893-9.00002-2

Woodroffe CD, Rogers K, McKee KL et al (2016) Mangrove sedimentation and response to relative sea-level rise. Annu Rev Mar Sci 8(1):243–266. https://doi.org/10.1146/annurev-marine-122414-034025

Worthington TA, zu Ermgassen PSE, Friess DA, et al (2020) A global biophysical typology of mangroves and its relevance for ecosystem structure and deforestation. Sci Rep 10(1):14652. https://doi.org/10.1038/s41598-020-71194-5

Zweng MM, Reagan JR, Seidov D et al (2018) World Ocean Atlas, volume 2: Salinity. In: Mishonov A (ed) vol NOAA Atlas NESDIS 82. NOAA/NESDIS, Silver Spring, MD. p 50. http://www.nodc.noaa.gov/OC5/indprod.html

Chapter 4
The Wave-Dominated Deltas of Brazil

José Maria Landim Dominguez

Abstract This chapter presents a synthesis of the existing knowledge on the wave-dominated deltas of Brazil (Parnaíba, São Francisco, Jequitinhonha, Doce, Paraíba do Sul). One of these, Parnaíba, is also strongly influenced by tides. These deltas are subject to different climatic zones, tidal regimes and wave climate, and different degrees of regulation and human occupation, which causes them to range from pristine deltas (Parnaíba) to human-influenced deltas (all others). Additionally, all these rivers flow in this area are controlled to greater or lesser extent controlled by the South American Monsoon System (SAMS). During the Holocene, the construction of these deltas was affected by variations in SAMS intensity associated with a progressive increase in Southern Hemisphere summer insolation. During the last decades, the reduction in river flows has triggered extensive erosion at the river mouth. Nevertheless, no significant losses in the delta plain area have been detected, and the eroded sediments are distributed to the regions lateral to the river mouth following patterns determined by the angular wave climate. Vulnerability to climate change is also examined, particularly concerning the rise in sea level. Parnaíba was determined to be the most vulnerable of the examined deltas, despite its almost pristine state.

Keyword Deltaic sedimentation · Climate change · Quaternary evolution · Vulnerability

4.1 Deltas—Hot Spots of Vulnerability to Climate Changes

Deltas are considered hot spots of vulnerability to climate change due to a number of factors (Day et al. 2019): (i) delta plains are lowland areas built under an approximately stable sea level, (ii) they experience high subsidence rates due to the progressive compaction of delta front muds, (iii) they have high-fertility soils and therefore are areas of intense agricultural activity and high population densities, (iv) they are very sensitive to variations in rainfall in their catchments, which affect river flows

J. M. L. Dominguez (✉)
Federal University of Bahia (UFBA), Salvador, Bahia, Brazil
e-mail: landim@ufba.br

J. M. Landim Dominguez et al. (eds.), *Tropical Marine Environments of Brazil*,
The Latin American Studies Book Series, https://doi.org/10.1007/978-3-031-21329-8_4

and therefore sediment supply to the delta plain, (v) their rivers have been extensively regulated due to the construction of dams, and (vi) they have high biodiversity and offer a variety of environmental services. Even in the absence of ongoing climate change, many deltas have already had their resilience to environmental changes irreparably compromised due to the magnitude of human interventions.

The construction of modern delta systems began more or less synchronously around the world due to the stabilization of sea level after 8000 years (Stanley and Warne 1994). Delta construction in association with climate stabilization during the Holocene allowed the emergence of great agricultural civilizations in many regions of the world (Day et al. 2007), such as in the Tigris-Euphrates delta (Kennett and Kennett 2006). It is estimated that more than 500 million people currently live in deltas (Giosan et al. 2014).

Man has intensively exploited the delta's natural resources for centuries and in some cases for more than a few thousand years. There is even a proposition that some Mediterranean deltas are human creations (Maselli and Trincardi 2013). Human population expansion in the catchments and the consequent deforestation have substantially increased the solid discharge of the Mediterranean rivers, causing delta progradation (Anthony et al. 2014; Besset et al. 2019). Moreover, because most delta plains are composed of fine sediments, deltas will only be geomorphologically sustainable if their long-term vertical build-up is greater than the magnitude of sea-level rise (Day et al. 2019). This vertical build-up is made possible by the input of mineral sediments during major floods, when the river overflows its bed and inundates the low areas of the delta plains. According to Giosan et al. (2014), almost all large deltas with areas greater than 10,000 km^2, and most deltas with areas between 1000 and 10,000 km^2, do not receive enough mineral sediment input to offset the 1 m sea-level rise by 2100. This is particularly true for deltas made up mainly of fine sediment, which experience high subsidence rates and have their rivers regulated. Smaller deltas made up of sandy sediments, such as wave-dominated deltas, have a larger portion of the delta plain located above sea level and therefore are less vulnerable to sea-level rise (Giosan et al. 2014). Notwithstanding, the ongoing climate changes and their effects on sea level, river discharge, and the increased frequency of extreme events have the potential to significantly alter the sustainability of all deltas.

On the coast of Brazil, in addition to the Amazon, there are 5 other deltas (Parnaíba, São Francisco, Pardo-Jequitinhonha—hereafter named Jequitinhonha, Doce, and Paraíba do Sul) (Fig. 4.1), with areas ranging from 2400 km^2 (Paraíba do Sul and Doce) to 1000–800 km^2 (Parnaíba, São Francisco, and Jequitinhonha). With the exception of the Parnaíba, which experiences an expressive influence of the tides, the São Francisco, Jequitinhonha, Doce, and Paraíba do Sul deltas are typical examples of wave-dominated deltas (Dominguez et al. 1987).

These 5 deltas extend from 22°S (Paraíba do Sul) to almost the equator (Parnaíba) and thus are subject to different climatic zones, tidal regimes, and wave climates and different degrees of regulation and human occupation, which causes them to range from pristine deltas (Parnaíba) to human-influenced deltas (all the others) (sensu Ibáñez et al. 2019). Additionally, all these rivers have their flows to a greater or lesser degree controlled by the South American Monsoon System (SAMS) (Marengo

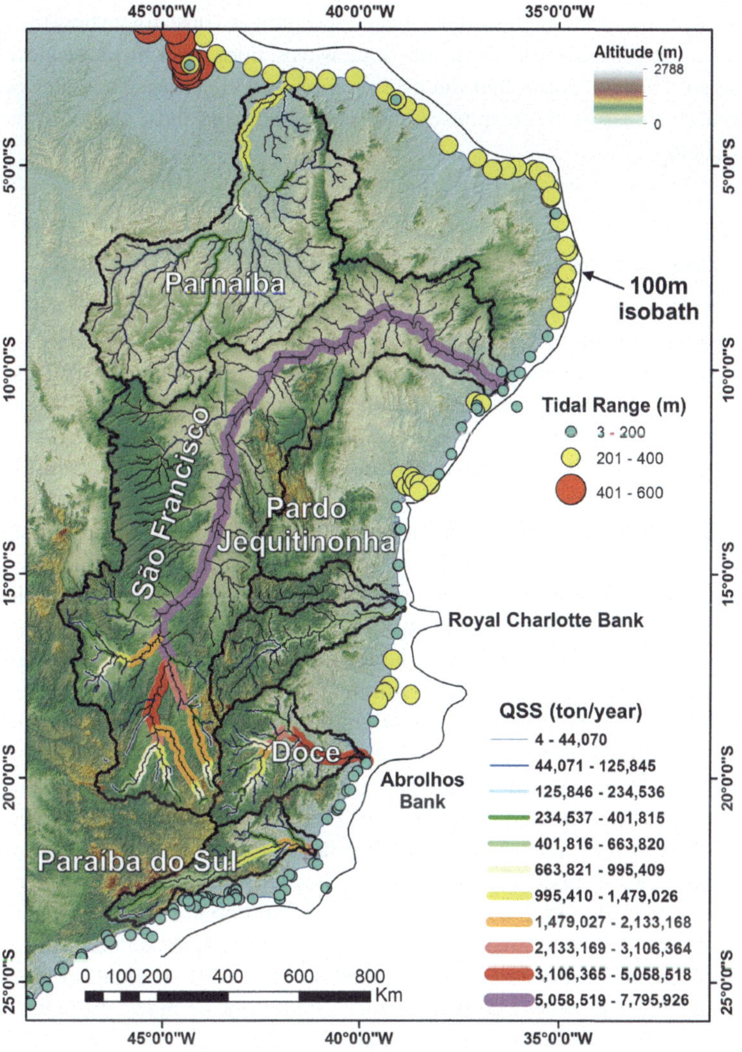

Fig. 4.1 Location of the five Brazilian deltas and their catchments, plotted on the Digital Elevation Model for eastern Brazil. Additionally, depicted are: Tidal range (Salles et al. 2020) and annual suspended sediment discharge (QSS) (Fagundes et al. 2021). The Pardo River catchment was also included in Figs. 4.2, 4.3 and 4.4, because it discharges in the Jequitinhonha delta

et al. 2012a, b), which has experienced variations in its intensity during the Holocene (Shimizu et al. 2020). These variations, documented through the study of speleothems (Cruz et al. 2009; Ward et al. 2019; Utida et al. 2020) and other paleoclimate proxies (Chiessi et al. 2021), have caused changes in precipitation in the river catchments and consequently in their flows, with direct repercussions on the construction of delta plains, as recently documented for São Francisco (Dominguez and Guimarães 2021).

In this way, these five deltas offer a great opportunity to investigate how the spatial variation in the delta controlling factors (e.g., waves, tides, river flows, sea level) and past climate changes controlled the evolution of these features. Such analysis can then be used to assess the responses and sustainability of these deltas to ongoing climate change.

4.2 Wave Versus Tide Dominance in the Brazilian Deltas

Deltas are very dynamic depositional environments sensitive to changes in marine and terrestrial environments and have been classified according to a triangular scheme where the vertices correspond to the relative dominance of fluvial inputs and wave and tidal energy in the receiving basin (Galloway 1975; Anthony 2015).

In the particular case of receiving basins with high wave energy, there is direct feedback between river inputs and the orientation of the coastline close to the river mouth. The greater the solid discharge from the river, the greater the deflection or reorientation of the shoreline in relation to its original orientation in the absence of the river. This deflection in turn increases the longitudinal transport that diverges from the mouth until a steady state situation is reached (Ashton and Giosan 2011). However, if the river discharge is very large, there will not be a coastline orientation large enough to allow the removal of sediments from the river mouth by the waves (Nienhuis et al. 2015). Based on this reasoning, these authors proposed an index R called the "fluvial dominance ratio", expressed by the ratio between the fluvial sediment flux (Q_r) and the maximum possible total coastal transport to both sides of the mouth (Q_{smax}) calculated from the wave climate for the region. When the value of $R < 1$, there will be a theoretical shoreline deflection that will allow the waves to remove sediments from the mouth as soon as they are introduced by the river. When $R > 1$, the delta is classified as fluvial dominated. An important implication derived from this reasoning is that wave-dominated deltas are usually associated with medium to small rivers, with reduced Q_r values. Nienhuis et al. (2015) calculated a value of $R = 0.3$ for the São Francisco delta, which should not be much different for the other wave-dominated deltas present on the east coast of Brazil, considering that the significant wave height along the stretch of coast investigated here is approximately 1–2 m and the solid discharges of the other rivers are lower than that of São Francisco (Fig. 1.4 and Table 1.4). There is thus a marked dominance of wave action in the Brazilian deltas. Of the five deltas investigated here, only the Parnaíba, in addition to wave action, is also highly influenced by tides.

The most striking morphological effect of the tides in river mouths is the increase in the width of the river channel (Langbein 1963) because when the tidally driven discharge is much larger than the river discharge, the channel width must increase downstream. Additionally, the tides tend to keep abandoned river channels on the delta plain open (Nienhuis et al. 2018). These authors proposed an index T called the "tide dominance ratio" expressed by the following relationship:

$$T = Q_{\text{tide}}/Q_{\text{river}}$$

where Q_{tide} is the characteristic tidal discharge amplitude at the river mouth (m^3/s) and Q_{river} is the mean annual fluvial discharge (m^3/s).

Thus, if the value of T is greater than 1, the delta is considered to be dominated by tides. In the absence of tides ($T = 0$), the discharge upstream of the delta is approximately equal to that at the mouth, resulting in a channel with an approximately constant width. When the value of T = infinite, the discharge at the mouth is completely controlled by the tides, and the channel cross-sectional area will depend on the tidal prism (Stive and Rakhorst 2008).

In fact, for the Brazilian deltas, the ratio between the channel width at the river mouth and at the point where it enters the delta plain increases from south to north following the increase in tidal amplitude from a microtidal to mesotidal regime (Fig. 4.1). This ratio varies from 1.25 at Paraíba do Sul, 1.16 at Doce, and 1.14 at Jequitinhonha to 2.36 at São Francisco and 4.5 at Parnaíba (Table 4.1).

Although we have not calculated the value of the T index for the Brazilian deltas, the significant increase in channel width at the river mouth, the absence of beach ridges and the dominance of mangrove swamp muds on the deltaic plain would justify classifying the Parnaíba as tide dominated, although it is also influenced by waves and is exposed to significant wind action.

These distinctions are important because the way these deltas will respond to ongoing climate change and other anthropogenic influences will depend on the relative dominance of river, wave and tidal action, among other aspects.

4.3 Regional Setting

4.3.1 Catchment

With the exception of the Parnaíba, whose catchment drains Paleozoic sediments of the Parnaíba sedimentary basin, the other rivers flow dominantly on Precambrian rocks (Fig. 4.2). The upper and middle courses of the São Francisco River run entirely within the São Francisco craton, a geotectonic unit of Archean-Paleoproterozoic age (Heilbron et al. 2017), while the other rivers and the lower course of the São Francisco River drain the Brasiliano (Neoproterozoic) fold belts that surround the craton (Fig. 4.2). Table 4.1 presents the main characteristics of these catchments. It is worth noting that a 1300 km-long sector of the São Francisco catchment in its middle portion acts as a terrestrial sink (Fig. 4.2) (Nyberg et al. 2018) and traps a considerable volume of alluvial and aeolian sediments that otherwise would be available for delta construction. This might also be the reason why the São Francisco delta is small compared to the other deltas.

In general, significant portions of the Parnaíba, São Francisco, and Jequitinhonha catchments are subjected to an arid to semiarid climate, mainly in their middle

Table 4.1 Major characteristics of the deltas and their catchments

DELTA	Catchment Area (km^2)	Catchment maximum altitude (m)	Catchment mean altitude (m)	Suspended sediment discharge (ton/year)	Population in catchment	Delta area (km^2)	Population in delta	Distance river mouth to shelf break (km)	Channel width at river mouth/channel width upstream	Tidal Range (m)	Delta plain area loss (km^2)	Delta plain area gain (km^2)
Parnaíba	333.163	975	353	1.277.532	4.466.744	1.000	80.760	64	4.50	2.80	7.00	6.00
São Francisco	636.005	1.931	634	7.795.926	15.303.971	856	55.865	26	2.36	1.74	7.43	2.61
Jequitinhonha/Pardo	102.444	2.005	685	600.776	1.875.385	838	29.754	22	1.14	1.73	3.40	5.18
Doce	82.967	2.574	567	5.045.658	3.719.820	2.316	75.339	30	1.16	1.16	1.49	2.07
Paraíba do Sul	57.218	2.711	612	2.239.843	6.636.908	2.468	396.484	85	1.25	1.06	1.51	4.82

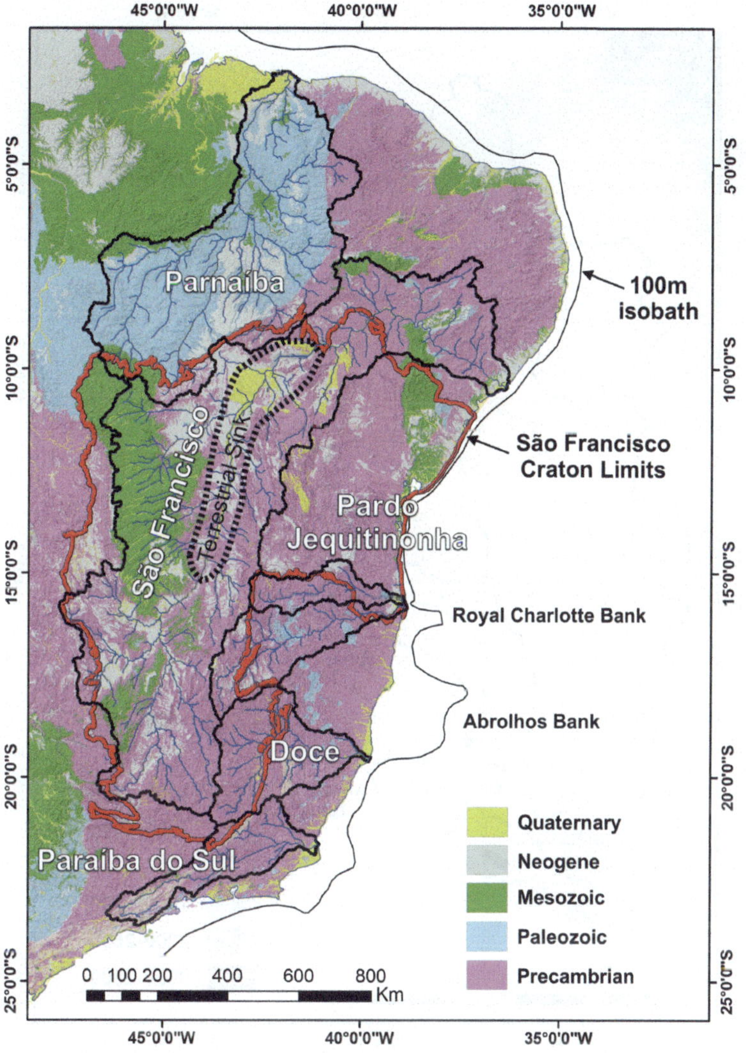

Fig. 4.2 Geology of the catchments of the five Brazilian deltas (from CPRM Brazilian Geological Service—http://www.cprm.gov.br/publique/Geologia/Geologia-Basica/Carta-Geologica-do-Brasil-ao-Milionesimo-298.html). Additionally, shown are the limits of the São Francisco craton and of the terrestrial sink located in the middle catchment of the São Francisco (Nyberg et al. 2018)

portions, but extending in some cases to the river mouth (Fig. 4.3). In the Parnaíba and São Francisco deltas, the combination of a climatic regime characterized by more than 3 dry months during the year, persistent trade winds and abundant sediment input favored the development of expressive coastal dune fields (Barbosa and Dominguez 2004).

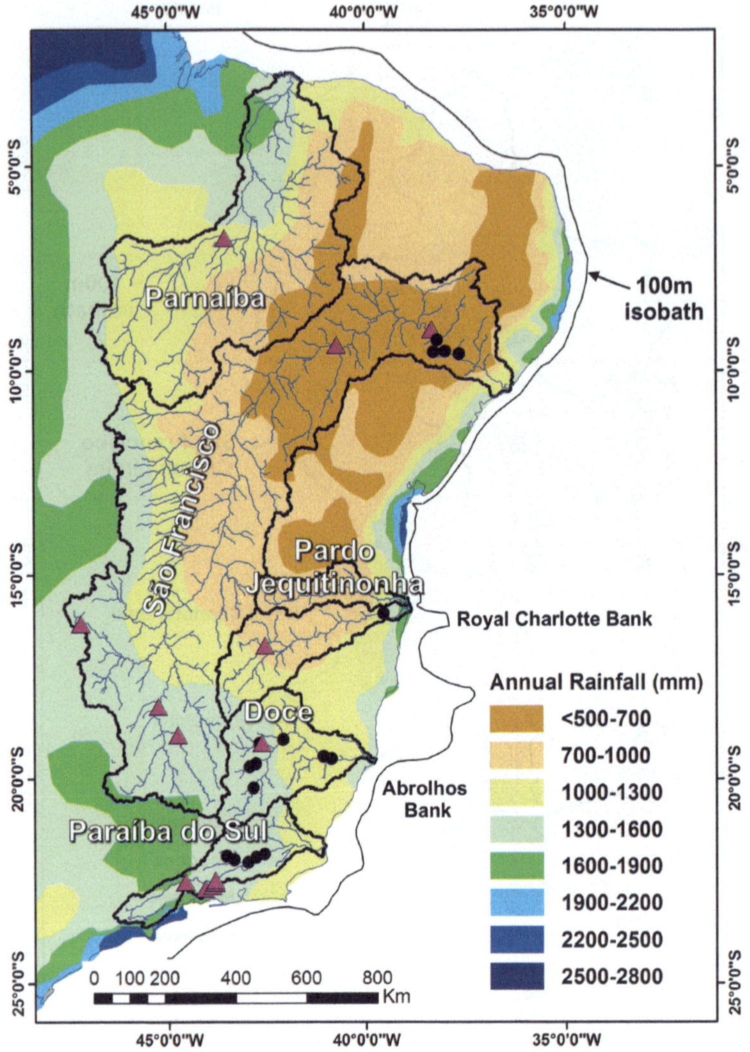

Fig. 4.3 Annual rainfall in the catchments of the five Brazilian deltas (Alvares et al. 2013). Additionally, depicted are the major dams built and their types: run of the river (black circles) and reservoir (magenta triangles) (From Agência Nacional de Águas: https://www.ana.gov.br/sar/sin)

The rains in these watersheds occur mainly in the austral summer (DJF) and early autumn (MAM—in north—northeastern Brazil) months and are controlled by the South American Monsoon System (SAMS), which is responsible for more than 70% of the annual precipitation over Tropical South America, associated with the southward migration of the ITCZ and formation of the South Atlantic Convergence Zone (SACZ) (Chiessi et al. 2021) (Fig. 4.4). These summer rains are mainly concentrated in the upper reaches of these rivers (Fig. 4.4). The SACZ is currently located at approximately 20–22°S coincident with the Doce and Paraíba do Sul watersheds (Fig. 4.4). Although the ITCZ is primarily a Northern Hemisphere phenomenon during extreme El Niño episodes, it extends to the Southern Hemisphere (Takahashi and Battisti 2007).

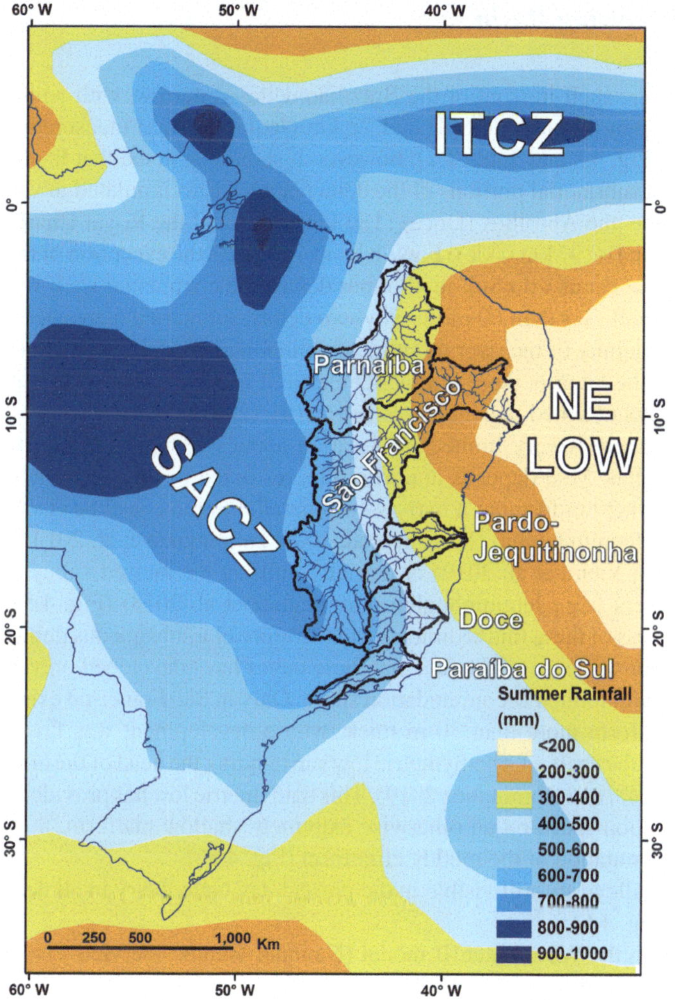

Fig. 4.4 Summer rainfall in the catchments of the five Brazilian deltas (based on Climate Prediction Center Merged Analysis of Precipitation (CMAP) data for the period 1979–2004) and the major features associated with the South America Monsoon System (SAMS). SACZ—South American Convergence Zone. ITCZ—Intertropical Convergence Zone. NE Low—Nordeste Low is the descending limb of the regional Walker circulation characterized by very low humidity

Overall, Paraíba do Sul (17 dams), Doce (8 dams), and São Francisco (8 dams) are the most regulated rivers when compared to Jequitinhonha (2 dams) and Parnaíba (1 dam) (Fig. 4.3) (Agência Nacional de Águas: https://www.ana.gov.br/sar/sin).

4.3.2 Receiving Basin

In general, the shelf in front of the Brazilian deltas is narrow, with widths ranging between 85 km (Paraíba do Sul) and 22 km (Jequitinhonha) measured in front of the mouth (Fig. 4.5 and Table 4.1); however, it should be noted that in almost all of these deltas, substantial portions of the delta plain have accumulated in very shallow areas, such as the Abrolhos (Doce) (189 km wide) and the Royal Charlotte banks (Jequitinhonha) (107 km wide) (Figs. 4.1 and 4.5). With the exception of the southern portion of the Paraíba do Sul, in all other deltas, the continental shelf in almost its entirety is shallower than 60 m. In only two deltas, submarine canyons occur in the immediate vicinity of the river mouths, Jequitinhonha and São Francisco (Fig. 4.5), but only in the latter is there a direct connection between the river mouth and the canyon, as documented in detail by Ribeiro et al. (2021) (Fig. 4.5).

Overall, there is very limited information about the submerged portion of the Brazilian deltas. Mud deposits are, as a rule, restricted to the inner shelf up to the isobath of 10–20 m forming a narrow strip parallel to the coastline (Doce, Paraíba do Sul, and Jequitinhonha) (Murillo et al. 2009; Dominguez et al. 2013; Lopes and Frazão 2018; Vieira et al. 2019) or partially infilling the incised valleys excavated by these rivers (e.g., Jequitinhonha) (Dominguez et al. 2013) (Fig. 4.5). The low expressiveness of these fine sediments and therefore of a subaqueous delta clinoform is a direct consequence of the fact that these deltas have developed on very shallow platforms, with limited accommodation space. Only in São Francisco exists a muddy deltaic clinoform more than 30 m thick, whose development was favored by the presence, on the shelf, of a bathymetric low surrounding the head of the homonymous canyon (Rangel and Dominguez 2019). This bathymetric low has provided additional accommodation space on an otherwise extremely shallow platform, thus allowing for the accumulation of the muddy clinoform (Fig. 4.5b).

Incised valleys are still visible in the present day bathymetry in all deltas, except for Doce (Fig. 4.5).

Data from the WaveWatchIII model (National Weather Service—NOAA) show that the significant wave height (Hs) at all deltas varies from 1 to 2 m. The wave angular distribution in relation to the river mouth is quite variable. In the Parnaíba, São Francisco and Paraíba do Sul deltas, the distribution is asymmetric, while in the Jequitinhonha and Doce Rivers, it is approximately symmetrical (Fig. 4.5). This angular distribution was fundamental in controlling various aspects of the Holocene evolution of these deltaic plains (e.g., presence/absence of a groyne effect) (Dominguez et al. 1987; Bhattacharya and Giosan 2003) and will exert a marked

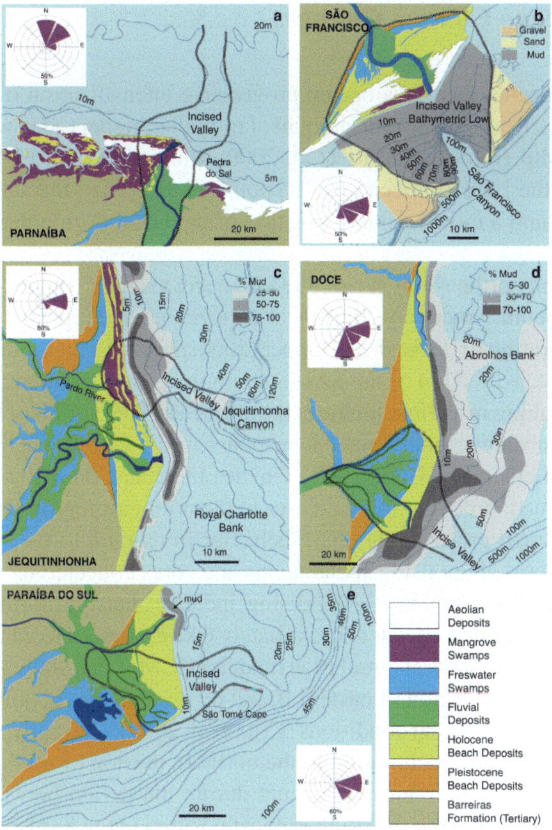

Fig. 4.5 Simplified geology of the Parnaíba, São Francisco, Jequitinhonha, Doce and Paraíba do Sul deltas modified from Dominguez et al. (1987) and Martin et al. (1993a, b). Additionally, shown are the major isobaths (from Brazilian Navy Admiralty Nautical Charts). Limits of incised valleys were inferred from bathymetry and from the following sources: São Francisco - Rangel and Dominguez (2019), Jequitinhonha—Lopes and Frazão 2018), Doce—Dessart (2009), and Paraíba do Sul—Carelli et al. (2018) and Carvalho et al. (2019). Distribution of fine grained sediments on the shelf are from: São Francisco—Araújo et al. (2018), Jequitinhonha—Dominguez et al. (2013), Doce—Vieira et al. (2019), and Paraíba do Sul—Murillo et al. (2009). No information on distribution of fine sediments is available to the Parnaíba. Insets show the angular wave climate (WaveWatchIII—National Weather Service—NOAA)

influence on the future evolution of these deltas in a scenario of progressive reduction of rainfall in the hydrographic basins (Milly et al. 2005; Marengo et al. 2010; Marengo et al. 2012a, b) (see Sect. 4.7 below).

4.3.3 Holocene Climate Changes

During the Holocene, there has been a progressive intensification of the SAMS in southeastern Brazil, where the total amount of rainfall is higher today when compared to the Early—Mid Holocene (Cruz et al. 2009; Shimizu et al. 2020). This intensification during the Late Holocene is a response to increased insolation during the austral summer and a more southern position of the ITCZ during this period (Haug et al. 2001).

Specifically, in northeastern Brazil, the long-term average climatic conditions are a result of the South American subtropical anticyclone and a persistent subsidence related to the Northeast Low in the upper troposphere (Lenters and Cook 1997) (Fig. 4.4). An intensification of the SAMS strengthens South America's subtropical anticyclone, resulting in a drier climate in the northeast. The Northeast low is linked to the large-scale subsidence maintained in response to condensational heating over the Amazon basin (Lenters and Cook 1997). On the other hand, during periods when the SAMS is weakened, precipitation increases in the northeast region. Thus, the northeast region of Brazil presents antiphase behavior in relation to the south-southeast region. Studies carried out using speleothems and cores collected on the continental margin complemented by climate modeling indicate that the northeast region of Brazil was considerably wetter at approximately 6 ka compared to what occurs today (Shimizu et al. 2020). General circulation models (GCMs) also show that around this time, the entire coastal region of eastern Brazil received a greater volume of rainfall when compared to the current conditions (Cruz et al. 2009). Currently, approximately 50% of the total annual precipitation in the northern portion of northeastern Brazil, where the Parnaíba delta is located, occurs during the period between March and April, when the ITCZ reaches its southernmost position in the northeastern region of South America (Chiessi et al. 2021), a period when the highest flows in the Parnaíba River occur.

Utida et al. (2020) also showed that during the Holocene, the abrupt climatic transition that occurred at approximately 4.2 ka and marks the base of the Meghalayan chronozone is characterized by the onset of drier conditions in northeastern Brazil (Oliveira et al. 1999; Zular et al. 2018). Speleothem records from the region show that the period extending from 9 to 5 ka was much wetter than today. Between 4.8 and 4.2 ka, there was an abrupt change in precipitation coinciding with a reversal in the insolation trend (increase) in South America at approximately 5 ka (Utida et al. 2020).

Chiessi et al. (2021) interpreted the upper core decrease in Fe/Ca and Ti/Ca values in a core collected in the continental margin off the Parnaíba River as a result of a decrease in the inflow of siliciclastic sediment from that river since the Mid Holocene. This decrease in the siliciclastic input was attributed by these authors to a latitudinal contraction of the ITCZ migration range in association with an intensification of the descending limb of the regional Walker circulation (Northeast Low) (Fig. 4.4).

Finally, during the last millennia, a number of palaeoprecipitation records point to an intensification of the SAMS, with a wider SACZ during the Little Ice Age (Campos et al. 2019).

These changes in the SAMS regime during the Holocene have high potential to have affected sedimentation in the Brazilian deltas, as exemplified by Dominguez and Guimarães (2021) for the São Francisco River delta.

4.3.4 Sea-Level History

The fact that the continental shelf in front of the Brazilian deltas is very shallow (<60 m) has two important implications: (i) during most of the last glacial cycle, the shelf was exposed, favoring the incision of valleys by these rivers, and (ii) flooding of the shelf only began after the Younger Dryas (12.9–11.7 ka), when the sea level was on average positioned 60 m below the current level (Bard et al. 2010). Thus, the stratigraphic record of delta sedimentation during the 125 m rise in sea level since the Last Glacial Maximum (LGM) was most likely not preserved due to the lack of accommodation space; otherwise, this record was potentially preserved only in the interior of the incised valleys excavated by these rivers on the shelf (Fig. 4.5). However, this preservation is probably incomplete because in most of these deltas, the incised valleys have not yet been completely infilled and still have a bathymetric expression. In this regard, the delta of São Francisco deserves attention because it was entirely built in a bathymetric depression that developed around the head of the homonymous canyon (Fig. 4.5b) (Rangel and Dominguez 2019). The bottom of this depression is situated at a depth of 120 m, providing additional accommodation space and thus allowing for the accumulation of a complete sedimentary record of sea-level rise since the LGM. For São Francisco, the main episode of delta construction during the last deglaciation occurred during the Younger Dryas. Afterward, this delta was completely drowned during the Melt Water Pulse 1B (Rangel and Dominguez 2019).

Finally, given that most of the Brazilian coast is located in a GIA far field region, a relative drop in sea level of approximately 3–4 m since the mid-Holocene is expected (Peltier 2007). In fact, numerous sea-level reconstructions for the region indicate a high sea level of this magnitude approximately 5700 years ago (Kenitiro et al. 1985; Martin et al. 2003; Angulo et al. 2006). The major exception is the Parnaíba delta, which is located in a region where the GIA far-field effects were insignificant (Peltier 2007; Hammond et al. 2021). This is corroborated by the fact that no higher-than-present Holocene sea-level records have yet been found for that region.

4.4 Geology-Geomorphology of the Deltas

The delta plain of a wave-dominated delta is characterized by well-sorted beach sand deposits that form terraces capped by beach dune ridges. These ridges indicate the successive positions occupied by the shoreline, left behind as a result of progradation. With the exception of the Parnaíba delta, in all other deltas on the east coast of Brazil, there are two generations of beach deposits (regressive littoral sands) (Fig. 4.5).

Pleistocene beach deposits that accumulated during the high sea level of the MIS5e isotopic stage (Martin et al. 1988). These terraces have altitudes of approximately 8 m above present sea level and usually occupy the innermost portion of the deltaic plains (Figs. 4.5 and 4.6). These altitudes agree reasonably well with similar deposits in other regions of the world (Polyak et al. 2018). They were deposited against currently inactive sea-cliffs carved into the Barreiras Formation of Early-Mid Miocene age (Rossetti et al. 2013). In the Paraíba do Sul delta, in particular, these deposits reach their greatest expression in the area and constitute the entire southern half of the delta plain, even reaching the present day coastline (Fig. 4.5e). MIS5e beach deposits have not yet been found in the Parnaíba delta. Likewise, inactive sea cliffs carved into the Barreiras Formation are absent in that region.

Holocene beach deposits associated with the current marine isotopic stage (MIS1) are found in all east coast deltas, but they are virtually absent in the Parnaíba delta, where they occur as sand spit deposits scattered on the western portion of the delta plain (Fig. 4.5a). This might be derived from the fact that Parnaíba has characteristics more akin to those of tidal-dominated deltas. An important morphological aspect of the Holocene beach deposits is that overall, their altitudes increase toward the interior of the delta plain as a result of the 3–4 m GIA far-field drop in relative sea level since the Mid Holocene (Peltier 2007) (Fig. 4.6). Additionally, as already mentioned, in the Parnaíba, transitional deposits indicative of a higher than present Holocene sea level have not yet been documented, which seems to derive from the fact that this delta is located in a region where the far-field effects of the GIA are significantly small when compared to the east coast (Peltier 2007).

In the outermost portions of the Paraíba do Sul, São Francisco and Parnaíba deltas, there is an asymmetry in facies distribution between the updrift (continuous sand sheet of beach deposits at the Paraíba do Sul and São Francisco, and aeolian deposits at the Parnaíba) and the downdrift (sandy spits separated by low lying regions occupied by mangroves) portions of the delta plain (Fig. 4.7), resulting from the so-called groyne effect (Dominguez et al. 1987; Bhattacharya and Giosan 2003). The groyne effect is common in river mouths subjected to an asymmetrical wave angle climate, where the solid and liquid discharges from the river act as a kind of hydraulic jetty, retaining in the updrift portion of the mouth the sediments transported by the regional littoral drift while the downdrift portion is fed dominantly by river sediments. As the fluvial input is episodic and concentrated in major flood events, the shoreline progradation downdrift of the river mouth is intermittent. After major floods, river-borne sediments are reworked by the waves, originating sandy spits that extend longitudinally and isolate portions of the shoreface. The resulting lowland is later infilled with fine sediments and colonized by mangroves. This process has been described in detail by Rodriguez et al. (2000) and Fraticelli (2006) for the Brazos River delta in the Gulf of Mexico. The formation and longitudinal extension of the sandy spits is facilitated by the high angle that the waves form with the downdrift

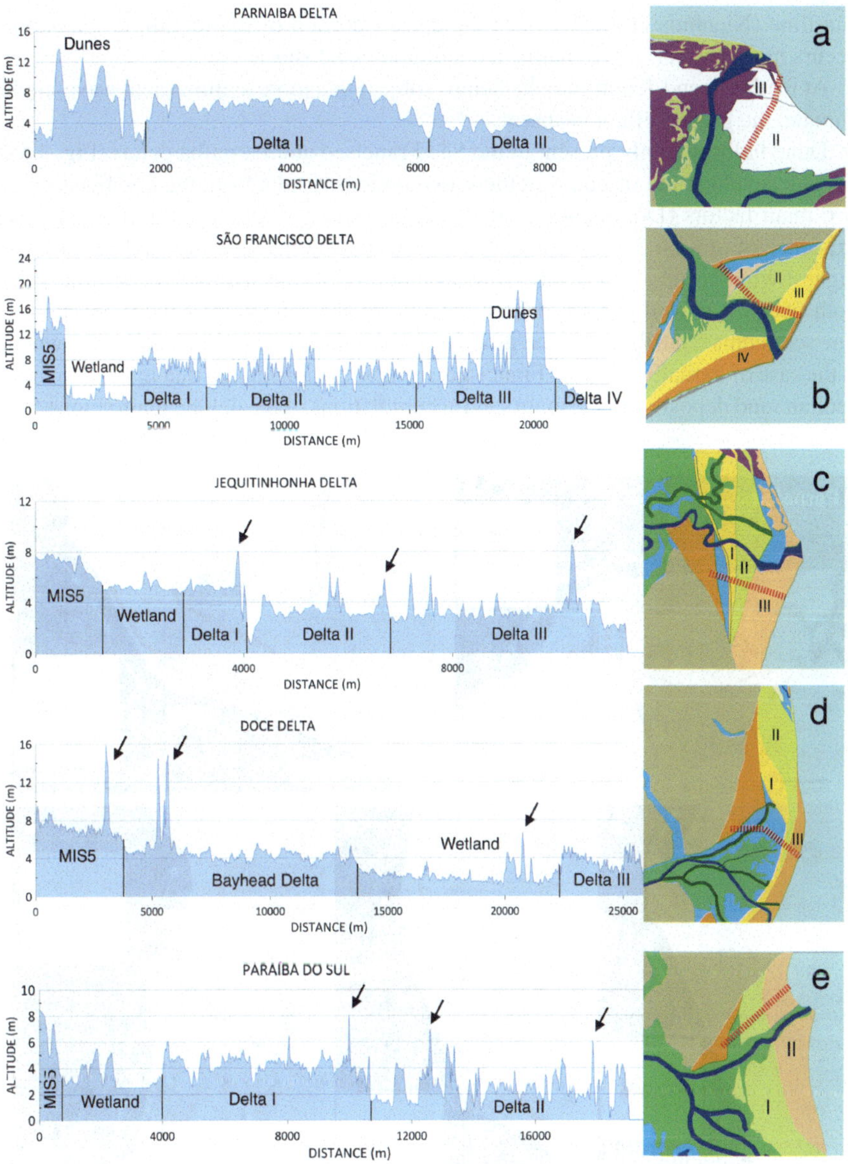

Fig. 4.6 Topographical profiles extracted from copernicus-DEM digital elevation models produced by the European Space Agency (absolute vertical accuracy: < 2 m for slopes below or equal 20%). Note that Copernicus-DEMs do not distinguish vegetation from bare soil, which might result in overestimation of elevation. We have tried to avoid this by extracting elevation from regions of the DEM with rarified vegetation cover, which was not always possible. Some spikes caused by arboreous vegetation are indicated by small black arrows. See Fig. 4.5 for legend of all units but for the Holocene beach deposits. Roman numerals indicate major stages of delta construction during the Holocene (see Figs. 4.8 through 4.12 and text for more details)

coastline (Nienhuis et al. 2012). In the updrift portion of the mouth, progradation occurs more regularly, originating a continuous sand sheet.

At the Doce and Jequitinhonha deltas, subject to a more symmetrical wave-angle climate, the groyne effect is absent.

Dune fields are only present in the São Francisco and Parnaíba deltas (Fig. 4.5a, b). The development of dunes in these two regions results from the combination of two main factors (Dominguez et al. 1992; Barbosa and Dominguez 2004): (i) the existence of at least 3 consecutive dry months during the year, coinciding with the highest wind speeds, and (ii) the wide availability of fine sediments contributed by both rivers. It is well known that the longer a river is, the greater the proportion of fine sediments contributed to the mouth (Gasparini et al. 2004; Frings 2008), which is the case of Parnaíba and São Francisco. It should also be noted that in the Parnaíba, aeolian sand deposits are the main sand accumulations of the deltaic plain (Fig. 4.5a).

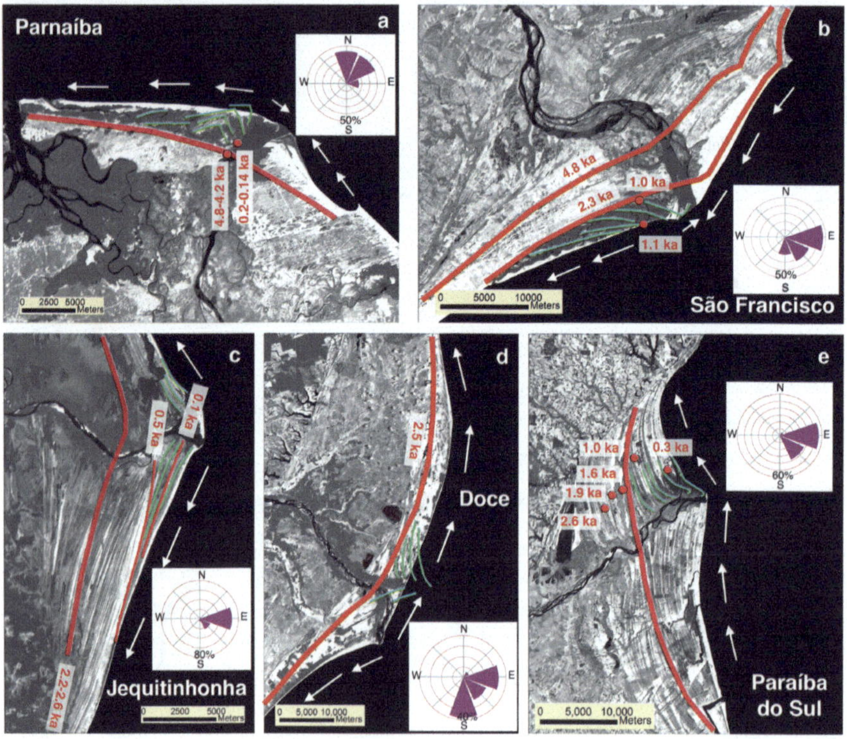

Fig. 4.7 Details of the delta plains emphasizing the groyne effect **a**, **b**, **e**, truncations in beach-dune ridge orientation **c**, **d**, **e**, and significant dates of episodes of delta construction. Additionally, shown are the angular wave climate (WaveWatchIII—National Weather Service—NOAA) and the regional longshore transport. Dates are from the following sources: Parnaíba—Szczygielski et al. (2015), São Francisco—Dominguez and Guimarães (2021), Jequitinhonha—Dominguez (1982) and Rodrigues (2018), Doce—Dominguez and Wanless (1991) and Martin et al. (1993a, b) and Paraíba do Sul—Rocha et al. (2019)

Fluvial deposits are usually of secondary importance in the Brazilian deltas and restricted to those areas marginal to river channels. This derives from the absence of distributaries in these deltas and from the extensive reworking by waves and tides of the riverine sediments. Fluvial deposits are also more restricted to the innermost portions of delta plains that were built around the mid-Holocene, suggesting that fluvial sedimentation was more important in the early phases of delta development. Thus, the vertical delta plain build-up, which is so important in fluvial-dominated deltas, is secondary in importance in the Brazilian deltas, with the exception of the Parnaíba, where a large portion of the delta plain is occupied by mangroves.

4.5 Holocene Evolutionary History

The first works on the Brazilian deltas consisted mainly of mapping delta plain deposits, with special emphasis on the geometry of the beach dune ridges, their truncation patterns and their dating by the radiocarbon method (Martin and Suguio 1992; Martin et al. 1993a, b). From these data, Holocene evolutionary models have been proposed, consistently emphasizing the changes in the relative sea level, mainly its lowering since the mid-Holocene (Suguio and Martin 1981; Dominguez et al. 1987). In these models, river-borne sediments were always considered secondary when compared to the sediment supply supposedly derived from the 3–4 m drop in the relative sea level, following the inverse of Bruun's rule (Dominguez and Wanless 1991). However, this hypothesis has never been examined in greater depth or effectively tested.

Additionally, in these evolutionary models, the low-lying zone that usually separates the Pleistocene from the Holocene beach deposits was interpreted as a paleolagoon (currently, they would be more appropriately classified as an estuary) formed during the maximum Holocene transgression by analogy with the coastal evolution models available for the North American eastern and Gulf coasts. In the Brazilian models, the rivers would have initially built intralagoonal deltas, and later with sea-level lowering since the Mid Holocene, the paleolagoons would become emergent, and the rivers would begin to flow directly into the open sea, constructing the Holocene portion of the delta plain (Suguio and Martin 1981; Dominguez et al. 1987, 1992). The best example of this model would be represented by the Doce delta, where several abandoned distributaries, interpreted as belonging to an intralagoonal delta, occur in the central portion of the delta plain (Suguio and Martin 1981; Dominguez and Wanless 1991) (Fig. 4.5e). These models would apply mostly to the wave-dominated deltas of eastern Brazil. For the tidal-dominated Parnaíba, to date, there are no proposed evolutionary models, although more recent works have advanced in this direction (Szczygielski et al. 2015).

Little or no attention was given in these earlier models to the role played by the incised valleys excavated by these rivers during lowstands. In fact, if we consider that the mean sea level during the last million years was positioned between 45 and 60 m below the current level (Blum et al. 2013), the eastern-northeastern continental shelf of Brazil was exposed to subaerial conditions most of the time. Thus, the development of expressive incised valley systems associated with these rivers is expected. The presence of these valleys must have played an important role in delta development during the post-LGM sea-level rise by generating additional accommodation space on an otherwise very shallow shelf. As already mentioned, this aspect has been very well documented for the São Francisco River (Rangel and Dominguez 2019) and can certainly be applied to the other deltas investigated here. Incised valleys are still visible in the present day bathymetry in all deltas except for Doce (Fig. 4.5) or have been documented in high-resolution seismic surveys (Jequitinhonha and São Francisco—Lopes and Frazão 2018; Rangel and Dominguez 2019) or through boreholes in the delta plain (Paraíba do Sul, Doce and São Francisco—Dessart 2009; Guimarães 2010; Carelli et al. 2018; Carvalho et al. 2019).

During the Holocene transgression, estuarine/deltaic sedimentation took place essentially inside these valleys. Only after they were infilled in their most proximal regions could the fluvial sediments be used to prograde the shoreline, which for the deltas discussed herein took place essentially in the shallow areas lateral to these incised valleys. The greater shoreline progradation in these shallow marginal areas is possibly due to the smaller accommodation space available in them. Thus, in the Doce and Jequitinhonha rivers, most of the Holocene delta plain was built in the adjacent shallow shelf areas represented by the Abrolhos and Royal Charlotte banks, respectively (Fig. 4.5c, d). In the Paraíba do Sul, the Holocene delta was built in the shallower shelf area located to the north of the cape of São Tomé (Fig. 4.5e). In the Parnaíba Delta, the continental shelf surrounding the incised valley is extremely shallow and characterized by large areas with depths less than 20 m. In the delta's updrift portion, there are a series of reefs very close to or along the present day coastline (e.g., at Pedra do Sal, where rocks from the crystalline basement outcrop) (Figs. 4.5a and 4.6a) influencing the coastline geometry and showing how reduced the water depths in this stretch are. The São Francisco delta is a special case since the entire delta (deltaic plain and subaqueous clinoform) was built inside a bathymetric depression (Fig. 4.5b) (Rangel and Dominguez 2019) that surrounds the head of the São Francisco canyon. Even so, the progradation was slightly more expressive in the northeastern portion of the plain where the antecedent topography of the bathymetric low was shallower (Dominguez and Guimarães 2021).

Based on these elements and their integration into the literature, Holocene evolutionary models for each of the Brazilian deltas are proposed and presented in Figs. 4.8, 4.9, 4.10, 4.11 and 4.12.

4.6 Holocene Climate Changes and Delta Construction

There is a vast body of literature discussing how changes in watershed precipitation influenced delta construction in different regions of the world, such as the Nile (Revel et al. 2010; Marriner et al. 2012), Volga (Overeem et al. 2003), Rhone (Fanget et al. 2013, 2014), and Ebro (Xing et al. 2014). More recently, Dominguez and Guimarães

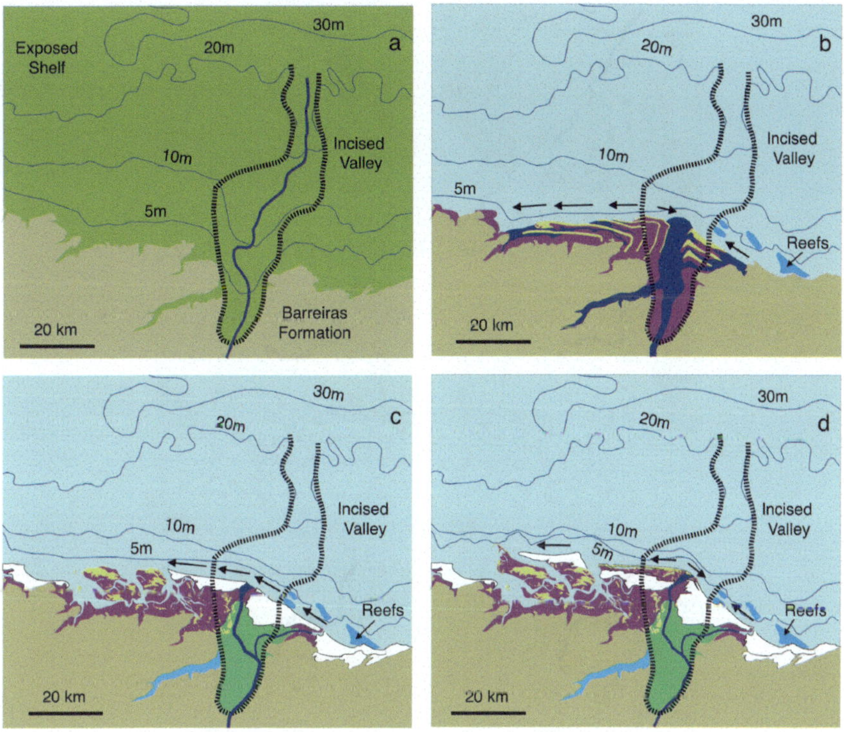

Fig. 4.8 Proposed evolutionary scheme for the Parnaíba delta: **a** Last Glacial Maximum: shelf exposed subaerially and inception of the incised valley. **b** Delta construction began sometime during the Mid-Holocene and extended up to 4.2 ka (Szczygielski et al. 2015). **c** By 4.2 ka, an abrupt climate transition took place marked by the onset of drier conditions in northeastern Brazil (Utida et al. 2020) associated with an intensification of the SAMS. This interruption in delta construction resulted in shoreline retreat and reworking by wind and formation of an aeolian sand sheet in both sides of the river mouth. **d** In more recent times, since at least 0.2 ka (Szczygielski et al. 2015) progradation restarted apparently coinciding with the southward migration of the ITCZ during the LIA. Small arrows indicate the regional longshore transport and local reversals. See Fig. 4.5 for legend

(2021) documented how changes in precipitation in the watershed controlled the main episodes of delta construction in São Francisco.

These variations in precipitation at various time scales and their effects on river flows are recorded in the delta plains in the form of the geometry of the sedimentary accumulations and the patterns of orientation and truncation of the beach dune ridges. In Brazil, beach-dune ridge geometries and truncation patterns have been extensively used as indicators of changes in regional littoral transport deriving from changes in wave climate (Dominguez et al. 1983; Martin et al. 1985; Dominguez et al. 1992). This interpretation, however, is questionable since it is very difficult to change the regional wave climate to the point of reversing the direction of regional transport. It is much easier to explain the beach dune ridge geometries and truncations as a result of variations in river discharge and associated morphodynamic adjustments. As already

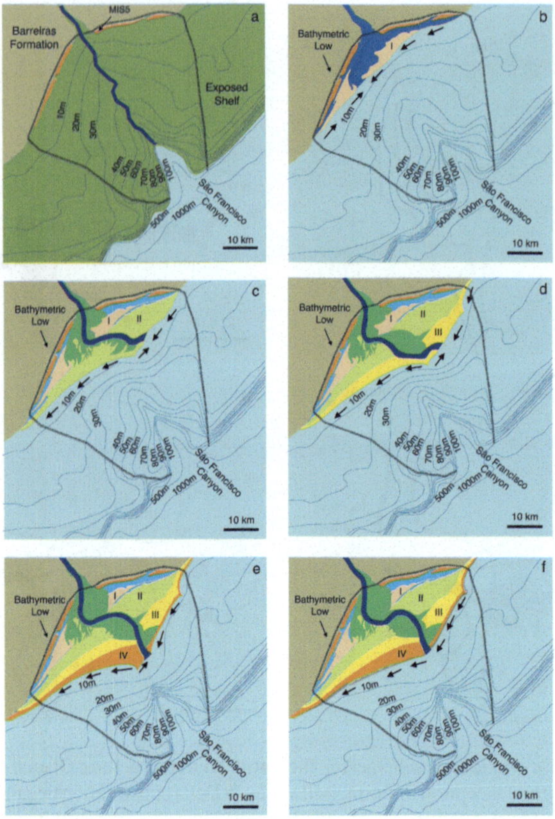

Fig. 4.9 Proposed evolutionary scheme for the São Francisco delta. **a** Last Glacial Maximum: shelf exposed subaerially and inception of the incised valley connecting river mouth and canyon. **b** Maximum of the Holocene transgression and beginning of episode I of delta construction 7.3–5.7 ka. **c**, **d** and **e** major episodes of delta progradation (II—5.7–4.8 ka, III—3.2–2.3 ka, and IV—1.1–1.0 ka) associated with periods of increased humidity in the catchment as recorded in speleothems (Stríkis et al. 2011; Novello et al. 2012) and coinciding with Bond events 4, 2 and 1. During episodes II and III the river mouth migrated updrift of the regional longshore transport. **f** After episode IV of delta construction the shoreline remained approximately stable. Small arrows indicate the regional longshore transport and local reversals. Based on Dominguez and Guimarães (2021). See Fig. 4.5 for legend of all units but for the Holocene beach deposits

mentioned, there is direct feedback between river inputs and the orientation of the coastline close to the river mouth. The greater the solid discharge from the river is, the greater the deflection or reorientation of the shoreline in relation to its original orientation in the absence of the river (Ashton and Giosan 2011).

Based on the evolutionary models presented in Figs. 4.8, 4.9, 4.10, 4.11 and 4.12, the major aspects important to understanding the impact of Holocene climate changes on delta construction are summarized below.

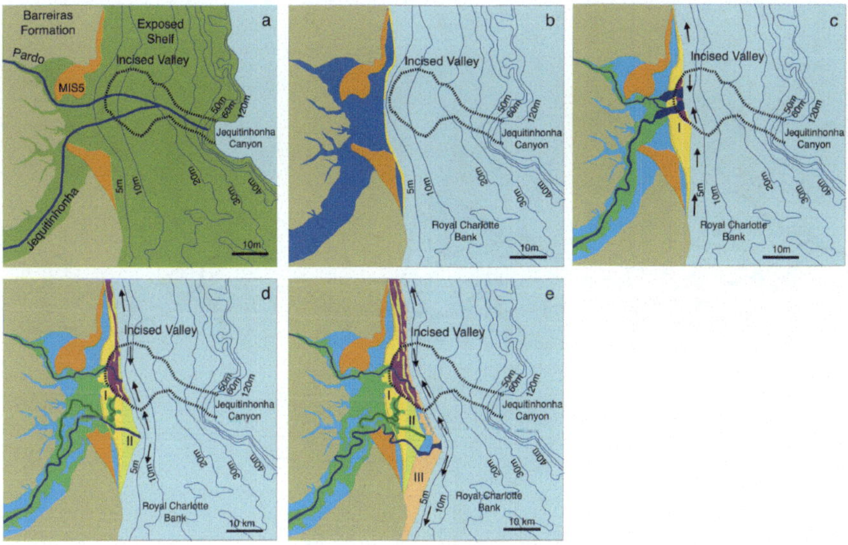

Fig. 4.10 Proposed evolutionary scheme for the Jequitinhonha delta: **a** Last Glacial Maximum: shelf exposed subaerially and inception of the incised valley connecting the river to the Jequitinhonha canyon. **b** Maximum of the Holocene transgression (5.8 ka). **c**, **d** and **e** During the Holocene three different episodes of delta construction (I, II, and III) occurred in association with the southward migration of the lower river course with the most recent episode beginning approximately 2.5 ka (Based on Dominguez et al. 1987). Note that most of the mangrove swamps in the delta are concentrated around the head of the incised valley which is also responsible for a slight concavity in the present day shoreline. Small arrows indicate the regional longshore transport and local reversals. See Fig. 4.5 for legend of all units but for the Holocene beach deposits

Parnaíba—based on the dates published by Szczygielski et al. (2015) at approximately 4.2 ka, a substantial portion of the delta plain had already been built (Figs. 4.7a and 4.8b). After this date, the coastline remained apparently stationary or even experienced an erosional retreat, which favored the development of dunes that migrated toward the interior of the plain (Fig. 4.8c). Apparently, only in a much more recent period (0.2 ka) (Szczygielski et al. 2015) did the shoreline restart, in association with a pronounced groyne effect (Figs. 4.7a and 4.8d).

São Francisco—according to Dominguez and Guimarães (2021), almost half of the delta plain was built up to 4.8 ka (Fig. 4.7b). Since then, there has been a progressive reduction in progradation rates, accompanying a progressive increase in the aridity in the middle-lower catchment. Major episodes of delta construction coincided with increased precipitation in the catchment, as indicated by speleothem studies (Dominguez and Guimarães 2021). The last major episode of shoreline progradation also took place at approximately 1.0 ka in association with a pronounced groyne effect (Figs. 4.7b and 4.9e). Apparently, the coastline has changed very little in the last 1000 years (Dominguez and Guimaraes 2021).

Jequitinhonha—a significant portion of the delta plain was built after 2.5 ka, when the lower river course reached its southernmost position (Dominguez 1982)

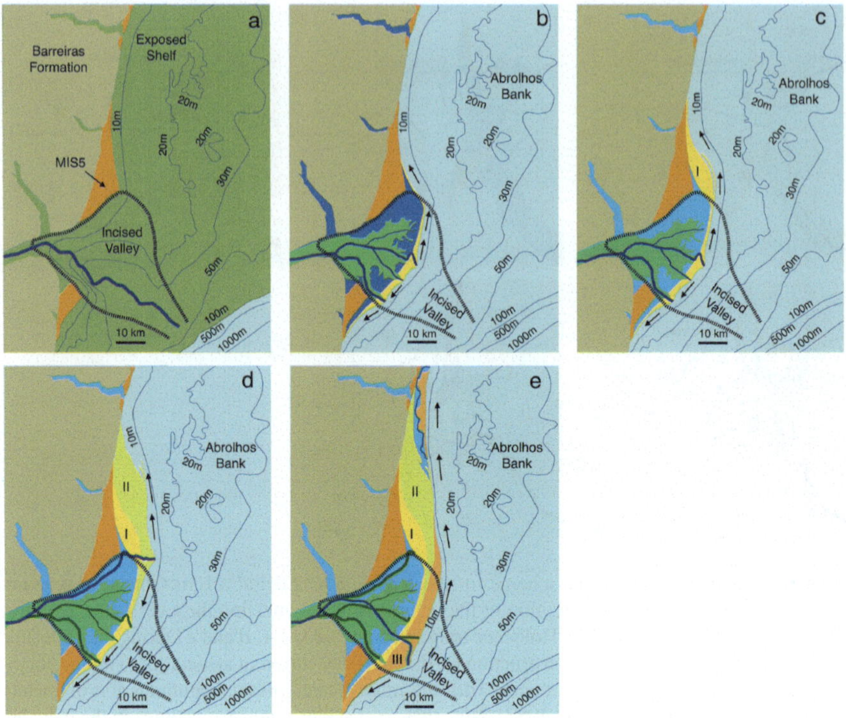

Fig. 4.11 Proposed evolutionary scheme for the Doce delta: **a** Last Glacial Maximum: shelf exposed subaerially and inception of the incised valley. **b** Maximum of the Holocene Transgression (5.8 ka) with beginning of the development of a large bay head delta in the most proximal portions of the incised valley (Dominguez and Wanless 1991). **c**, **d** and **e** Three major episodes of delta construction (I, II, and III) have been documented by Dominguez and Wanless (1991) and Martin et al. (1993a, b). The last episode of delta construction began approximately 2.5 ka. Note that during most of the time riverborne sediments were preferentially transported northward. Small arrows indicate the regional longshore transport and local reversals. See Fig. 4.5 for legend of all units but for the Holocene beach deposits

(Figs. 4.7c and 4.10e). The large area occupied by the third delta may be the result of reduced accommodation space, as it was built on the Royal Charlotte bank, or may derive from an increase in river input. In fact, during the construction of delta 3 (Figs. 4.7c and 4.10e), there were important variations in the fluvial flow that produced systematic truncations in the beach dune ridges (Fig. 4.8c). Of these truncations, the most important appears to have occurred approximately 500 years ago according to LOE ages reported by Rodrigues (2018) (Fig. 4.7c). After this episode of severe erosional retreat, caused by a decrease in river flow, the coastline prograded approximately 2 km (Fig. 4.7c). This progradation was once more interrupted by another episode of severe erosion that occurred in the beginning of the nineteenth century (Fig. 4.7c). Afterward, the shoreline was more prograded by almost 1 km.

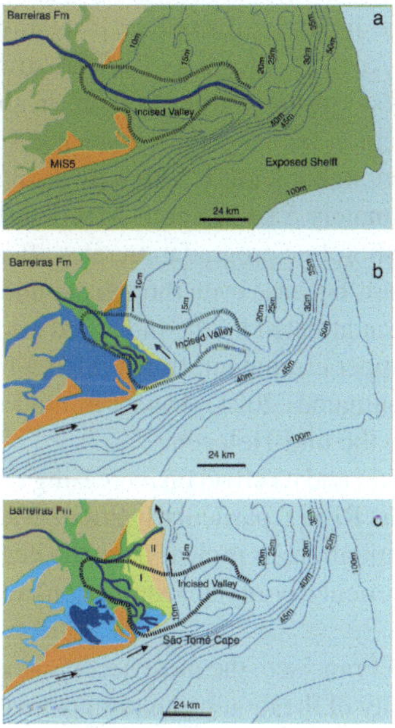

Fig. 4.12 Proposed evolutionary scheme for the Paraíba do Sul delta. **a** Last Glacial Maximum: shelf exposed subaerially and inception of the incised valley. **b** Maximum of the Holocene transgression (5.8–4.3 ka) (Martin et al. 1993a, b)—Construction of a bay head delta in the head of the incised valley. **c** At least two episodes (I, II) of delta construction have been documented for the Holocene. Episode I from 4.1 to 1.6 ka, and episode II starting approximately 1.0 ka (Rocha et al. 2019). Small arrows indicate the regional longshore transport and local reversals. See Fig. 4.5 for legend of all units but for the Holocene beach deposits

Doce—the most important shoreline progradation associated with the current river course also took place after 2.5 ka (Dominguez and Wanless 1991; Martin et al. 1993a, b; Martin et al. 1996). Unfortunately, good chronological control is not available for this last episode of progradation. However, the inspection of historical maps from the seventeenth and eighteenth centuries shows that the Doce River mouth protruded more pronounced seaward and was referred to as the Doce cape. This is corroborated by the beach-ridge geometries to the north of the present river mouth, which marks the position of a subrecent mouth that was later abandoned and truncated (Fig. 4.7d).

Paraíba do Sul—in this delta, the most expressive progradation of the coastline during the Holocene occurred after 1.5 ka according to LOE ages reported by Da Rocha et al. (2019) (Fig. 4.7e). During this period of rapid progradation, a pronounced groyne effect came into action, resulting in more deflected beach dune ridge sets separated by truncation discontinuities indicative of river flow variations (Fig. 4.7e).

In summary, it can be said that while in the São Francisco and Parnaíba deltas, the most expressive progradation took place before 4 Ka, in the Jequitinhonha, Doce and Paraíba do Sul, there was an increase in the progradation after 2.5 ka. Perhaps this behavior can be explained by the intensification of the SAMS since the Mid

Holocene as a result of the increase in summer insolation in the Southern Hemisphere (Utida et al. 2020). An intensification of SAMS strengthens South America's subtropical anticyclone, resulting in a drier climate in the northeast. Speleothem records in the Northeast region indicate that between 4.8 and 4.2 ka, there was an abrupt change in precipitation, coinciding with a reversal in the summer insolation trend (increase) at approximately 5 ka in South America (Utida et al. 2020). In São Francisco, the decrease in shoreline progradation rates after 4.8 ka was more gradual than that in Parnaíba, perhaps because even though the middle and lower portions of its catchment may have been affected by the increased aridification that occurred in the Northeast region, the upper catchment, from which 50% of the river's flow originates (Dominguez and Guimarães 2021), has its rainfall controlled by the SAMS, which has intensified since the mid-Holocene.

This SAMS intensification and its effect on increasing river flows in the Paraíba do Sul, Doce and Jequitinhonha Rivers, particularly after 2.5 ka, were also accompanied by great variability in precipitation, as is clear from the beach dune ridge geometries and truncation patterns, particularly in the Jequitinhonha and Paraíba do Sul deltas (Fig. 4.7c, e).

In all deltas, but for São Francisco, the most accentuated deflections of the coastline in the immediate vicinity of the mouths, implying greater fluvial flows, occurred in a much more recent period (last few centuries) and could be associated with the great intensification of the SAMS during the Little Ice Age (LIA) (Bird et al. 2011; Vuille et al. 2012; Novello et al. 2018). The absence of a more detailed chronological control, however, does not allow for a more in-depth testing of this hypothesis.

The concomitant increase in progradation in the Parnaíba delta during the same period is apparently contradictory to the speleothem records, according to which, during the LIA, extreme aridity dominated in the northeast region of Brazil. However, as already mentioned, between March and April, the ITCZ reaches its southernmost position in the northeast region of South America, a period in which the largest outflows occur in the Parnaíba River. Thus, a more southern position of the ITCZ during the LIA could also explain the marked progradation experienced in this delta during the last 200 years. However, this more southern ITCZ position apparently did not imply increased precipitation in the middle-lower catchment of São Francisco, which is more affected by the intensification of the descending limb of the regional Walker circulation (NE Low) (Cruz et al. 2009). This could explain why, at least from 1857 to 1978, the coastline in this delta remained apparently stable (Dominguez and Guimarães 2021).

4.7 Recent Coastline Changes

The comparison of satellite images from different dates allowed us to evaluate the losses and gains in the delta plain area, whose results are summarized in Table 4.1 and Fig. 4.13. Only in São Francisco did the losses significantly surpass the gains in the delta plain area, likely due to sediment removal from the shoreline by wind action. In the other extreme is the Paraíba do Sul, where gains did surpass losses, although this is one of the more regulated rivers and has had its mouth even temporarily sealed by a sand spit in October 2019 (https://www.jornalterceiravia.com.br/2020/01/12/mudanca-da-foz-retrata-problemas-enfrentados-pelo-paraiba-ha-decadas/). In the other deltas, gains and losses cancelled each other approximately.

In all deltas, the greatest losses in area occurred in the immediate vicinity of the river mouth with the eroded sediments redistributed to the lateral areas, which in this way experienced progradation. In deltas subject to a greater degree of angular asymmetry of the wave climate, the eroded sediments were preferentially transported to one side of the deltaic plain (e.g., São Francisco and Paraíba do Sul) (Fig. 4.13).

Nienhuis et al. (2013) examined the effect of the angular asymmetry of the wave climate onshoreline decay in wave-dominated deltas after river abandonment. They

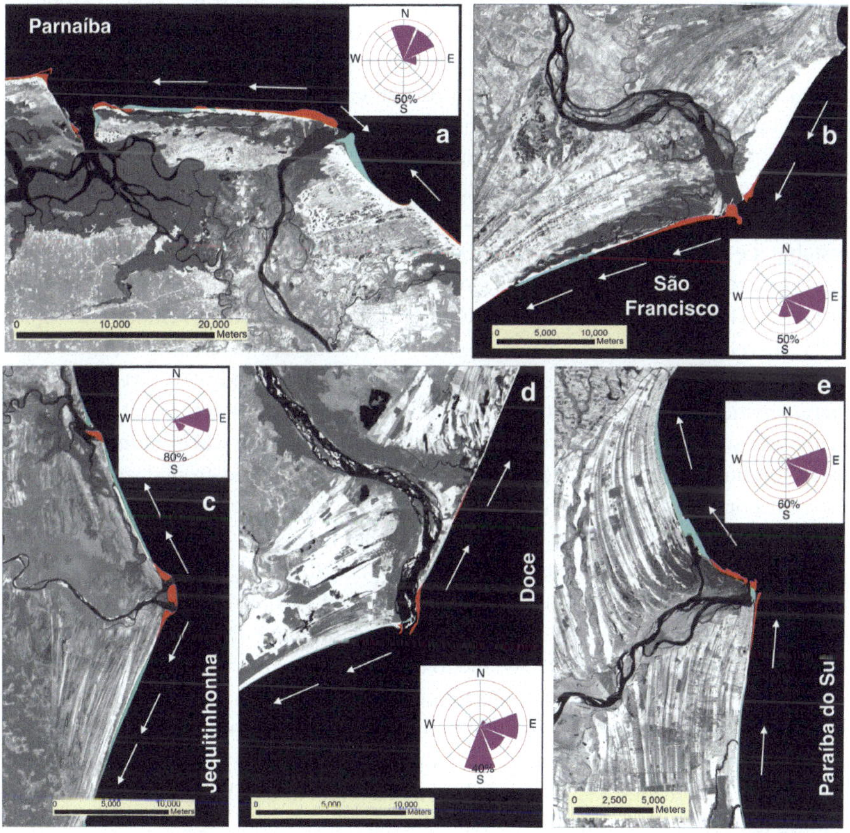

Fig. 4.13 Area losses and gains at the delta plains determined by comparison of Landsat infrared spectral band imagery, for the following time intervals: **a** Parnaíba (1987–2020), **b** São Francisco (1984–2019), **c** Jequitinhonha (1986–2020), **d** Doce (1985–2020), **e** Paraíba do Sul (1984–2021). See Table 4.1, for quantitative information. Insets show the angular wave climate (WaveWatchIII—National Weather Service—NOAA)

identified four distinct shoreline decay modes according to their geomorphological expression: (i) smooth diffusive shoreline, (ii) discontinuous shoreline, (iii) growing spit, and (iv) decaying shoreline sand waves. Although these models have been developed for a delta abandonment scenario, with a complete cut off of the sediment supply for the river mouth, they still help us to understand the current and future behavior of the coastline in the Brazilian deltas in a scenario of decreasing flows.

The Parnaíba, Jequitinhonha, and Doce deltas show a "smooth diffusive" type of behavior (Fig. 4.13a, c, d). This mode occurs when, before the abandonment or reduction of sediment input, the delta had a cuspate geometry where both the updrift and the downdrift sides had stable coastlines, that is, when the diffusivity is greater than zero, meaning that the waves reach the coastline at an angle of less than 45°. In this case, the abandoned delta will maintain its general shape, with erosion at the mouth and deposition on the lateral flanks. In the specific case of the Parnaíba, the above statement seems at first sight contradictory, since the delta exhibits a pronounced groyne effect typical of an asymmetric angular wave climate. However, since the coastline downdrift of the mouth has significantly prograded before the onset of the current erosional retreat, it became more susceptible to the action of waves coming from the north, promoting a significant redistribution of sediments to updrift of the regional longshore transport.

The São Francisco delta fits better in the "discontinuous shoreline" type of behavior (Fig. 4.13b). This mode appears when the downdrift shoreline is marginally unstable, i.e., the diffusivity is close to zero. After abandonment, a quick rearrangement occurs at the mouth with the eroded sediments transported in the downdrift direction. The discontinuity represented by the delta cusp will eventually dissipate as the shoreline flattens.

Finally, Paraíba do Sul presents a "decaying shoreline sand wave"-type behavior characteristic of very asymmetric angular wave climates, which result in a highly unstable downdrift coastline, triggering the formation of sand waves along the downdrift shoreline (Fig. 4.13e). Here, after leaving the mouth, the eroded sediments are also transported in the downdrift direction. The discontinuity represented by the delta cusp tends to persist longer compared to the other modes.

Briefly, in all deltas analyzed, the degree of shoreline deflection in the immediate vicinity of the mouth was much more pronounced in the recent past (decades to centuries), with deltaic cusps much more prominent than currently. Today, all these delta cusps are experiencing decay due to the combined effect of rainfall decrease in the catchments with river regulation and sediment retention in dam reservoirs. For the São Francisco River, Dominguez and Guimarães (2021) demonstrated that erosion at the river mouth is the result of a combination of these two factors. A reduction in flows in recent decades has affected all of these rivers and will be exacerbated in the coming decades due to ongoing climate change (Arias et al. 2021).

4.8 Vulnerability to Ongoing Climate Change

There is extensive literature published over the last decades discussing the impact of climate change on deltas (Tessler et al. 2015; Day et al. 2019; Prado et al. 2019; Hoitink et al. 2020). Much of this literature is focused on the great deltas of Southeast Asia and the Mediterranean Sea due to their high population densities and intensive use for agricultural purposes. In addition to the reduction in rainfall in the catchments, another major impact of climate change on deltas is sea-level rise and its implications for delta plain inundation and erosional shoreline retreat (Ibáñez et al. 2014). These aspects have been aggravated in recent decades by the construction of dams and artificial dikes that have reduced the ability of deltas to grow vertically (aggradation) and to feed the coastline with sediments (Vörösmarty et al. 2003; Syvitski et al. 2009; Day et al. 2016).

In the particular case of the Brazilian deltas located on the east coast, a few aspects must be considered when analyzing their vulnerability to climate change, since most of the delta plain is made up of beach dune deposits. This results in some important implications: (i) the soils are very poor in nutrients, therefore with limited use for agriculture which results in low population contingents in the delta plain, concentrated as a rule in a few urban centers (Table 4.1), (ii) the top of these beach-dune deposits is significantly higher than mean sea level due to wave run-up (berm height is a function of the wave height at breaking), and aeolian deposition (actually most of the delta plain beach ridges were formed as dune ridges), which further contributes to increase the average elevation of the terrain (Fig. 4.6). In the particular case of the São Francisco delta, the development of dune fields has caused some regions of the delta plain to have altitudes higher than 20 m (Fig. 4.6b), and (iii) because they were built in a GIA far-field region, the more internal beach deposits are approximately 3–4 m higher compared to those located in the outermost region of the delta plain (Fig. 4.6). These three factors contribute to increasing the resilience of these deltas to a rise in sea level (Fig. 4.14). In these deltas, the areas of greatest vulnerability to sea level rise are those occupied by alluvial or freshwater swamps (São Francisco and Paraíba do Sul) (Fig. 4.14b, e) and mangrove swamps (São Francisco and Jequitinhonha) (Fig. 4.14b, c). In the particular case of Jequitinhonha and Doce, the vertical build-up of the river's natural levees makes the alluvial plain less susceptible to inundation (Fig. 4.14c, d). At Doce, in particular, the northern portion of the delta plain is very prone to inundation by sea level rise, although situated in an area of beach-dune ridge deposits (Fig. 4.14d). A possible reason for the existence of such an extensive low-lying area may be because these beach ridges were deposited during a small fall in sea level between 4.6 and 3.8 cal ka and later invaded by the sea at approximately 3.5 ka, as hypothesized by Dominguez and Wanless (1991).

For the tidal-dominated delta of Parnaíba, the vulnerability to sea level rise is much higher than that of the wave-dominated east coast deltas because most of this delta plain is occupied by mangrove swamps (Fig. 4.14). The survivability of the Parnaíba do sea-level rise will depend on the capacity of these mangrove swamps

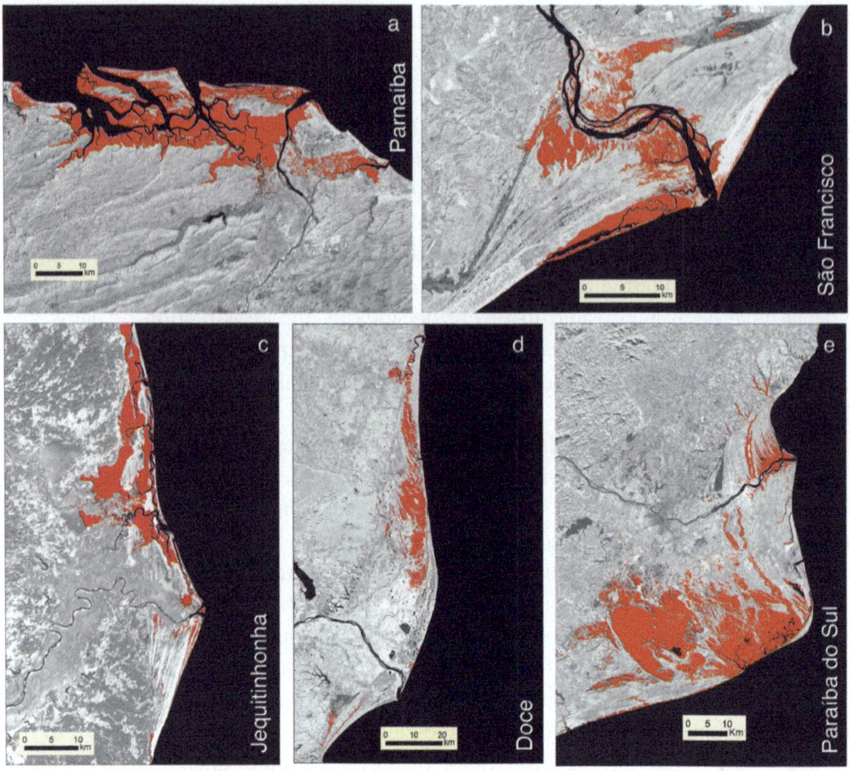

Fig. 4.14 Depiction of delta plain regions (red) more susceptible to inundation by a rise in sea level. In preparing these figures we have used Copernicus DEMs to simulate a 2 m rise in sea level, to emphasize the existing low-lying areas. We are aware that sea level rise projections by the end of this century are much lower but since Copernicus DEMs do not differentiate between vegetation from bare soil, a 1 m rise in sea level produced irrelevant results. Additionally, areas occupied by mangrove swamps were traced manually in Landsat 8 images and later incorporated in these figures

to build up in response to a rising sea level. However, this capacity will certainly be compromised by future changes in rainfall in the catchment over the next decades and by further river regulation (Rudorff et al. 2021). The least vulnerable areas of the plain are the sandy terraces that are made up of aeolian deposits. These deposits are almost exclusively interdune in origin and were left behind by barcan dune migration. The vertical aggradation during dune migration has produced sandy terraces with elevations as high as 6 m above current sea level (Fig. 4.6a). Therefore, Parnaíba, although still considered pristine, is the most vulnerable delta to climate change in Brazil.

For all Brazilian deltas, the results of climate models point to a reduction in precipitation in the watersheds, implying a reduction in river flows and, consequently, in the contribution of sediments to the mouth (Arias et al. 2021). This indicates that the decay process of the deltaic cusps presently taking place in all deltas will continue in the coming decades. It is possible that a significant decrease in the delta plain area will not occur; instead it there might be just a redistribution of sediments to the sides of the delta as the shoreline bulge decays.

4.9 Research Opportunities

In the last decade, a wealth of paleoclimatic data was produced in Brazil. These proxies are concentrated in two categories: (i) $\delta^{18}O$ and $^{87}Sr/^{86}Sr$, measured in speleothems with a good latitudinal distribution from northern to southern Brazil, complemented by records collected in other regions of South America, have allowed us to investigate variations in the intensity of SAMS, responsible for most of the rainfall in Brazil and therefore in the catchments of the deltas investigated herein (Cruz et al. 2009; Stríkis et al. 2011; Novello et al. 2018; Ward et al. 2019; Utida et al. 2020) and (ii) bulk sediment Ca, Ti and Fe elemental intensities, $\delta^{2}H$, TOC, and other sediment properties in cores collected in the Brazilian continental margin near major river mouths (Mulitza et al. 2017; Chiessi et al. 2021; Bahr et al. 2021) or in lakes in the continent (Utida et al. 2019) to reconstruct rainfall changes in the catchment of major rivers.

The next step would be to integrate these proxies with the evolutionary history of the Brazilian deltas, as done by Dominguez and Guimarães (2021) for the São Francisco delta.

Specifically, for the Paraiba do Sul and Jequitinhonha deltas, this application would be very promising due to the following aspects: (i) the catchments of these deltas are located in different climatic zones. While the Paraíba do Sul basin is located directly in the SACZ, Jequitinhonha is located further north in an area marginal to the SACZ (Bahr et al. 2021). (ii) In the outermost portions of these plains, built during the last 2.5 ka, there is a series of beach-dune ridge sets, with apparently cyclical truncation patterns indicative of variations in river flows. The mapping and establishment of a detailed chronology of these beach dune ridges using LOE would be essential to recover the palaeohydrology of these two rivers during the last 2.5 ka. Currently, existing age determinations are very limited and practically restricted to the radiocarbon method, which imposes several limitations in establishing an accurate chronology.

Another aspect that deserves attention is the study of subaqueous delta clinoforms due to their importance as archives of environmental signals in the watershed (changes in climate, sea level variations, anthropogenic activities, etc.) (Cattaneo et al. 2004; Liu et al. 2004; Bianchi and Allison 2009; Fanget et al. 2014). The delta clinoforms are characterized by high deposition rates, allowing for the preservation of different environmental records, unlike what happens, for example, in regions further away from the mouths, where the lower sediment input often results in the dominance of erosion, producing incomplete environmental records (Cattaneo and Steel 2003; Zecchin et al. 2011; Green et al. 2014). Bahr et al. (2021) present one of the few documented cases of environmental information reconstruction (rainfall variation) off the coast of Brazil using a core collected in a muddy accumulation on the continental shelf, located northeast of the Jequitinhonha River mouth.

The two most promising areas for carrying out studies of this nature are the São Francisco delta, which has the most expressive muddy delta clinoform of all Brazilian deltas (Rangel and Dominguez 2019), and the Jequitinhonha delta, which, although not presenting a well-developed subaqueous clinoform, has an expressive mud accumulation partially infilling its incised valley in the shelf (Fig. 6.5e).

In addition, other equally important research opportunities include (i) detailed assessment of the vulnerability of these deltas to climate change and (ii) integration of their evolutionary history with paleoclimate proxy records.

4.10 Concluding Remarks

The Brazilian deltas present an excellent opportunity to assess how the spatial variability in physiographic, oceanographic, and climatic factors determined their characteristics, Holocene evolution, and vulnerability to climate change. Of the five deltas investigated here, the four located on the east coast (Paraíba do Sul, Doce, Jequitinhonha and São Francisco) can be classified as wave-dominated, while the Parnaíba on the north coast is tide-dominated, although it is also strongly influenced by waves. This results in contrasting characteristics in the types of deposits present in the deltaic plains, with that of the Parnaíba consisting predominantly of mangrove swamps, while in the other four, sandy beach-dune ridge deposits dominate. This makes Parnaíba the most vulnerable delta to climate change due to the combined effects of projected sea-level rise and decrease in precipitation in the watershed, although of all the Brazilian deltas, it is the only delta that can be considered pristine.

Finally, these deltas and their clinoforms are important archives of variations in environmental signals in their catchments (e.g., precipitation, sediment retention in dams), which until now have been only incipiently explored.

Acknowledgements The following research grants have made this study possible: InctAmb-Tropic (CNPq/FAPESB No. 565054/2010-4, 8936/2011, 465634/2014-1, and inc004/2019), CAPES/PNPD (No. 2983/2010), and CNPQ (No. 405873/2016-6).

References

Angulo RJ, Lessa GC, Souza MC (2006) A critical review of mid- to late-Holocene sea-level fluctuations on the eastern Brazilian coastline. Quat Sci Rev 25:486–506. https://doi.org/10.1016/j.quascirev.2005.03.008

Anthony EJ (2015) Wave influence in the construction, shaping and destruction of river deltas: a review. Mar Geol. https://doi.org/10.1016/j.margeo.2014.12.004

Anthony EJ, Marriner N, Morhange C (2014) Human influence and the changing geomorphology of mediterranean deltas and coasts over the last 6000 years: from progradation to destruction phase? Earth-Sci Rev 139:336–361. https://doi.org/10.1016/j.earscirev.2014.10.003

Alvares CN, Stape JL, Sentelhas PC et al (2013) Köppen's climate classification map for Brazil. Meteorol Z 22:711–728. https://doi.org/10.1127/0941-2948/2013/0507

Araújo HAB, Dominguez JML, Machado AJ et al (2018) Benthic foraminifera distribution in a deltaic Clinoform (São Francisco delta, Eastern Brazil): a reference study. J Mar Sys 186:1–16. https://doi.org/10.1016/j.jmarsys.2018.05.004

Arias PA, Sathyendranath SN, Bellouin E et al (2021) Technical summary. In: Zhou MS, Delmotte V, Zhai P et al (eds) Climate Change 2021: the physical science basis. Contribution of working group I to the sixth assessment report of the intergovernmental panel on climate change. Cambridge University Press. https://www.ipcc.ch/report/ar6/wg1/downloads/report/IPCC_AR6_WGI_TS.pdf

Ashton AD, Giosan L (2011) Wave-Angle control of delta evolution. Geophys Res Lett 38:1–6. https://doi.org/10.1029/2011GL047630

Bahr A, Kaboth-Bahr S, Jaeschke A et al (2021) Late Holocene precipitation fluctuations in South America triggered by variability of the North Atlantic overturning circulation. Paleoceanogr Paleoclimatol. https://doi.org/10.1029/2021pa004223

Barbosa LM, Dominguez JML (2004) Coastal Dune Fields at the São Francisco River Strandplain, Northeastern Brazil: morphology and environmental controls. Earth Surf Process Landf 29:443–456. https://doi.org/10.1002/esp.1040

Bard E, Hamelin B, Delanghe-Sabatier D (2010) Deglacial Meltwater Pulse 1B and younger Dryas sea levels revisited with boreholes at Tahiti. Science 327:1235–1237. https://doi.org/10.1126/science.1180557

Besset M, Anthony EJ, Bouchette F (2019) Multi-Decadal variations in delta shorelines and their relationship to river sediment supply: an assessment and review. Earth-Sci Rev 193:199–219. https://doi.org/10.1016/j.earscirev.2019.04.018

Bhattacharya JP, Giosan L (2003) Wave-Influenced deltas: geomorphological implications for facies reconstruction. Sedimentology 50:187–210. https://doi.org/10.1046/j.1365-3091.2003.00545.x

Bianchi TS, Allison MA (2009) Large-River delta-front estuaries as natural "recorders" of global environmental change. PNAS 106:8085–8092. https://doi.org/10.1073/pnas.0812878106

Bird BW, Abbott MB, Vuille M et al (2011) A 2300-Year-long annually resolved record of the South American summer monsoon from the Peruvian Andes. PNAS 108:8583–8588. https://doi.org/10.1073/pnas.1003719108

Blum M, Martin J, Milliken K et al (2013) Paleovalley systems: insights from quaternary analogs and experiments. Earth-Sci Rev 116:128–169. https://doi.org/10.1016/j.earscirev.2012.09.003

Campos MC, Chiessi CM, Prange M et al (2019) A new mechanism for millennial scale positive precipitation anomalies over tropical South America. Quat Sci Rev. https://doi.org/10.1016/j.quascirev.2019.105990

Carelli TG, Plantz JB, Borghi L (2018) Facies and Paleoenvironments in Paraíba Do Sul Deltaic Complex Area, North of Rio de Janeiro State, Brazil. J S Am Earth Sci 86:431–446. https://doi.org/10.1016/j.jsames.2018.07.005

Carvalho MA, Plantz JB, Carelli TG et al (2019) The impact of quaternary sea-level changes on the sedimentary organic matter of the Paraíba Do Sul Deltaic Complex Area, Southeastern Brazil. J S Am Earth Sci. https://doi.org/10.1016/j.jsames.2019.102274

Cattaneo A, Steel RJ (2003) Transgressive deposits: a review of their variability. Earth-Sci Rev 62:187–228. https://doi.org/10.1016/S0012-8252(02)00134-4

Cattaneo A, Trincardi F, Langone L et al (2004) Clinoform generation on mediterranean margins. Oceanography 17:104–117. https://doi.org/10.5670/oceanog.2004.08

Chiessi CM, Mulitza S, Taniguchi NK et al (2021) Mid- to Late Holocene contraction of the intertropical convergence zone over Northeastern South America. Paleoceanogr Paleoclimatol 36:1–20. https://doi.org/10.1029/2020PA003936

Cruz FW, Vuille M, Burns SJ et al (2009) Orbitally driven East-West Antiphasing of South American precipitation. Nature Geosci 2:210–214. https://doi.org/10.1038/ngeo444
Day JW, Guun JD, Folan WJ et al (2007) Emergence of complex societies after sea level stabilized. Eos 88:169–176. https://doi.org/10.1029/2007EO420011
Day JW, Agboola J, Chen Z et al (2016) Approaches to defining deltaic sustainability in the 21st Century. Estuar Coast Shelf Sci 183:275–291. https://doi.org/10.1016/j.ecss.2016.06.018
Day JW, Ramachandran R, Giosan L et al (2019) Delta winners and losers in the anthropocene. Coasts and Estuaries: The Future. https://doi.org/10.1016/B978-0-12-814003-1.00009-5
Dessart RL (2009) Caracterizaçao Estratigrafica Do Preenchimento Do Vale Inciso Do Baixo Rio Doce Durante o Pleistoceneo Superior-Holoceno. Doctoral Dissertation. Unisinos, São Leopoldo, Brazil
Dominguez JML (1982) Evolucao Quaternaria da Planicie Costeira Associada à Foz do Rio Jequitinhonha (BA): Influência das Variações do Nível do Mar e da Deriva Litorânea de Sedimentos. Master Thesis. Federal University of Bahia, Brazil
Dominguez JML, Guimarães JK (2021) Effects of Holocene climate changes and anthropogenic river regulation in the development of a wave-dominated delta: the São Francisco River (Eastern Brazil). Mar Geol. https://doi.org/10.1016/j.margeo.2021.106456
Dominguez JML, Wanless HR (1991) Facies Architecture of a Falling Sea-Level Strandplain, Doce River Coast, Brazil. In: Swift DJP, Oertel GF, Tillman RW et al (eds) Shelf Sand and Sandstone Bodies. Wiley, New York, pp 257–281. https://doi.org/10.1002/9781444303933.ch7
Dominguez JML, Bittencourt ACSP, Martin L (1983) O Papel da Deriva Litorânea de Sedimentos Arenosos na Construção das Planícies Costeiras Associadas às Desembocaduras dos rios São Francisco (SE, AL), Jequitinhonha (BA), Doce (ES) e Paraíba Do Sul (RJ). Rev Bras Geoc 13:98–105. https://doi.org/10.25249/0375-7536.198313298105
Dominguez JML, Bittencourt ACSP, Martin L (1992) Controls on quaternary coastal evolution of the east-northeastern coast of Brazil: roles of sea-level history, trade winds and climate. Sediment Geol 80:213–232. https://doi.org/10.1016/0037-0738(92)90042-P
Dominguez JML, Martin L, Bittencourt ACSP (1987) Sea-Level history and quaternary evolution of river mouth-associated beach-ridge plains along the east-southeast brazilian coast: a summary. In: Nummedal D, Pilkey DH, Howard JD (eds) Sea-Level fluctuation and coastal evolution. Soc. Econ. Paleontol. Mineral. Special Publication 41, Tulsa, Oklahoma pp 115–127. https://doi.org/10.2110/pec.87.41.0115
Dominguez JML, Silva RP, Nunes AS et al (2013) The Narrow, shallow, low-accommodation shelf of Central Brazil: sedimentology, evolution, and human uses. Geomorphology 203:46–59. https://doi.org/10.1016/j.geomorph.2013.07.004
Edward AJ (2015) Wave influence in the construction, shaping and destruction of river deltas: a review. Mar Geol 361:53–78. https://doi.org/10.1016/j.margeo.2014.12.004
Fagundes HO, Fan FM, Paiva RCD et al (2021) Sediment flows in South America supported by daily hydrologic-hydrodynamic modeling. Water Resour Res 57:1–26. https://doi.org/10.1029/2020WR027884
Fanget AS, Bassetti MA, Arnaud M et al (2013) Historical evolution and extreme climate events during the last 400 years on the Rhone Prodelta (NW Mediterranean). Mar Geol 346:375–391. https://doi.org/10.1016/j.margeo.2012.02.007
Fanget AS, Berné S, Jouet G et al (2014) Impact of relative sea level and rapid climate changes on the architecture and lithofacies of the Holocene Rhone subaqueous delta (Western Mediterranean Sea). Sediment Geol 305:35–53. https://doi.org/10.1016/j.sedgeo.2014.02.004
Fraticelli CM (2006) Climate forcing in a wave-dominated delta: the effects of drought-flood cycles on delta progradation. J Sedimentary Res 76:1067–1076. https://doi.org/10.2110/jsr.2006.097
Frings RM (2008) Downstream fining in large sand-bed rivers. Earth-Sci Rev 87:39–60. https://doi.org/10.1016/j.earscirev.2007.10.001
Galloway WE (1975) Process framework for describing the morphologic and stratigraphic evolution of delta depositional systems. In: Broussard ML (ed) Deltas: Models for Exploration. Houston Geological Society, Houston, pp 87–98

Gasparini NM, Gregory ET, Bras RL (2004) Network-Scale dynamics of grain-size sorting: implications for downstream fining, stream-profile concavity, and drainage basin morphology. Earth Surf Process Landf 29:401–421. https://doi.org/10.1002/esp.1031

Giosan L, Syvitski J, Constantinescu S et al (2014) climate change: protect the world's deltas. Nature 516:31–33. https://doi.org/10.1038/516031a

Green AN, Cooper JAG, Salzmann L (2014) Geomorphic and stratigraphic signals of postglacial meltwater pulses on continental shelves. Geology 42:151–154. https://doi.org/10.1130/G35052.1

Guimarães JK (2010) Evolução Do Delta Do Rio São Francisco - Estratigrafia Do Quaternário e Relações Morfodinâmicas. Doctoral Dissertation. Federal University of Bahia, Brazil.

Hammond WC, Blewitt G, Kreemer C et al (2021) GPS imaging of global vertical land motion for studies of sea level rise. J Geophys Res Solid Earth 126:1–26. https://doi.org/10.1029/2021JB022355

Haug GH, Hughen KA, Sigman DM et al (2001) Southward migration of the intertropical convergence zone through the Holocene. Science 293:1304–1308. https://doi.org/10.1126/science.1059725

Heilbron M, Cordani UG, Alkmim FF (eds) (2017) São Francisco Craton, Eastern Brazil. Regional Geology Reviews: São Francisco Craton, Eastern Brazil: Tectonic Genealogy of a Miniature Continent. Springer International Publishing. https://doi.org/10.1007/978-3-319-01715-0

Hoitink AJF, Nittrouer JA, Passalacqua P et al (2020) Resilience of River deltas in the anthropocene. J Geophys Res Earth Surface 125:1–24. https://doi.org/10.1029/2019JF005201

Ibáñez C, Day JW, Reyes E (2014) The response of deltas to sea-level rise: natural mechanisms and management options to adapt to high-end scenarios. Ecol Eng 65:122–130. https://doi.org/10.1016/j.ecoleng.2013.08.002

Ibáñez C, Alcaraz C, Caiola N et al (2019) Basin-Scale land use impacts on world deltas: human vs natural forcings. Glob Planet Change 173:24–32. https://doi.org/10.1016/j.gloplacha.2018.12.003

Kenitiro S, Martin L, Bittencourt ACSP et al (1985) Flutuações Do Nível Relativo Do Mar Durante o Quaternário Superior Ao Longo Do Litoral Brasileiro e Suas Implicacões Na Sedimentação Costeira. Rev Bras Geoc 15:273–86. http://www.ppegeo.igc.usp.br/index.php/rbg/article/view/12026/11570

Kennett DJ, Kennett JP (2006) Early State formation in Southern Mesopotamia: sea levels, shorelines, and climate change. J Isl Coast Archaeol 1:67–99. https://doi.org/10.1080/15564890600586283

Langbein WB (1963) The hydraulic geometry of a shallow estuary. Intern Assoc Scient Hydrol Bull 8:84–94. https://doi.org/10.1080/02626666309493340

Lenters JD, Cook KH (1997) On the origin of the Bolivian high and related circulation features of the South American climate. J Atmos Sci 54:656–677. https://doi.org/10.1175/1520-0469(1997)054%3c0656:otootb%3e2.0.co;2

Liu JP, Milliman JD, Gao S et al (2004) Holocene Development of the Yellow River's subaqueous delta, North Yellow Sea. Mar Geol 209:45–67. https://doi.org/10.1016/j.margeo.2004.06.009

Lopes VHR, Frazão EP (2018) Mapa Batimétrico Da Plataforma Rasa Da Bahia—Paleocanais Da Foz Do Rio Jequitinhonha e Rio Pardo-Salobro—CPRM, 2018, Scale 1:100.000 (Programa Mar, Zona Costeira e Antártica. Folha de Referência: S A.24-Y -A-V)

Marengo JA, Tomasella J, Nobre CA (2010) Mudanças Climáticas e Recursos Hídricos. In: Bicudo CEM, Tundisi JG, Scheuenstuhl MCB (eds) Águas do Brasil Análises Estratégicas. SIMA, São Paulo, Brazil 199–215.

Marengo JA, Chou SC, Kay G et al (2012a) Development of regional future climate change scenarios in South America Using the Eta CPTEC/HadCM3 climate change projections: climatology and regional analyses for the Amazon, São Francisco and the Paraná River Basins. Clim Dyn 38:1829–1848. https://doi.org/10.1007/s00382-011-1155-5

Marengo JA, Liebmann B, Grimm AM et al (2012b) Recent developments on the South American Monsoon System. Int J Climatol 32:1–21. https://doi.org/10.1002/joc.2254

Marriner N, Flaux C, Kaniewski D et al (2012) ITCZ and ENSO-like pacing of Nile delta hydrogeomorphology during the Holocene. Quat Sci Rev 45:73–84. https://doi.org/10.1016/j.quascirev.2012.04.022

Martin L, Suguio K (1992) Variation of coastal dynamics during the last 7000 years recorded in beach-ridge plains associated with river mouths: example from the Central Brazilian coast. Palaeogeogr Palaeoclimatol Palaeoecol 99:119–140. https://doi.org/10.1016/0031-0182(92)90010-3

Martin L, Suguio K, Flexor JM (1988) Hauts Niveaux Marins Pleistocenes Du Littoral Bresilien. Palaeogeogr Palaeoclimatol Palaeoecol 68:231–239. https://doi.org/10.1016/0031-0182(88)90042-9

Martin L, Fournier M, Mourguiart P et al (1993a) Southern oscillation signal in South American Palaeoclimatic data of the Last 7000 years. Quat Res 39:338–346. https://doi.org/10.1006/qres.1993.1040

Martin L, Suguio K, Flexor JM (1993b) As Flutuações Nível do Mar Durante o Quaternário Superior e a Evoluçao Geológica dos Deltas Brasileiros. Boletim IG-USP - Publicação Especial 15, São Paulo, Brazil pp 1–186

Martin L, Suguio K, Flexor JM et al (1996) Quaternary sea-level history and variation in dynamics along the central Brazilian coast: consequences on coastal plain construction. An Acad Bras Ci 68:303–354

Martin L, Dominguez JML, Bittencourt ACSP (2003) Fluctuating Holocene Sea Levels in Eastern and Southeastern Brazil: Evidence from Multiple Fossil and Geometric Indicators. J Coast Res 19:101–124. https://www.jstor.org/stable/4299151

Maselli V, Trincardi F (2013) Man made deltas. Sci Rep. https://doi.org/10.1038/srep01926

Milly PCD, Dunne KA, Vecchia AV (2005) Global pattern of trends in streamflow and water availability in a changing climate. Nature 438:347–350. https://doi.org/10.1038/nature04312

Mulitza S, Chiessi CM, Schefuß E et al (2017) Synchronous and proportional Deglacial changes in Atlantic meridional overturning and northeast Brazilian precipitation. Paleoceanography 32:622–633. https://doi.org/10.1002/2017PA003084

Murillo VC, Silva CG, Fernandez GB (2009) Nearshore Sediments and Coastal Evolution of Paraíba Do Sul River Delta, Rio de Janeiro, Brazil. J Coast Res, Si 56:650–654. https://www.jstor.org/stable/25737658

Nienhuis JH, Ashton AD, Roos PC et al (2012) Modeling plan—form deltaic response to changes in fluvial sediment supply. In: NCK-days 2012: crossing borders in coastal research. March 2012, Enschede, the Netherlands. https://doi.org/10.3990/2.192

Nienhuis JH, Ashton AD, Roos PC et al (2013) Wave reworking of abandoned deltas. Geophys Res Lett 40:5899–5903. https://doi.org/10.1002/2013GL058231

Nienhuis JH, Ashton AD, Giosan L (2015) What makes a delta wave-dominated? Geology 43:511–514. https://doi.org/10.1130/G36518.1

Nienhuis JH, Hoitink AJF, Törnqvist TE (2018) Future change to tide-influenced deltas. Geophys Res Lett 45:3499–3507. https://doi.org/10.1029/2018GL077638

Novello VF, Cruz FW, Karmann I et al (2012) Multidecadal climate variability in Brazil's Nordeste during the last 3000 years based on speleothem isotope records. Geophys Res Lett 39:1–6. https://doi.org/10.1029/2012GL053936

Novello VF, Cruz FW, Moquet JS et al (2018) Two Millennia of South Atlantic convergence zone variability reconstructed from isotopic proxies. Geophys Res Lett 45:5045–5051. https://doi.org/10.1029/2017GL076838

Nyberg B, Gawthorpe RL, Helland-Hansen W (2018) The distribution of rivers to terrestrial sinks: implications for sediment routing systems. Geomorphology 316:1–23. https://doi.org/10.1016/j.geomorph.2018.05.007

Oliveira PE, Barreto AMF, Suguio K (1999) Late Pleistocene/Holocene climatic and vegetational history of the Brazilian Caatinga: the fossil dunes of the middle Sao Francisco River. Palaeogeogr Palaeoclimatol Palaeoecol 152:319–337. https://doi.org/10.1016/S0031-0182(99)00061-9

Overeem I, Kroonenberg SB, Veldkamp A et al (2003) Small-Scale stratigraphy in a large ramp delta: recent and Holocene sedimentation in the Volga Delta, Caspian Sea. Sediment Geol 159:133–157. https://doi.org/10.1016/S0037-0738(02)00256-7

Peltier WR (2007) Postglacial coastal evolution: Ice-Ocean-solid earth interactions in a period of rapid climate change. In: Harff J, Hay WW, Tetzlaff DM (eds) Coastline changes: interrelation of climate and geological processes: geological society of America special paper 426, pp 5–28. https://doi.org/10.1130/2007.2426(02)

Polyak VJ, Onac BP, Fornós JJ et al (2018) A highly resolved record of relative sea level in the Western Mediterranean Sea during the last interglacial period. Nature Geosci 11:860–864. https://doi.org/10.1038/s41561-018-0222-5

Prado P, Alcaraz C, Benito X et al (2019) Pristine vs. Human-Altered Ebro Delta habitats display contrasting resilience to RSLR. Sci Total Environ 655:1376–1386. https://doi.org/10.1016/j.scitotenv.2018.11.318

Rangel AGAN, Dominguez JML (2019) Antecedent topography controls preservation of latest Pleistocene-Holocene transgression record and Clinoform development: the case of the São Francisco Delta (Eastern Brazil). Geo-Mar Lett 40:935–947. https://doi.org/10.1007/s00367-019-00609-8

Revel M, Ducassou E, Grousset FE et al (2010) 100,000 Years of African Monsoon variability recorded in sediments of the Nile Margin. Quat Sci Rev 29:1342–1362. https://doi.org/10.1016/j.quascirev.2010.02.006

Ribeiro RF, Dominguez JML, Santos AA et al (2021) Continuous Canyon-River connection on a passive margin: the case of São Francisco Canyon (Eastern Brazil). Geomorphology. https://doi.org/10.1016/j.geomorph.2020.107549

Rocha TB, Vasconcelos SC, Pereira TG et al (2019) Datação Por Luminescência Opticamente Estimulada (LOE) nas Cristas de Praia do Delta do Rio Paraíba do Sul (RJ): Considerações Sobre a Evolução Geomorfológica entre o Pleistoceno Superior e o Holoceno. Rev Bras Geomorfol 20:563–80. https://doi.org/10.20502/rbg.v20i3.1620

Rodrigues GB (2018) Geocronologia das Coberturas Superficiais dm Baixos Terraços e na Foz da Bacia Hidrográfica do Rio Jequitinhonha (BA) relacionada com as Pulsações Climáticas do Holoceno. Master Thesis. State University of Campinas, Brazil

Rodriguez AB, Hamilton MD, Anderson JB (2000) Facies and evolution of the modern Brazos Delta, texas: wave versus flood Influence. J Sediment Res 70:283–295. https://doi.org/10.1306/2dc40911-0e47-11d7-8643000102c1865d

Rossetti DF, Bezerra FHR, Dominguez JML (2013) Late Oligocene-Miocene transgressions along the equatorial and eastern margins of Brazil. Earth-Sci Rev 123:87–112. https://doi.org/10.1016/j.earscirev.2013.04.005

Rudorff C, Sparrow S, Guedes MRG et al (2021) Event attribution of Parnaiba River floods in Northeastern Brazil. Climate Resil Sustain. https://doi.org/10.1002/cli2.16

Salles FJP, Bentes FCM, Santos JA (2020) Catálogo de Estações Maregráficas. Fundação de Estudos do Mar, Rio de Janeiro

Shimizu MH, Sampaio G, Venancio IM et al (2020) Seasonal changes of the South American Monsoon system during the Mid-Holocene in the CMIP5 simulations. Clim Dyn 54:2697–2712. https://doi.org/10.1007/s00382-020-05137-1

Stanley DJ, Warne AG (1994) Worldwide initiation of Holocene marine deltas by deceleration of sea-level rise. Science 265:228–231. https://doi.org/10.1126/science.265.5169.228

Stive MJF, Rakhorst RD (2008) Review of empirical relationships between inlet cross-section and Tidal Prism. Int J Water Resour Environ Eng 23:89–95

Stríkis NM, Cruz FW, Cheng H et al (2011) Abrupt variations in South American Monsoon rainfall during the Holocene based on a Speleothem record from Central-Eastern Brazil. Geology 39:1075–1078. https://doi.org/10.1130/G32098.1

Suguio K, Martin L (1981) Significance of quaternary sea-level fluctuations for delta construction along the Brazilian coast. Geo-Mar Lett 1:181–185. https://doi.org/10.1007/BF02462431

Syvitski JPM, Kettner AJ, Overeem I et al (2009) Sinking deltas due to human activities. Nat Geosci 2:681–686. https://doi.org/10.1038/ngeo629
Szczygielski A, Stattegger K, Schwarzer K et al (2015) Evolution of the Parnaíba Delta (NE Brazil) during the late Holocene. Geo-Mar Lett 35:105–117. https://doi.org/10.1007/s00367-014-0395-x
Takahashi K, Battisti DS (2007) Processes controlling the mean tropical pacific precipitation pattern. Part II: the SPCZ and the southeast pacific dry zone. J Clim 20:5696–5706. https://doi.org/10.1175/2007JCLI1656.1
Tessler ZF, Vörösmarty CJ, Grossberg M et al (2015) Profiling risk and sustainability in coastal deltas of the world. Science 349:638–643. https://doi.org/10.1126/science.aab3574
Utida G, Cruz FW, Etourneau J et al (2019) Tropical South Atlantic influence on Northeastern Brazil precipitation and ITCZ displacement during the past 2300 years. Sci Rep. https://doi.org/10.1038/s41598-018-38003-6
Utida G, Cruz FW, Santos RV et al (2020) Climate changes in Northeastern Brazil from deglacial to meghalayan periods and related environmental impacts. Quat Sci Rev. https://doi.org/10.1016/j.quascirev.2020.106655
Vieira FV, Bastos AC, Quaresma VS et al (2019) Along-Shelf changes in mixed carbonate-siliciclastic sedimentation patterns. Cont Shelf Res. https://doi.org/10.1016/j.csr.2019.103964
Vörösmarty CJ, Meybeck M, Fekete B et al (2003) Anthropogenic sediment retention: major global impact from registered river impoundments. Glob Planet Change 39:169–190. https://doi.org/10.1016/S0921-8181(03)00023-7
Vuille M, Burns SJ, Taylor BL et al (2012) A review of the South American Monsoon history as recorded in stable isotopic proxies over the past two Millennia. Clim Past 8:1309–1321. https://doi.org/10.5194/cp-8-1309-2012
Ward BM, Wong CI, Novello VF et al (2019) Reconstruction of Holocene coupling between the South America Monsoon system and local moisture variability from Speleothem δ 18O and 87Sr/86Sr records. Quat Sci Rev 210:51–63. https://doi.org/10.1016/j.quascirev.2019.02.019
Xing F, Kettner AJ, Ashton A et al (2014) Fluvial response to climate variations and anthropogenic perturbations for the Ebro River, Spain in the Last 4000 years. Sci Total Environ 473–474:20–31. https://doi.org/10.1016/j.scitotenv.2013.11.083
Zecchin M, Ceramicola S, Gordini E et al (2011) Cliff overstep model and variability in the geometry of transgressive erosional surfaces in high-gradient shelves: the case of the Ionian Calabrian margin (Southern Italy). Mar Geol 281:43–58. https://doi.org/10.1016/j.margeo.2011.02.003
Zular A, Utida G, Cruz FW et al (2018) The effects of mid-holocene fluvio-eolian interplay and coastal dynamics on the formation of dune-dammed lakes in NE Brazil. Quat Sci Rev 196:137–153. https://doi.org/10.1016/j.quascirev.2018.07.022

Chapter 5
Reefs of the Western Tropical South Atlantic Ocean: Distribution, Environmental Impacts and Trends on Environmental Suitability Due to Climate Changes

Ruy Kenji Papa de Kikuchi, Zelinda Margarida de Andrade Nery Leão, Marilia de Dirceu Machado de Oliveira, Marcelo Oliveira Soares, Maria Elisabeth De Araújo, João Lucas Leão Feitosa, Caroline Vieira Feitosa, Carlos Eduardo Rocha Duarte Alencar, and Fulvio Aurélio Morais Freire

Abstract Reefs that grow in the western tropical South Atlantic are known as marginal reefs because they thrive in environmental conditions (e.g., elevated turbidity) far from those considered to be the optimal conditions for framework builders (calcareous skeleton secreting organisms, such as corals). This is assumed to be one reason for the lower taxonomic richness in coral species and relatively higher endemism compared to reefs in other regions of the world, such as the Caribbean and the Pacific. These reefs, considered hot spots of biodiversity and home to turbidity-tolerant corals, are increasingly affected by local anthropogenic-driven impacts (pollution, dredging-related sedimentation, overfishing, unregulated tourism, and bioinvasions), global climate changes (warming, heatwaves, acidification, and sea-level rise) and their synergic effects. As a consequence, ecosystem goods and services are severely affected. Reef location in relation to the coastline and to urbanized areas are key points of vulnerability to the impacts of climate change.

R. K. P. de Kikuchi (✉) · Z. M. de A. N. Leão · M. de D. M. de Oliveira
Federal University of Bahia (UFBA), Salvador, Bahia, Brazil
e-mail: kikuchi@ufba.br

M. O. Soares · C. V. Feitosa
Federal University of Ceará (UFC), Fortaleza, Ceará, Brazil

M. O. Soares
Leibniz Center for Tropical Marine Research (ZMT), Bremen, Germany

M. E. De Araújo · J. L. L. Feitosa
Federal University of Pernambuco (UFPE), Recife, Pernambuco, Brazil

C. E. R. D. Alencar
Southwestern Bahia State University (UESB), Jequié, Bahia, Brazil

C. E. R. D. Alencar · F. A. M. Freire
Federal University of Rio Grande do Norte (UFRN), Natal, Rio Grande do Norte, Brazil

J. M. Landim Dominguez et al. (eds.), *Tropical Marine Environments of Brazil*,
The Latin American Studies Book Series, https://doi.org/10.1007/978-3-031-21329-8_5

In this chapter, we synthesize the main characteristics of the distribution of reef and coralline ecosystems of the tropical southwestern Atlantic. Finally, we use three endemic coral species (*Mussismilia hispida*, *Mussismilia harttii* and *Mussismilia braziliensis*) as proxies of the reef ecosystem to evaluate the trend of environmental suitability across the Brazilian Tropical Marine region in the RCP8.5 scenario.

Keywords Marginal reefs · Coralline environments · Coral distribution · Fish diversity · Global change

5.1 Introduction

Reefs that grow in the southwestern Atlantic thrive in environmental conditions that are far from optimal. They are subject to turbidity and sedimentation higher than what is considered the normal condition for reef builders (calcareous skeleton secreting organisms, such as corals). This is why they are known as marginal reefs (Kleypas et al. 1999; Suggett et al. 2012). This condition is assumed to be one reason for the lower diversity in coral species and relatively higher endemism compared to reefs in other ecoregions, such as in the Caribbean and in the Pacific (Laborel 1969; Leão and Ginsburg 1997; Leão et al. 2003). Thus, despite being hot spots of biodiversity and the home of turbid-tolerant corals, a synergy of local anthropogenic-driven impacts (pollution, dredging-related sedimentation, overfishing, unregulated tourism, and bioinvasions) and of globally controlled climate changes (warming, heatwaves, acidification, and sea-level rise) have enhanced their vulnerability. As a consequence of this increase in impacts, ecosystem goods and services are severely affected (Elliff and Kikuchi 2017). Reef location in relation to the coastline and to urbanized areas are key points of vulnerability to the impacts of climate change (Kikuchi et al. 2010). Adding to the current stress on some selected sites, an orphan oil spill occurred in 2019/2020, affecting many unique reef sites, previously pristine or already affected by local impacts (Soares et al. 2020).

In this chapter, we synthesize the main characteristics of the distribution and variability of reef and coralline ecosystems of the Western Tropical South Atlantic Ocean. We also synthesize the distribution of coral richness and fish diversity and the main impacts that affect the reef ecosystem along the Brazilian coast and shelf. Finally, we use three endemic coral species as proxies of the reef ecosystem growth to evaluate the trend of environmental suitability across the Brazilian Tropical Marine region in the RCP8.5 climate change scenario.

5.2 Reef Distribution Along the Continental Shelf and Seamounts

Reef ecosystems occur along the continental shelf in most of the Tropical and Subtropical Western Atlantic, comprising shallow water (euphotic), mainly in the Tropical region, to the deep realm, mapped mostly on the Subtropical Atlantic. The transition of the shallow water reef realm to the mesophotic realm is set at 30 m depth. The mesophotic realm ranges down to 150 m deep, from which the deep reef realm begins (Lesser et al. 2009) (Fig. 5.1).

In this chapter, we will focus on the shallow reefs of the Western Tropical South Atlantic realm. We use a regional classification proposed by Kikuchi et al. (2022) (Fig. 5.1) that assumed the definition of an ecoregion as the smallest scale unit in the marine system area where species composition is relatively homogenous and clearly distinct from adjacent systems (Spalding et al. 2007). The present classification was proposed on the basis of reef morphology distribution, seasonal migration of the South Equatorial Current (SEC) and the biogeography of fish and of milleporids (Pinheiro et al. 2023). Following this classification, Brazilian shallow coral reefs occur in seven ecoregions (Table 5.1) grouped into two marine provinces: the North Brazil Shelf and the Tropical Southwestern Atlantic.

The North Brazil Shelf province corresponds to the Amazon (AMZ) ecoregion, whereas the Tropical Southwestern Atlantic province includes five ecoregions: the São Pedro and São Paulo islands (ASPSP), the Fernando de Noronha islands and Atol das Rocas (FNAR), North Northeastern Brazil (NNEB), East Northeastern Brazil (ENEB), Eastern Brazil (EB) and Trindade and Martin Vaz islands (TMV).

As a general trend, shallow reefs are rare in the AMZ and in the NNEB ecoregions. Shallow reefs occupy larger areas in the ENEB ecoregion and are widely distributed in the EB ecoregion. The mesophotic reefs megahabitat was widely mapped in the AMZ (Moura et al. 2016) and NNEB (Carneiro et al. 2022) ecoregions and in the Abrolhos region, situated in the EB ecoregion, approximately 17°S (Fig. 5.1).

In the AMZ ecoregion, known shallow reef structures are the Parcel de Manuel Luiz reefs located 90 km off the state of Maranhão coastline (Fig. 5.1) (Coura 2016). They grow as giant pinnacles to depths of approximately 25–30 m, and their tops reach up to 2 m of water depth. During spring low tides, some of these reefs are at sea-level (Amaral et al. 2007; Coura 2016).

In the FNAR ecoregion, the Fernando de Noronha Archipelago (03° 5′ S—32° 25′ W) comprises 21 islands and islets of volcanic origin. It is located approximately 350 km off the coast of Rio Grande do Norte state (Fig. 5.1). These islands are part of the Fernando de Noronha volcanic mountain chain (Almeida 1955). Abundant coral fauna grows on some of their rocky shores (Maida and Ferreira 1997; Amaral et al. 2009), and coralline algae crusts occur on the eastern side of the main island. The Rocas atoll (03° 51′ S—33° 49′ W) developed on a flat top of a seamount. It has an elliptical shape with the larger axis (3.5 km long) oriented E-W and the minor axis (approximately 2.5 km long) oriented N-S. The reef ring is open on its western and northern faces. Despite its small dimensions, all of the characteristic features

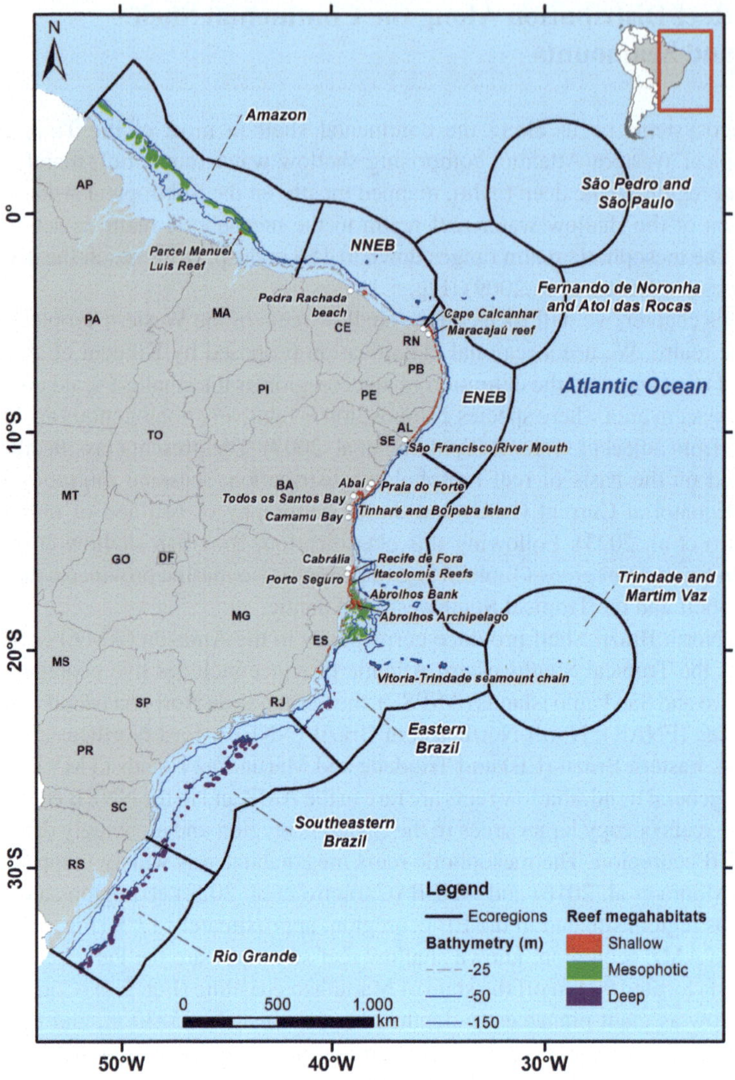

Fig. 5.1 Reef distribution on ecoregions of the Brazilian Economic Exclusive Zone (EEZ), according to Kikuchi et al. (2023)

of a reef can be distinguished, such as the reef front, reef flat and a shallow lagoon (Kikuchi and Leão 1997).

The NNEB ecoregion (from approximately 4–5° S) is oriented east–west and extends from the mouth of the Parnaíba River (Piauí State) to Cape Calcanhar (Rio Grande do Norte State) (Fig. 5.1). Patch (Fig. 5.3a, b) and bank reefs (Fig. 5.3c, d) have been mapped on their inner (Laborel 1969; Laborel-Deguem 2019; Santos et al. 2007), middle and outer shelves (Soares et al. 2016; Gomes et al. 2020a, b;

Table 5.1 Ecoregions of the Southwestern Atlantic region according to Kikuchi et al. (2023). Ecoregions are drawn in Fig. 5.1

Realm	Province	Acronym	Ecoregion	Code
Tropical Atlantic	North Brazil Shelf	AMZ	Amazon	72
	Tropical Southwestern Atlantic	SPSP	São Pedro and São Paulo Islands	73
		FNAR	Fernando de Noronha and Atol das Rocas	74
		NNEB	North Northeastern Brazil	75
		ENEB	East Northeastern Brazil	75
		EB	Eastern Brazil	76
		TMV	Trindade and Martin Vaz Islands	77
Temperate Atlantic	Warm Temperate Southwestern Atlantic	SEB	Southeastern Brazil	180
		RG	Rio Grande	181

Code refers to a numeric identification of the ecoregion according to Spalding et al. (2007)

Carneiro et al. 2022). More recently, De Araújo et al. (2020) and Carneiro et al. (2021) provided revisions of the occurrences of discontinuous shallow substrates found in the intertidal zone that constitute a reef morphotype that was named cemented terraces (Fig. 5.2a, b).

The ENEB ecoregion extends from Cape Calcanhar at approximately 5°S to the São Francisco River mouth, at approximately 10°S the borderline between the Alagoas and Sergipe states (Fig. 5.1). The major reef types found in this ecoregion are fringing (Fig. 5.2c, d) and patch reefs (Fig. 5.3a, b). Fringing and bank reefs that originate from lines of beach rocks (Fig. 5.2e, f) are common as well (Dominguez et al. 1990; De Araújo et al. 2020). Some of these linear reefs are attached to and fringe the coastline, and others are several kilometers offshore, generally aligned parallel to the coast at depths of approximately 5–10 m, but can also be found at 20–25 m depth (Testa 1997; Santos et al. 2007) and may constitute former coastlines (Gomes et al. 2020a).

The EB ecoregion extends from the São Francisco River mouth to Arraial do Cabo mouth at approximately 23°S. This is the largest and richest coral reef area along the entire Brazilian coast. In this ecoregion, most of the reef types are found. Shallow isolated bank reefs (Fig. 5.3c, d) of various sizes occur between the beaches of Abaí and Praia do Forte (Leão et al. 2003). Deeper reef structures occur toward the continental shelf break (Kikuchi and Leão 1998). In the area from Todos os Santos Bay to Camamu Bay, shallow bank reefs are observed inside these bays (Leão et al. 2003). Near Camamu Bay, a line of shallow fringing reefs borders the shores of the Tinharé and Boipeba Islands (Fig. 5.2c, d), and a series of lines of elongated bank reefs (approximately north–south) (Fig. 5.2e, f) occur down to depths of 20–30 m

Fig. 5.2 Schematic illustrations of major reef types: **a** and **b** cemented terrace exemplified by the Pedra Rachada beach, state of Ceará, in the North Northeastern Brazil ecoregion, **c** and **d** fringing reefs exemplified by Itaparica island reefs, state of Bahia, in Eastern Brazilian ecoregion, and **e** and **f** linear bank reef exemplified by the bank reef of Arembepe beach, state of Bahia, Eastern Brazil ecoregion; arrows indicate two trends of linear bank reefs. Illustrations **a**, **c** and **e** are reprinted from Marine Environmental Research, 160, De Araújo et al. (2020), Diversity patterns of reef fish along the Brazilian tropical coast, Appendix A, Copyright (2020), with permission from Elsevier

(Kikuchi et al. 2008). The area of Cabralia/Porto-Seguro is characterized by the presence of bank reefs of various shapes and dimensions in water no deeper than 20 m, running mostly parallel to the coastline (Leão et al. 2003). The most studied reef in this area is the Recife de Fora, an isolated bank reef a few kilometers off the coastline (Seoane et al. 2012). The elongated reefs may have grown on submerged strings of beach rocks (Fig. 5.2e, f). Southward, there are Itacolomis reefs, which

Fig. 5.3 Schematic illustrations of major reef types **a** and **b** patch reefs, exemplified by the Maracajaú reefs, Rio Grande do Norte State, East Northeastern Brazil ecoregion, **c** and **d** bank reefs exemplified by the reefs off Tinharé Island in Bahia State, Eastern Brazil ecoregion and **e** and **f** pinnacles exemplified by the Itacolomis reefs. They are known locally as "chapeirão" meaning "big hat" (plural "chapeirões", "big hats", Bahia State, Eastern Brazil ecoregion. In this last picture, the broad reef tops of several "chapeirões" can be seen. Illustrations **a**, **b**, **c** and **e** are reprinted from Marine Environmental Research, 160, De Araújo et al. (2020), Diversity patterns of reef fish along the Brazilian tropical coast, Appendix A, Copyright (2020), with permission from Elsevier

are the northernmost limit of the occurrence of the Brazilian giant "chapeirões" and isolated bank reefs (Fig. 5.3e, f) separated from one another by deep irregular channels (Cruz et al. 2008). Patch reefs with varied dimensions are generally found in the vicinity of fringing (Fig. 5.2c, d) and bank reefs (Fig. 5.3a, b).

At approximately 17°S, the continental shelf widens to form the Abrolhos Bank, in which the richest and the most well-known coral reefs of the Eastern Brazil ecoregion occur (Laborel 1969; Hartt 1870; Leão 1996). These reefs form two arcs: the coastal arc composed of bank reefs of various shapes and dimensions and the outer arc east of the islands of the Abrolhos Archipelago, which is formed by isolated "chapeirões" (Fig. 5.3e, f) in water deeper than 20 m. This distinctive reef type ranges in height above the sea floor from 5 to 25 m, and the horizontal extension of their tops varies from 5 to 50 m. Incipient fringing reefs border the shores of two of the five islands of the archipelago. More recently, in the southern portion of the Abrolhos Bank, shallow pinnacles ("chapeirão") were described by Mazzei et al. (2017). Milliman and Barreto (1975) documented the occurrence of drowned reefs at the shelf break, and mesophotic reefs are described across the mid and outer shelves on the Abrolhos Bank at depths from 25 to 90 m (Moura et al. 2013).

The TMV ecoregion (20° 30′ S—29° 20′ W) is formed by rocky islands in the Vitória—Trindade Seamount Chain. They are located approximately 1200 km off the coast of Espírito Santo state. No true reefs exist, but coral species have been reported on these islands (Pereira-Filho et al. 2011).

5.3 Coral and Fish Communities (Along the Tropical Shelf) and Impact Variability

5.3.1 Coral Distribution

The zooxanthellate scleractinian corals provide most of the structural complexity in the reef ecosystem. These corals produce colonies secreting calcium carbonate that are the basic structure of the coral reefs (Kinzie and Buddemeier 1996). Encrusting organisms such as coralline algae, bryozoans, vermetid gastropods, and foraminifera contribute to stabilizing sediment produced by organisms such as mollusks, corals, hydroids, and octocorals, among other carbonate-secreting organisms. The reef-building Scleractinia coral fauna of Brazil has the following three distinctive characteristics (Laborel 1969; Leão et al. 2003): (i) low taxonomic richness (number of coral species) compared with that of the North Atlantic reefs; (ii) the major reef builders are species endemic to Brazilian waters, and (iii) they are predominantly composed of massive growth forms.

It is a low diversity fauna because it has 20 indigenous coral species and five calcareous hydroids (Table 5.2). However, in the last three decades, nonindigenous scleractinian corals were introduced in these waters. They are corals that bear no symbiosis with the microalgae known as zooxanthellae and were described as *Tubastraea coccinea* and *T. tagusensis* (De Paula and Creed 2004). A recent study, based on genetic and morphological data, indicated that the species known as *T. tagusensis* was mistakenly named and recommended the use of *Tubastraea* sp. and that there is a possibility of the occurrence of a third species (Bastos et al. 2022). On the other hand, five coral species are endemic. Three of the endemic coral species, *Siderastrea stellata*, *Mussismilia hispida*, and *Mussismilia hartii*, are widely distributed along the coast and shelf, from the AMZ to the SEB ecoregions. The other two endemic species have a more restricted distribution: *Favia leptophylla* occurs in the AMZ and EB ecoregions, and *Mussismilia braziliensis* is restricted to the EB ecoregion (Table 5.2). Three calcareous hydroids are endemic to the Tropical Southwestern Atlantic, *Millepora brasiliensis*, *Millepora nitida* and *Millepora laboreli.* The first one occurs in the AMZ to the EB ecoregions. *Millepora nitida*, on the other hand, occurs in the ENEB ecoregion, and *Millepora laboreli* is confined to the AMZ ecoregion.

The eastern shelf of the Brazilian continental margin, comprising the ENEB and EB ecoregions (Table 5.1, Fig. 5.1), is where a greater number of hermatypic coral and calcareous hydroid species thrive. More than twenty species are found in this region. A high variety of reef forms and a greater number of hermatypic and calcareous hydroids are found in the EB ecoregion. This pattern of species richness distribution was originally described by Laborel (1969). As of the nonindigenous species of the *Tubastraea genus*, the SEB ecoregion is where they were first introduced, and two

Table 5.2 Distribution of zooxanthellate shallow water corals and hidrocorals (grey squares) across the marine ecoregions and nonindigenous species (black squares) occurrences

	Ecoregion	State/island	*Millepora alcicornis* (h)	*Siderastrea stellata* (c, e)	*Madracis decactis* (c)	*Mussismilia hispida* (c, e)	*Favia gravida* (c, e)	*Porites branneri* (c)	*Agaricia agaricites/humilis* (c)	*Meandrina brasiliensis* (c)	*Montastraea cavernosa* (c)	*Porites astreoides* (c)	*Millepora braziliensis* (h, e)	*Mussismilia harttii* (c, e)	*Scolymia wellsi* (c)	*Agaricia fragilis* (c)	*Siderastrea radians* (c)	*Stephanocoenia michelini** (c)	*Astrangia braziliensis* (c)	*Phyllangia americana* (c)	*Astrangia rathbuni* (c)	*Favia leptophylla** (c, e)	*Millepora nitida* (h, e)	*Mussismilia braziliensis* (c, e)	*Stylaster roseus* (h)	*Millepora laboreli* (h, e)	*Scolymia cubensis* (c)	*Siderastrea siderea* (c)	*Tubastraea tagusensis** (nic)	*Tubastraea coccinea* (nic)	*Total no. species/state*	*Total no. species/ecoregion*
Oceanic islands	ASPSP	SPSP			grey										grey				grey												3	**3**
	FNAR	FN	grey	grey		grey	grey	grey	grey		grey		grey	grey											grey						10	**12**
		AR	grey	grey		grey	grey		grey	grey	grey	grey																			8	
	TMV	TR	grey	grey				grey																							3	**3**
Continental margin	AMZ	MA	grey	grey	grey	grey	grey	grey	grey	grey	grey	grey	grey			grey						grey				grey					14	**14**
	NNEB	PI	grey	grey						grey																					3	
		CE	grey	grey			grey	grey		grey	grey									grey									black		9	**16**
		RN	grey	grey	grey	grey	grey	grey	grey	grey	grey	grey	grey	grey		grey		grey													14	
	ENEB	PB	grey	grey	grey	grey	grey	grey	grey	grey	grey	grey	grey	grey			grey	grey													14	
		PE	grey	grey	grey	grey	grey	grey	grey	grey	grey	grey	grey	grey	grey	grey	grey	grey	grey	grey											18	**21**
		AL	grey	grey	grey	grey	grey	grey	grey	grey	grey	grey	grey	grey	grey	grey	grey	grey	grey				grey					grey	black		20	
	EB	SE		grey	grey	grey	grey		grey		grey	grey		grey	grey												grey		black	black	12	
		BA	grey	grey	grey	grey	grey	grey	grey	grey	grey	grey	grey	grey	grey	grey	grey	grey	grey			grey	grey	grey	grey				black	black	23	**26**
		ES	grey	grey	grey	grey	grey	grey	grey	grey			grey		grey		grey					grey		grey							13	
		RJ (N)	grey	grey	grey	grey		grey												grey	grey								black	black	9	
	SEB	RJ (S)			grey	grey														grey	grey								black	black	6	
		SP	grey		grey	grey						grey									grey								black	black	7	**10**
		SC			grey															grey	grey									black	4	

Indigenous species are organized from the most widely distributed (left) to the most restricted (right). Species are listed as quoted in the papers surveyed. Species with (*) present taxonomic classification issues explained in the text. (c) coral; (h) calcareous hydroid; (e) endemic species; (nic) nonindigenous coral species. Ecoregions: AMZ = Amazon, ASPSP = São Pedro and São Paulo Islands, FNAR = Fernando de Noronha and Atol das Rocas, NNEB = North northeastern Brazil, ENEB = East northeastern Brazil, EB = eastern Brazil, TMV = Trindade and Martin Vaz Islands, SEB = southeastern Brazil. Sources of information: Laborel (1969); Leão et al. (2003); De Paula and Creed (2004); Amaral et al. (2007, 2009); Ferreira and Maida (2006); Neves et al. (2006, 2008, 2010); Correia (2011); Lindner et al. (2017); Mantelatto et al. (2011); Pereira-Filho et al. (2011, 2019); Sampaio et al. (2012); Bastos et al. (2022).

species occur (Bastos et al. 2022). They migrate gradually north (Sampaio et al. 2012) and south (Mantelatto et al. 2011; Lindner et al. 2017). The range of *Tubastraea coccinea* extends further south, whereas the species called *Tubastraea tagusensis* (see the explanation identification issue, below) reached the NNEB ecoregion to the north.

Among the species listed in Table 5.2, there are three indigenous and two nonindigenous coral species with taxonomic classification issues. *Agaricia agaricites* used to have numerous subspecies, with *Agaricia agaricites humilis* as one of them. Laborel (1969), in the most comprehensive survey of the coral species along Brazilian waters, mentions *Agaricia agaricites humilis* in his text, but on the illustrations captions and on the list of species, he only wrote the naming "*Agaricia agaricites*", as remarked by Nunes et al. (2019). The subspecies *Agaricia agaricites humilis* was accepted as a species by van Morsel in 1986 and became *Agaricia humilis*, but some papers on species occurrences and reef monitoring still cite *Agaricia agaricites*. Nunes et al. (2019) considered that the species that occur in the Tropical Southwestern Atlantic is *Agaricia humilis*.

Another issue reported by Nunes et al. (2019) refers to *Favia/Mussismilia leptophylla*. Although *Favia leptophylla* is a name frequently used in the literature, Nunes et al. (2008) stated that based on genetic data it is closer to the three endemic *Mussismilia* species. Later, Budd et al. (2012) accepted the name *Mussismilia leptophylla* as a new designation of the species based on the molecular and microstructural characteristics of its skeleton. However, as there are similarities in living tissue characteristics between this species and *Favia gravida*, the name *Favia leptophylla* is still used (Nunes et al. 2019).

Nunes et al. (2019) list another naming change of *Stephanocoenia michelini*. In fact, as listed in the World Register of Marine Species (WoRMS), it refers to Hoeksema and Cairns (2022). *Stephanocoenia intersepta* is the accepted name of the species.

Last, but not least, the designation by de De Paula and Creed (2004) of the two species of nonindigenous coral, known in Brazil as "sun-coral", *Tubastraea coccinea* and *Tubastraea tagusensis*, has been recently questioned. Bastos et al. (2022), in a study based on skeletal morphology and molecular characteristics, confirmed the occurrence of *Tubastraea coccinea*. However, they indicate that *Tubastraea tagusensis* is not a valid classification and call it *Tubastraea sp.*

As this is a general revision of the distribution of species along Brazilian waters, we have used the classification found in the papers surveyed in Table 5.2.

5.3.2 Fish Diversity Variability

Reef fish, i.e., fish that depend on hard bottom features at some point in their lives, and all reef-dependent species are somewhat disjunct at a local scale, as the distribution of reefs along the coast is usually patchy (Jones et al. 2002). Subpopulations are more connected by larvae dispersed by currents than by swimming adults

(Mora and Sale 2002); parental populations located hundreds of kilometers away can influence local populations. Moreover, reef fish recruitment often follows stochastic patterns, and the resulting communities are highly dependent on post-recruitment processes, such as regional and local habitat characteristics (e.g., temperature, depth, wave surge, currents, turbidity, reef morphology) and biotic drivers (e.g., population connectivity, larval supply, competition). Beyond the processes shaping these apparently unpredictable communities, patterns emerge from both historical biogeographies determining local species pools and major drivers interacting with species preferences.

The Southwestern Atlantic reef fish fauna is limited to the north by the Amazon River plume, to the south by the Malvinas Current and to the east by the Mid-Atlantic barrier; more than 400 resident reef fish species are registered in this area, of which 27% are endemics and nearly 20% are threatened with extinction (Pinheiro et al. 2018). Brazilian reef fishes have been increasingly investigated due to their relevance (Araújo et al. 2018); currently, they are the second most studied group in tropical Brazilian reef systems behind corals only.

As mentioned above, reef fish communities are highly influenced by a myriad of environmental drivers. Some species are generalist enough to cope with the variation range of these factors, and these species are widespread along the Brazilian tropical coast and generally dominate communities in numbers (e.g., *Acanthurus coeruleus*, *Haemulon plumierii* and *Sparisoma axillare*). However, they are a minority when considering total species richness; approximately a quarter of Brazilian tropical reef fish fauna can be regarded as common or dominant, while the bulk of reef fish diversity is composed of rare species (e.g., *Chaetodon ocellatus* and *Sphoeroides testudineus*). Those that occur in low numbers also appear to be more sparsely distributed along the coast and are specialists in the habitats they selected to live in (e.g., *Bathygobius soporator* and *Diodon holacanthus*). Some rare species can be typical of particular reef formations, i.e., disproportionately more abundant and/or more frequently found at reefs of a given type (De Araújo et al. 2020).

The number of rare species increases with latitude and reaches a maximum at the southernmost limit of the 'coral reefs' distribution in Brazil (south of Bahia and north of Espírito Santo States, EB ecoregion) (Fig. 5.1). In this region, we find a greater diversity of reef morphologies (De Araújo et al. 2020) than in any other area along the Brazilian coast. Additionally, more complex reef formations show greater numbers of rare species than more homogeneous reefs, which means reef morphology and fish diversity are connected.

Although Brazilian reefs share some similarities with other reef systems, the presence of a highly endemic reef-building coral fauna renders some of these reefs unique. Reef fishes associated with the five types of reef formations depicted above are addressed in De Araújo et al. (2020). The cemented terraces (Fig. 5.2a, b), which are predominantly located in the NNEB ecoregion (3° 20′ S—6°13′ S), are the most homogeneous and thus present the most distinctive reef fish fauna with a higher number of rare associated species. Conversely, the bank reefs (Fig. 5.3c, d) are the most widespread, variable and heterogeneous reefs along the Brazilian coast, and the associated fish fauna is equally the most variable. Pinnacle reefs ('chapeirão',

Fig. 5.3e, f), with their distinctive form unique to Brazilian reefs, are limited in distribution to the south of Bahia State (EB ecoregion, Fig. 5.1) and show just a fraction of the variability in fish communities associated with bank reefs. In summary, Brazilian reef fish communities can be shaped by reefs' geomorphology and their distribution; more homogenous reefs select species with a greater degree of habitat specificity, while high latitudes show a greater diversity of reef morphologies, increasing the overall diversity of rare species (De Araújo et al. 2020).

Despite the intrinsic relationship between reef morphology/formation and fish diversity, this diversity can also be affected by local (e.g., fishing, water pollution, habitat degradation) and global (e.g., sea-level rise and global warming) stressors that act concomitantly, cause adverse effects on the species and interfere with the ecological state of these reefs (Cruz et al. 2018). For example, sea-level rise may also affect food webs and reef assemblages that inhabit reef flats by increasing the time spent foraging by the larger species in this region (Harbone 2013). Global warming affects reef fish since they have a limited ability to acclimate to high temperatures as adults (Munday et al. 2012). The diversity of reef fish species decreases as the size of the local human population increases (Drew et al. 2015). Likewise, it is known that fishing causes changes in species richness, relative abundance, and patch occupancy (Tittensor et al. 2007).

As previously stated, a significant part of the species richness of Brazilian reefs is composed of rare species. Thus, this trait can favor low functional redundancy in reefs, which makes them easier to collapse if one or more species are lost in situations of anthropogenic impacts (e.g., habitat degradation and destructive fishing practices) (Soares et al. 2021). Therefore, the unbalanced distribution of rare species exposes the increased vulnerability of these reef communities to density fluctuations and human stressors (De Araújo et al. 2020).

5.4 Main Global and Local Impacts and Bleaching

Western Tropical South Atlantic reefs (Fig. 5.1) are affected by a plethora of human stressors, including long-term warming (Principe et al. 2021), marine heatwaves (Duarte et al. 2020), ocean acidification (Cotovicz et al. 2020), sea-level rise (Muehe 2018 documents erosive trends along the Brazilian coast), mismanaged touristic and industrial activities (Feitosa et al. 2012; Giglio et al. 2016), higher nutrient inputs (Costa et al. 2000; Mies et al. 2020), increased macro- and microplastic pollution (De Carvalho-Souza et al. 2018; Azevedo-Santos et al. 2021), environmental disasters (e.g., oil spills and mining dam collapses) (Coimbra et al. 2020; Soares et al. 2020; Cardoso et al. 2022), overfishing of reef species (Morais et al. 2017; Pereira et al. 2021), and invasive species such as lionfish (*Pterois* spp.) (Luiz et al. 2021) and sun corals (*Tubastraea* spp. with three morphotypes, Table 5.2) (Creed et al. 2017; Capel et al. 2019; Bastos et al. 2022). These human stressors impact these tropical shallow-water reefs as cumulative and synergistic pressures (Fig. 5.4) and consequently undermine their long-term persistence (Soares et al. 2021). In this regard, these unique reefs present a large number of phase-shift events (detailed by Cruz et al. 2018) and indicators of reef decline (e.g., decreasing richness, loss of suitable habitats, loss of reef fish biomass, and decreased architectural complexity) (Leão et al. 2016; Aued et al. 2018; Beneli et al. 2020; Principe et al. 2021), which suggests significant impacts of their functional roles, such as reef framework production and sediment generation (Costa et al. 2016; Elliff and Silva 2017), the stability of reef habitat complexity, nursery functions, and carbonate reef growth potential. Moreover, the uneven spatial distribution of rare reef fishes and decreased biomass also drive a high vulnerability to anthropogenic pressures (Bender et al. 2013; De Araújo et al. 2020).

Overall, ~ 16% of Southwestern Atlantic reefs have a high to very high risk of experiencing cumulative impacts when assuming equal sensitivity of these reefs across human pressures (Magris et al. 2018). Moreover, when variable sensitivities across pressures are included, 37% of these reefs have a high to very high risk of experiencing cumulative human impacts (Magris et al. 2018). In this regard, the spatial distribution of SW Atlantic reefs with high or very high risk is largely coincident with localities where the human population is higher along the Brazilian coast, such as in the ENEB ecoregion (Fig. 5.1). To date, 42% of Brazilian reefs have a low risk of cumulative impacts from both local and global stressors (Magris et al. 2018). Among the major human pressures, climate change and mismanagement of fishing activities account for the two main values of exposure (Magris et al. 2018, 2020). Warming extended across over 95% of the SW Atlantic reef region, while the area impacted by land-based human activities was much smaller, ranging from 7.6% (e.g., sewage pollution) to 22.6% (e.g., pesticides) due to Brazilian agriculture (Magris et al. 2020). In addition, approximately 82.8% of the Brazilian maritime area is under pressure from pelagic longlines, the most widespread fishery activity, followed by pelagic driftnet (61.6%) and bottom trawl (60.1%) (Magris et al. 2018, 2020), which affect key species on SW Atlantic reefs. In this way, mismanagement of fisheries, increased global warming and heatwaves (Duarte et al. 2020) and land-based activities are severe human threats (Magris et al. 2020) to the stability of SW Atlantic reefs, shifts their distribution on reef-building corals (Principe et al. 2021), and increased bleaching (Bleuel et al. 2021).

Recent data indicate that coral bleaching events affected at least 26 species of scleractinians, hydrocorals, octocorals, and zoantharians in tropical SW Atlantic reefs over 26 years (1994–2020). Although no species suffered postbleaching mass mortality between 1994 and 2018 (Mies et al. 2020; Soares et al. 2021), there has been an increase in the sea surface temperature in all Brazilian reef ecoregions in the last 39 years (Soares et al. 2022). A clear trend of 0.2 °C increase per decade was detected in low-latitude equatorial coastal and oceanic reefs (FNAR and NNEB ecoregions) (Soares et al. 2022). An increase of 0.13 °C in the Abrolhos Bank (EB ecoregion) and 0.1 °C in the Coral Coast marine protected area located in the ENEB ecoregion has been reported (Soares et al. 2021, 2022). Moreover, the marine heat wave frequency was close to three events per year during the 1982–2020 period in the coral reef ecoregions, reaching values larger than 10 events per year in oceanic islands such as the Rocas Atoll (FNAR ecoregion) and in the ENEB and NNEB

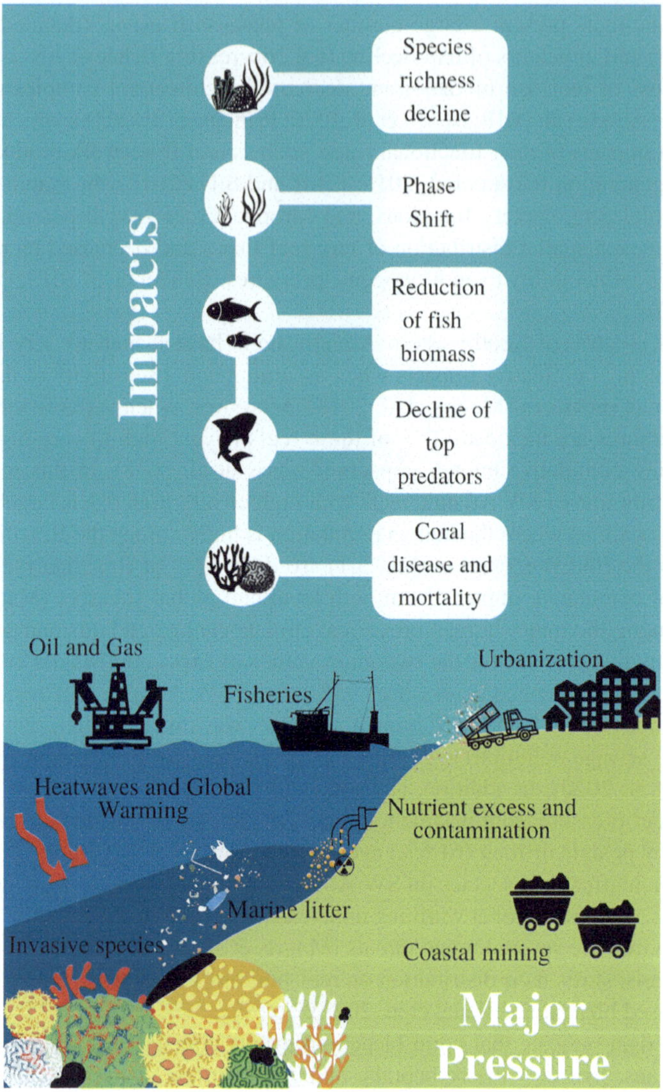

Fig. 5.4 Major pressures and impacts have been detected on shallow-water (<30 m depth) reefs (Tropical Southwestern Atlantic, Brazil). *Source* Emanuelle Rabelo

ecoregions (Soares et al. 2021, 2022). In this regard, marine heatwaves have become more frequent and longer lasting, which may lead to more mass bleaching events, as observed in 2010 (Soares et al. 2019), 2019 (Duarte et al. 2020) and again in 2020 (Soares et al. 2022).

These recent and intense heatwaves of 2019 and 2020 caused higher mortality rates in several key foundation corals (e.g., *Millepora alcicornis*, *Millepora braziliensis*, and *Mussismilia harttii*), showing that the SW Atlantic reefs are under severe threat from climate change scenarios (Leão et al. 2016; Pereira et al. 2022; Soares et al. 2021). This is especially important considering the low coral diversity, endemic species, and low functional redundancy of reef assemblages in marginal reefs (Soares 2020). The repeated occurrence of intense and recent marine heatwaves (2019 and 2020) affecting SW Atlantic reefs is modern evidence of the detected regional trends without a significant reduction in carbon emissions (e.g., RCP 8.5—business as usual scenario), which will drive increased coral bleaching (Bleuel et al. 2021) and habitat loss in major reef-building corals such as *Siderastrea* spp., *Mussismilia hispida*, and *Montastraea cavernosa* in this century (2100) (Principe et al. 2021). Despite the resilience of Brazilian coral assemblages to specific suboptimal conditions, such as turbidity and resuspension of sediments (Loiola et al. 2019; Mies et al. 2018, 2020), it is likely that the cumulative effects of local, regional, and global pressures may lead to the large scale degradation of the unique tropical SW Atlantic reefs (Principe et al. 2021; Soares et al. 2021, 2022). The high phenotypic plasticity of some Brazilian corals and their adaptations to short-term environmental fluctuations (Mies et al. 2020; Godoy et al. 2021) do not imply that these species will be successful under ongoing global environmental change (Bleuel et al. 2021; Principe et al. 2021; Soares et al. 2021). In addition to climate change, other direct and indirect human pressures may threaten these tropical SW Atlantic reefs (Fig. 5.4), as we reviewed in this section.

5.5 Projections/Tendencies of Endemic Coral Species Distribution

This modeling experiment was performed using three endemic coral species as proxies of potential species distribution trends of the reef ecosystem. *Mussismilia hispida* is the most widespread reef building species, occurring across all ecoregions from the AMZ ecoregion through the tropical southeastern Atlantic province to the SEB ecoregion. *Mussismilia harttii* occurs in the tropical southeastern Atlantic province, and *Mussismilia braziliensis* is a species confined to the eastern Brazil ecoregion. In this way, we try to have an indication of the change in reef environment suitability expected with a business as usual future RCP 8.5 scenario (high-emissions or business as usual scenario; Riahi et al. 2011).

5.5.1 Methods

We used the maximum entropy model (Maxent) to estimate the probability of species suitability in response to estimated global climate change for the coming decades under the premises of the RCP8.5 scenario (global carbon increase of 30%, fossil fuels generating more than 60% of global electricity, sea-level rise trajectory between 0.5 and 1.0 m). This model estimates the widest possible distribution (maximum entropy) from species presence and environmental abiotic factors available about the target species distribution (Phillips and Dudík 2008; Elith et al. 2011). For each target species, an extensive review of the occurrence of the species in both primary and secondary sources (systematic review) was performed according to the procedures described in Alencar et al. (2017) and França et al. (2020). Occurrences recorded within a 30 arc-second resolution grid (grid cell of approximately 1 km; Davies and Guinotte 2011) were considered as duplicate presence, thus constituting a binary character for each cell (Tittensor et al. 2009).

The present and future scenarios under climate and marine environmental factors used in the species distribution model were obtained from the Bio-ORACLE database (http://www.oracle.ugent.be; Tyberghein et al. 2012; Assis et al. 2017). Bio-ORACLE is a larger spectrum dataset in the water column containing geophysical information and climatic (cloud cover, temperature, diffuse attenuation, dissolved oxygen) and marine (photosynthetic radiation, silica, phosphate, nitrate, calcite, and chlorophyll *a*) information. The most recent environmental layers is the benthic stratum available from Bio-Oracle (Assis et al. 2017) inferred through ocean bathymetric charts (GEBCO 2015) considering the depth in each grid cell and when one uses geographic position (Assis et al. 2016; Boavida et al. 2016). Three scenarios were evaluated: (1) present scenario, data survey between 2000 and 2014; (2) future scenario 1 for the layers of the years 2040 and 2050; and (3) future scenario 2 for the years 2090 and 2100 based on the representative concentration pathway (Moss et al. 2010) of gas emissions over time leading to a high greenhouse gas increase (RCP8.5 Scenario) with the highest impact on marine biodiversity.

The investigation and removal of collinear variables were performed by observing the correlation coefficient and variance inflation value (VIF > 3, Pearson's $r > 0.8$; Zuur et al. 2010; Davies and Guinotte 2011; Fourcade et al. 2018). In accordance with Tittensor et al. (2009), we assumed satisfactory spatial independence between grid cells, considering these effects to be minimal. Each model was built based on twenty-five bootstrap replications (Phillips et al. 2006). The remaining model parameters were kept as the default (a convergent threshold of 10^{-5}, maximum iteration value of 500, and a regularization multiplier of 1) (Phillips and Dudík 2008). Potential distribution maps on an environmental suitability scale for each species were plotted according to the procedures described in Dudík and Phillips (2008) and Tittensor et al. (2009). Models were validated by analyzing the ROC ("Receiver Operating Characteristics"; Phillips et al. 2006) and AUC ("area under the curve"; Fielding and Bell 1997) parameters by the concepts of sensitivity and specificity, using a subset of 30% of the total sample data (assessment data) and the remaining 70% for model calibration (calibration data) (Manel et al. 2001; Tittensor et al. 2009). The AUC ranges from zero to one, with values close to one (1) indicating high model performance, whereas values below 0.5 indicate low model performance (Allouche et al. 2006; Elith et al. 2006). The final models for each species relied on a procedure to remove variables with a contribution of less than 1% in the model.

For each environmental layer considered, response curves were estimated, in addition to a jacknife permutation analysis to detail the importance of the contribution of each variable in the model, individually and synergistically (Davie and Guinotte 2011; Tittensor et al. 2009). The percentage areas of suitability along future projections from the present scenario were calculated for each of the species according to the marine ecoregion delimitations proposed by Spalding et al. (2007) and using a threshold level of suitability of 0.8. Statistical procedures, environmental suitability models and distribution maps were generated using R software (R Development Core Team 2018) through the package 'raster' (raster data treatment and standardization, Hijmans et al. 2017), 'usdm' (Variable inflation factor, Naimi et al. 2014) and 'dismo' (Environmental suitability models, Hijmans et al. 2017).

5.5.2 Suitability Areas (Modern and Future Scenarios)

The suitability models for predicting future scenarios showed high performance, with AUC values above 0.9 (*Mussismilia braziliensis* = 0.997 + 0.001; *M. hartii* = 0.997 + 0.005 and *M. hispida* = 0.993 + 0.001). In all cases, the reduced models showed better performance (higher AUC) than the global models. For *M. braziliensis*, the jackknife permutation procedures suggest that the mean salinity (80.3%, jackknife permutation = 38.8%), salinity amplitude (9.9%, jackknife permutation = 46.3%), and mean temperature (8.6%, jackknife permutation = 10.3%) are variables that explain its distribution, in descending order of importance. The other variables showed a lower contribution to the model. Mean salinity is the environmental layer with a higher gain when variables are evaluated alone. For *M. harttii*, mean salinity (76.6%, jackknife permutation = 51.2%), salinity amplitude (14.9%, jackknife permutation = 42.6%), and mean temperature (6.1%, jackknife permutation = 0.1%) are variables that explain its distribution, also in descending order of importance. The environmental layer with the highest gain, when evaluated in isolation, was mean salinity. In contrast, it showed the lowest gain in the model when one omits Salinity Amplitude from the analysis. For *M. hispida*, salinity amplitude (55.5%, jackknife permutation = 18.4%), mean salinity (12.8%, jackknife permutation = 0.4%), mean temperature (11.9%, jackknife permutation = 36.5%) and mean current velocity (11.4%, jackknife permutation = 24%) are variables that explain its distribution. Mean temperature was the environmental layer with the highest gain when evaluated in isolation. In contrast, it showed the lowest gain in the model when the analysis omitted the mean current velocity. In the discussion that follows the ecoregion nomenclature of Spalding et al. (2007) will be used.

The outstanding common regions with high suitability prediction for the three species evaluated were the northern portion of the 76 Ecoregion (Eastern Brazil) (Todos os Santos Bay until Camamu Bay), including the southeastern portion of the 75 Ecoregion (Northeastern Brazil) and the central portion of the 76 Ecoregion (Eastern Ecoregion) (Abrolhos Archipelago and vicinities) (Figs. 5.5, 5.6 and 5.7). For *M. hispida*, high-suitability regions were still detected in the central portion of the

75 Ecoregion (northeastern Brazil) (Fig. 5.7). The predicted suitability was associated with a particular set of abiotic variables, with differences for each species analyzed. For *M. hispida*, the highest suitability values were inversely related to the amount of silica in the water column and directly related to the amount of photosynthetic radiation. On the other hand, these environmental layers did not contribute significantly to the model for the species *M. braziliensis* and *M. hartii*. High silica levels and low light penetration can be associated with high sedimentation/turbid conditions that characterize the marginal reef environment (Leão and Ginsburg 1997; Leão et al. 2006; Suggett et al. 2012). Within this environment, *M. braziliensis* and *M. hartti* are closely related in their photophysiology and capacity to adapt to a highly variable light penetration environment (Suggett et al. 2012). This result, however, contrasts with the indication that the photosynthesis to respiration ratio (P/R) of *M. braziliensis* is lower than the performance of *M. hispida* in the same experimental conditions (Freitas et al. 2019). Thus, a comparison of the relative influence of environmental variables measured on a regional scale, such as the one used in the present modeling, on the performance of the species is risky. The influence of microhabitat characteristics is decisive to the understanding of resource use by species and will bias any comparison between these two different scale studies.

Salinity and temperature were variables with a high influence on the species evaluated in this study. Salinity was the major contributing variable for *M. braziliensis* and *M. hartii*, whereas it was a low contributing variable to *M. hispida*. Both variables are closely related to water masses and are influenced by the proximity to the coast (consequently to depth). Thus, at the regional scale, it seems that oceanographic

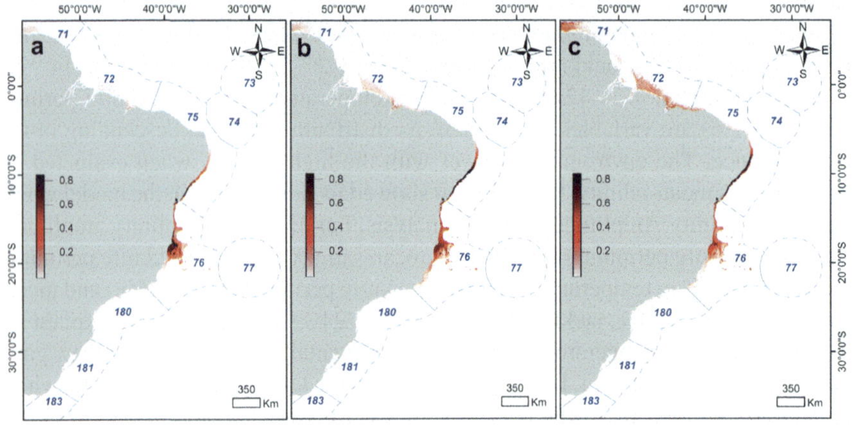

Fig. 5.5 Predicted environmental suitability for the species *Mussismilia braziliensis* using the maximum entropy model for the **a** modern 2000–2014, **b** RCP85 2040–2050 and **c** RCP85 2090–2100 scenarios. Darker colors indicate more suitable environments for the species. Delimitation and numbering of marine ecoregions according to that proposed by Spalding et al. (2007): 71. Guianan, 72. Amazonia, 73. São Pedro and São Paulo Islands, 74. Fernando de Noronha and Atoll das Rocas, 75. Northeastern Brazil, 76. Eastern Brazil, 77. Trindade and Martin Vaz Islands, 180. Southeastern Brazil, 181. Rio Grande, 183. Uruguay–Buenos Aires Shelf. See also Table 5.1

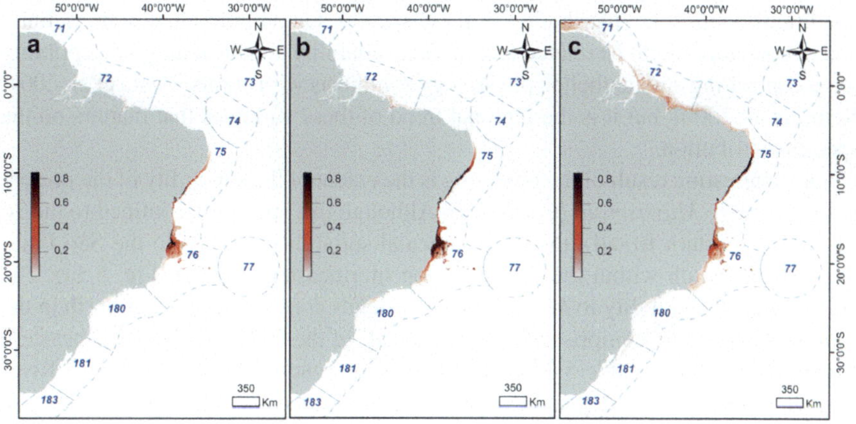

Fig. 5.6 Predicted environmental suitability for the species *Mussismilia hartii* using the maximum entropy model for the **a** modern 2000–2014, **b** RCP8.5 2040–2050 and **c** RCP8.5 2090–2100 scenarios. Darker colors indicate more suitable environments for the species. Delimitation and numbering of marine ecoregions according to that proposed by Spalding et al. (2007): 71. Guianan, 72. Amazonia, 73. São Pedro and São Paulo Islands, 74. Fernando de Noronha and Atoll das Rocas, 75. Northeastern Brazil, 76. Eastern Brazil, 77. Trindade and Martin Vaz Islands, 180. Southeastern Brazil, 181. Rio Grande, 183. Uruguay–Buenos Aires Shelf. See also Table 5.1

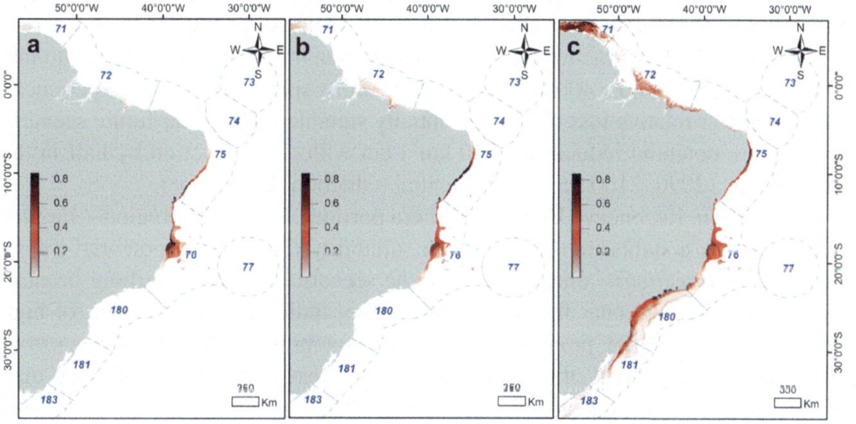

Fig. 5.7 Predicted environmental suitability for the species *Mussismilia hispida* using the maximum entropy model for the **a** modern 2000–2014, **b** RCP85 2040–2050 and **c** RCP85 2090–2100 scenarios. Darker colors indicate more suitable environments for the species. Delimitation and numbering of marine ecoregions according to that proposed by Spalding et al. (2007): 71. Guianan, 72. Amazonia, 73. São Pedro and São Paulo Islands, 74. Fernando de Noronha and Atoll das Rocas, 75. Northeastern Brazil, 76. Eastern Brazil, 77. Trindade and Martin Vaz Islands, 180. Southeastern Brazil, 181. Rio Grande, 183. Uruguay–Buenos Aires Shelf. See also Table 5.1

conditions control the distribution of these species. On an individual scale, salinity and temperature might be related to nutrient availability and bleaching susceptibility (e.g., controlling zooxanthellae physiology symbiosis with corals, Costa et al. 2004; Kemp et al. 2006), but it is the regional trend of these variables that matters on the species distribution.

One interesting result of the modeling is the extent of the suitability of the present environment to *Mussismilia braziliensis*. Although this species is confined to the 76 Ecoregion (eastern Brazil), the environmental suitability extends to the São Francisco River mouth within the 75 Ecoregion (northeastern Brazil) (Fig. 5.5a). The environmental suitability to *M. braziliensis* extends consistently further north in the future scenarios and comprises the southern half of the 75 Ecoregion (Northeastern Brazil) (Fig. 5.5a–c, Table 5.3). If the decreasing trend of the São Francisco River discharge during the last 1 ka (Dominguez and Guimarães 2021; Dominguez this book) persists in the future, this projection may be confirmed.

The 76 ecoregion (eastern Brazil) shows reduced values of the predicted probability of occurrence of all three species evaluated in future scenarios (Figs. 5.5b, c 5.6b, c and 5.7b, c, Table 5.3). In these scenarios, mean temperature and mean salinity were the variables with the greatest impact on environmental suitability. Therefore, in a future temporal mesoscale, several community (spatiality) and population intrinsic aspects (food availability, photosynthetic capacity of zooxanthellae, diseases) and, in addition survival and occurrence aspects should be compromised. The relative area of high suitability of species in the entire area of this ecoregion in the modern scenario, which accounted for approximately 3%, fell to close to 1% over a period of 100 years, which in practice represents a reduction in the area potentially suitable for the species by almost 400 km^2. Among the three species, *M. hartii* experienced a slightly lower relative loss of area potentially suitable for the near future scenario (2040–2050, potential reduction of 40 km^2) but with area reduction by half in the next 50 years (2090–2100 scenario, potential reduction of 210 km^2).

For *M. hartii*, Recôncavo Baiano (northern portion of the 76 Ecoregion—Eastern Brazil) may be a stable unit area of high suitability (Fig. 5.6), a potential refuge zone. For *M. braziliensis* and *M. hispida*, the scenario is more disturbing because the predictions in the near future scenario do not indicate potential areas of high suitability, although they point to signs of high suitability in the future scenario of 2090–2100. This interval of absence of suitable areas within this ecoregion may indicate that the next decades will be decisive for these species in terms of tolerance to new abiotic and biotic environmental conditions. The areas adjacent to Recôncavo Baiano exhibited high suitability for all species except *M. hispida* (Fig. 5.7b, c), while the region of the Abrolhos Archipelago maintained areas of high suitability for all species. This raises the importance of deepening knowledge of these regions that can be in the future a refuge for this and other species that share the same environmental envelope.

In the 75 ecoregions (northeastern Brazil), the trend of increase in the potential relative area of high suitability was similar for the three species in the future scenarios, with the exception of *M. hispida* for the 2090–2100 scenario. Shifts of the potential area to the south of those predicted by the modern scenario, such as what we see

Table 5.3 Relative suitability area for South Atlantic Ocean Ecoregions for present and future scenarios under RCP8.5 for *Mussismilia brasiliensis*, *M. hartii* and *M. hispida* according to Bio-Oracle benthic datasets and maximum entropy algorithm

Species/Scenarios	Marine Ecoregion (Spalding et al. 2007)				
	Amazonia (72)	NE Brazil (75)	E Brazil (76)	SE Brazil (180)	Rio Grande (181)
M. brasiliensis					
Present					
High suitability area (km^2)	–	40	540	–	–
Relative suitability area (%)	–	0.40	3.87	–	–
2040–2050					
High suitability area (km^2)	–	1050	–	–	–
Relative suitability area (%)	–	10.61	–	–	–
2090–2100					
High suitability area (km^2)	–	1280	140	–	–
Relative suitability area (%)	–	12.93	1.00	–	–
M. hartii					
Present					
High suitability area (km^2)	–	140	440	–	–
Relative suitability area (%)	–	1.41	3.15	–	–
2040–2050					
High suitability area (km^2)	–	1000	400	–	–
Relative suitability area (%)	–	10.10	2.86	–	–
2090–2100					
High suitability area (km^2)	–	1170	190	–	–
Relative suitability area (%)	–	11.82	1.36	–	–

(continued)

Table 5.3 (continued)

Species/Scenarios	Marine Ecoregion (Spalding et al. 2007)				
	Amazonia (72)	NE Brazil (75)	E Brazil (76)	SE Brazil (180)	Rio Grande (181)
M. hispida					
Present					
High suitability area (km^2)	–	40	540	–	–
Relative suitability area (%)	–	0.40	3.87	–	–
2040–2050					
High suitability area (km^2)	–	1050	–	–	–
Relative suitability area (%)	–	10.61	–	–	–
2090–2100					
High suitability area (km^2)	80	200	100	650	20
Relative suitability area (%)	0.23	2.02	0.72	3.22	0.13

Delimitation of ecoregions according to Spalding et al. (2007). High suitability is defined by maximum entropy levels greater than 0.8

for *M. hispida* (Fig. 5.7 and Table 5.3), are regularly shown in studies of climate variability (Frieler et al. 2012; Freeman et al. 2013). What strikes us in the present modeling is that all three species showed potential areas of high suitability under the modern scenario in the 75 Ecoregion (Northeastern Brazil) in areas close to the boundary of the 76 Ecoregion (Eastern Brazil). The potential area expansion in the 75 Ecoregion and to the North in future predictions is conspicuous. Overall, the three species showed an increase from modern scenario prediction estimates of 40 to 140 km^2 (0.4 to 1.4%) of potential high suitability areas to approximately 1000 to 1280 km^2 (10.1 to 12.93%) in future scenarios. Finally, *M. hispida* did not follow this pattern in the 2090–2100 scenario. Estimates indicate a retraction of potential high suitability areas after the 2040–2050 scenario followed by the estimated record of potential high suitability areas to the 72 Ecoregion (Amazonia) and to the 181 Ecoregion (Rio Grande) (Fig. 5.7c) (Table 5.3).

Our results indicate that climate change (temperature elevation and increased ocean acidification) will profoundly impact the distribution conditions of suitable habitats of these species under RCP8.5 for both the 2040–2050 and 2090–2100 scenarios. The new suitability areas predicted by the future scenarios, however, are only potential expansion areas because the actual colonization of new sites will depend on substrate availability, larval transport, competition, and predation, among other factors that regulate the process of new substrate colonization. However, these predictions should be considered as guidance for future studies that can support public and environmental policy.

5.6 Concluding Remarks

The EB and ENEB ecoregions are reef areas that concentrate a higher diversity of habitats (reef forms) and major reef organisms, such as coral and fish communities. The southern portion of the Eastern Brazil ecoregion has a wide shallow shelf (Abrolhos bank) that hosts one of the most important shallow water reef complexes. It is also one of the most studied reef regions in the country. Extensive mesophotic reef areas also occur in this region. To the north and to the south of this region, the availability of shallow water areas decreases, and mesophotic and deep water reefs become important and widespread interconnected seascapes.

Shallow water coral richness follows a similar pattern. It is higher in the 76 (eastern Brazil) and 75 (northeastern Brazil) ecoregions, where *Mussismilia* species and milleporids are important constituents of the reef-building fauna. Projections of future habitat suitability show that the most restricted species (*M. harttii* and *M. braziliensis*) can migrate north, a trend that can be enhanced by the increase in aridity and possibly an increase in temperature and salinity in the shallow shelf areas. On the other hand, turbidity might play an opposite role in these prognostics. *M. hispida*, on the contrary, might benefit from the tropicalization of the temperate realm and migrate south.

Tropical Southwestern Atlantic reefs are affected by a plethora of human stressors, including large-scale (global climate change-related processes, such as long-term warming, marine heatwaves, ocean acidification, and sea-level rise) or local but widespread impacts, such as mismanaged touristic and industrial activities, higher nutrient inputs, increased macro- and microplastic pollution, environmental disasters (e.g., oil spills and mining dam collapses), overfishing of reef species, and invasive species such as lionfish (*Pterois spp.*) and sun corals (*Tubastraea* spp.). Human stressors are impacting these tropical shallow-water reefs as cumulative and synergistic pressures and, consequently, undermining their long-term persistence.

Global climate change shifts limits that control the survival of reef-building and dwelling organisms, and regional shifts of optimal conditions also occur. However, the environmental conditions that control suitability for the survival of reef and coralline ecosystems are changing to sites where substrate availability is not guaranteed. Thus, we must apply all available efforts to develop research that supports conservation policies and environmental management actions to help ameliorate and reduce the impacts and degradation of reef and coralline ecosystems in the Tropical Western South Atlantic Ocean.

Acknowledgements This work is a result of the research led by the authors in the network established and funded through the inctAmbTropic - National Institute on Science and Technology for Tropical Marine Environments, CNPq/FABESB (565054/2010-4, 8936/2011, 465634/2014-1 and inc004/2019). RKPK receives Research Productivity Fellowship No. 311449/2019-0 from CNPq. MOS thanks the Conselho Nacional de Desenvolvimento Científico e Tecnológico (Research Productivity Fellowship No. 311449/2019-0 and 313518/2020-3, respectively) PELD Costa Semiárida do Brasil-CSB (No. 442337/2020-5), CAPES-PRINT, Alexander Von Humboldt Foundation and Fundação Cearense de Apoio ao Desenvolvimento Científico e Tecnológico (Chief Scientist Program) for their financial support. We would also like to thank Valéria F. Vale (LABEEC/UFRN) and Walter D. Oliveira (UFPE) for their invaluable help with systematic reviews on coral species occurrence and Alex B. Moraes (LABEEC/UFRN) for his support with the modeling procedures of the relative suitability algorithms. The picture that appears in Fig. 5.4 was drawn by Emanuelle Rabelo.

References

Alencar CERD, Vale VF, Moraes SASN et al (2017) New record of the Six-holed Keyhole Urchin, *Leodia sexiesperforata* (Leske, 1778) (Clypeasteroida, Mellitidae), from the Brazilian coast, with an updated distribution map. Check List 13(5):597–603. https://doi.org/10.15560/13.5.597

Allouche O, Tsoar A, Kadmon R (2006) Assessing the accuracy of species distribution models: prevalence, kappa and the true skill statistic (TSS). J Appl Ecol 43:1223–1232. https://doi.org/10.1111/J.1365-2664.2006.01214.X

Almeida FFM (1955) Geologia e Petrologia do Arquipélago de Fernando de Noronha. DNPM/DGM, Rio de Janeiro, Brazil

Amaral FMD, Hudson MM, Steiner AQ et al (2007) Corals and calcified hydroids of Manuel Luiz Marine Park (State of Maranhão, Northeast Brazil). Biota Neotrop 3:73–82. https://doi.org/10.1590/S1676-06032007000300008

Amaral FMD, Ramos CAC, Leão ZMAN et al (2009) Checklist and morphometry of benthic cnidarians from the Fernando de Noronha Archipelago, Brazil. Cah Biol Mar 50:277–290

Araújo ME, Mattos FGM, Feitosa CV et al (2018) Capítulo 6—Histórico do conhecimento dos 'peixes ósseos'. In: Araújo, ME de, Feitosa, CV, Mattos SMG (eds) Ecologia de peixes recifais de Pernambuco, 1st edn, UFPE, Recife, Brazil, pp 193–228

Assis J, Tyberghein L, Bosh S et al (2017) Bio-ORACLE v2.0: extending marine data layers for bioclimatic modelling. Glob Ecol Biogeogr 27:277–284. https://doi.org/10.1111/geb.12693

Assis J, Coelho NC, Lamy T et al (2016) Deep reefs are climatic refugia for genetic diversity of marine forests. J Biogeogr 43:833–844. https://doi.org/10.1111/jbi.12677

Aued AW, Smith F, Quimbayo JP et al (2018) Large-scale patterns of benthic marine communities in Brazilian Province. PLoS ONE 13(6):e0198452. https://doi.org/10.1371/journal.pone.0198452

Azevedo-Santos VM, Marques LM, Teixeira CR et al (2021) Digital media reveal negative impacts of ghost nets on Brazilian marine biodiversity. Mar Pollut Bull 172

Bastos N, Calazans SH, Altvater L et al (2022) Western Atlantic invasion of sun corals: incongruence between morphology and genetic delimitation among morphotypes in the genus *Tubastraea*. Bull Mar Sci. https://doi.org/10.5343/bms.2021.0031

Bender MG, Floeter SR, Mayer FP et al (2013) Biological attributes and major threats as predictors of the vulnerability of species: a case study with Brazilian reef fishes. Oryx 47:259–265. https://doi.org/10.1017/S003060531100144X

Beneli TM, Pereira PHC, Nunes JACC et al (2020) Ghost fishing impacts on hydrocorals and associated reef fish assemblages. Mar Environ Res 161. https://doi.org/10.1016/j.marenvres.2020.105129

Bleuel J, Pennino MG, Longo G (2021) Coral distribution and bleaching vulnerability areas in Southwestern Atlantic under ocean warming. Sci Rep 11. https://doi.org/10.1038/s41598-021-92202-2

Boavida J, Assis J, Silva I et al (2016) Overlooked habitat of a vulnerable gorgonian revealed in the Mediterranean and Eastern Atlantic by ecological niche modelling. Sci Rep 6:36460. https://doi.org/10.1038/srep36460

Budd AF, Fukami H, Smith ND et al (2012) Taxonomic classification of the reef coral family Mussidae (Cnidaria: Anthozoa: Scleractinia): Classification of Reef Corals. Zool J Linn Soc 166:465–529. https://doi.org/10.1111/j.1096-3642.2012.00855.x

Capel KCC, Creed JC, Kitahara MV et al (2019) Multiple introductions and secondary dispersion of *Tubastraea* spp. in the Southwestern Atlantic. Sci Rep 9. https://doi.org/10.1038/s41598-019-50442-3

Cardoso GO, Falsarella LN, Chiroque-Solano PM et al (2022) Coral growth bands recorded trace elements associated with the Fundão dam collapse. Sci Total Environ 807. https://doi.org/10.1016/j.scitotenv.2021.150880

Carneiro PB de M, Ximenes Neto AR, Feitosa CV et al (2021) Marine hardbottom environments in the beaches of Ceará State, Equatorial Coast of Brazil. ACMAR 54:120–153. https://doi.org/10.32360/acmar.v54i2.61440

Carneiro PBM, Ximenes Neto AR, Jucá-Queiroz B, Teixeira CEP, Feitosa CV, Barroso CX, Matthews-Cascon H, de Morais JO, Freitas JEP, Santander-Neto J, de Araújo JT, Monteiro LHU, Pinheiro LS, Braga MDA, Cordeiro RTS, Rossi S, Bejarano S, Salani S, Garcia TM, Lotufo TMC, Smith TB, Faria VV, Soares MO (2022) Interconnected marine habitats form a single continental-scale reef system in South America. Sci Rep 12(1):17359. https://doi.org/10.1038/s41598-022-21341-x

Coimbra KTO, Alcântara E, de Souza Filho CR (2020) Possible contamination of the Abrolhos reefs by Fundao dam tailings, Brazil—new constraints based on satellite data. Sci Total Environ 733. https://doi.org/10.1016/j.scitotenv.2020.138101

Correia MD (2011) Scleractinian corals (Cnidaria: Anthozoa) from reef ecosystems on the Alagoas coast, Brazil. J Mar Biolog Assoc U K 91:659–668. https://doi.org/10.1017/S0025315410000858

Costa OS Jr, Leão ZMAN, Nimmo M et al (2000) Nutrification impacts on coral reefs from northern Bahia, Brazil. Hydrobiologia 440:307–315. https://doi.org/10.1023/A:1004104118208

Costa CF, Sassi R, Amaral FD (2004) Population density and photosynthetic pigment content in symbiotic dinoflagellates in the Brazilian scleractinian coral *Montastrea cavernosa* (Linnaeus, 1767). Braz J Oceanogr 52(2):93–99. https://doi.org/10.1590/S1679-87592004000200001

Costa MBSF, Araújo M, Araújo TCM et al (2016) Influence of reef geometry on wave attenuation on a Brazilian coral reef. Geomorphology 253:318–327. https://doi.org/10.1016/j.geomorph.2015.11.001

Cotovicz LC, Chielle R, Marins RV (2020) Air-sea CO2 flux in an equatorial continental shelf dominated by coral reefs (Southwestern Atlantic Ocean). Cont Shelf Res 204:104175. https://doi.org/10.1016/j.csr.2020.104175

Coura MF (2016) Contribuição ao plano de manejo do Parque Estadual Marinho do Parcel de Manuel Luís/MA: atualização e avanços da unidade após 25 anos. Dissertation, Fundacion Universitaria Iberoamericana, Spain

Creed JC, Fenner D, Sammarco PW et al (2017) The invasion of the azooxanthellate coral *Tubastraea* (Scleractinia: Dendrophylliidae) throughout the world: history, pathways and vectors. Biol Invasions 19:283–305. https://doi.org/10.1007/s10530-016-1279-y

Cruz ICS, Kikuchi RKP, Leão ZMAN (2008) Use of the video transect method for characterizing the Itacolomis Reefs, Eastern Brazil. Braz J Oceanogr 56:271–280

Cruz ICS, Waters LG, Kikuchi RKP et al (2018) Marginal coral reefs show high susceptibility to phase shift. Mar Pollut Bull 135:551–561. https://doi.org/10.1016/j.marpolbul.2018.07.043

Davies AJ, Guinotte JM (2011) Global Habitat Suitability for Framework-Forming Cold-Water Corals. PLoS ONE 6. https://doi.org/10.1371/journal.pone.0018483

De Araújo ME, Mattos FMG, Melo FPL et al (2020) Diversity patterns of reef fish along the Brazilian tropical coast. Mar Environ Res 160:105038–105080. https://doi.org/10.1016/j.marenvres.2020.105038

De Carvalho-Souza GF, Tinôco ML, Medeiros DV et al (2018) Marine litter disrupts ecological processes in reef systems. Mar Pollut Bull 133:464–471. https://doi.org/10.1016/j.marpolbul.2018.05.049

De Paula AF, Creed JC (2004) Two species of the coral Tubastraea (Cnidaria, Scleractinia) in Brazil: a case of accidental introduction. Bull Mar Sci 74:175–183

Dominguez JML, Bittencourt ACSP, Leão ZMAN, Azevedo AEG (1990) Geologia do Quaternário costeiro do Estado de Pernambuco. Rev Bras Geoc 20:208–215

Dominguez JML, Guimarães JK (2021) Effects of Holocene climate changes and anthropogenic river regulation in the development of a wave-dominated delta: The São Francisco River (eastern Brazil). Mar Geol 435. https://doi.org/10.1016/j.margeo.2021.106456

Drew JA, Amatangelo KL, Hufbauer RA (2015) Quantifying the human impacts on Papua New Guinea reef fish communities across space and time. PLoS ONE 10. https://doi.org/10.1371/journal.pone.0140682

Duarte GAS, Villela HDM, Deocleciano M et al (2020) Heat waves are a major threat to turbid coral reefs in Brazil. Front Mar Sci 7. https://doi.org/10.3389/fmars.2020.00179

Elliff CI, Kikuchi RKP (2017) Ecosystem services provided by coral reefs in a Southwestern Atlantic Archipelago. Ocean Coast Manag 136:49–55. https://doi.org/10.1016/j.ocecoaman.2016.11.021

Elliff CI, Silva IR (2017) Coral reefs as the first line of defense: shoreline protection in face of climate change. Mar Environ Res 127:148–154. https://doi.org/10.1016/j.marenvres.2017.03.007

Elith J, Graham CH, Anderson RP et al (2006) Novel methods improve prediction of species' distributions from occurrence data. Ecography 29:129–151. https://doi.org/10.1111/j.2006.0906-7590.04596.x

Elith J, Phillips SJ, Hastie T et al (2011) A statistical explanation of MaxEnt for ecologists. Diversity Distrib 17(1):43–57. https://doi.org/10.1111/j.1472-4642.2010.00725.x

Feitosa CV, Chaves LDCT, Ferreira BP et al (2012) Recreational fish feeding inside Brazilian MPAs: impacts on reef fish community structure. Marine biological association of the United Kingdom. J Mar Biolog Assoc U K 92. https://doi.org/10.1017/S0025315412000136

Ferreira BP, Maida M (2006) Monitoramento dos recifes de coral do Brasil. MMA, Brasilia, Brazil

Fielding AH, Bell JF (1997) A review of methods for the assessment of prediction errors in conservation presence/absence models. Environ Conserv 24:38–49. https://doi.org/10.1017/S0376892997000088

Floeter SR, Gasparini JL (2001) The Brazilian endemic reef fishes. Coral Reefs 19. https://doi.org/10.1007/s003380000097

Fourcade Y, Besnard AG, Secondi J (2018) Paintings predict the distribution of species, or the challenge of selecting environmental predictors and evaluation statistics. Glob Ecol Biogeogr 27:245–256

França NFC, Alencar CERD, Mantelatto FL et al (2020) Filling biogeographic gaps about the shrimp *Farfantepenaeus isabelae* Tavares & Gusmão, 2016 (Decapoda: Penaeidae) in South America. Zootaxa 4718(4):497–508. https://doi.org/10.11646/zootaxa.4718.4.4

Freeman LA, Kleypas JA, Miller AJ (2013) Coral reef habitat response to climate change scenarios. PLoS ONE 8. https://doi.org/10.1371/journal.pone.0082404

Freire KMF, Carvalho-Filho A (2009) Richness of common names of Brazilian reef fishes. Pan-Am J Aquat Sci 4(2):96–145

Freitas LM, Oliveira M de DM, Leão ZMAN et al (2019) Effects of turbidity and depth on the bioconstruction of the Abrolhos reefs. Coral Reefs 38:241–253. https://doi.org/10.1007/s00338-019-01770-3

Frieler K, Meinshausen M, Golly A et al (2012) Limiting global warming to 2 °C is unlikely to save most coral reefs. Nat Clim Chang 3:165–170. https://doi.org/10.1038/nclimate1674

GEBCO (2015) General bathymetric chart of the oceans. https://www.gebco.net

Giglio VJ, Luiz OJ, Schiavetti A (2016) Recreational diver behavior and contacts with benthic organisms in the Abrolhos National Marine Park. Brazil. Environ Manag 57(3):637–648. https://doi.org/10.1007/s00267-015-0628-4

Godoy L, Mies M, Zilberberg C et al (2021) Southwestern Atlantic reef-building corals *Mussismilia* spp. are able to spawn while fully bleached. Mar Biol 168:1–8. https://doi.org/10.1007/s00227-021-03824-z

Gomes MP, Vital H, Droxler AW (2020a) Terraces, reefs, and valleys along the Brazil northeast outer shelf: deglacial sea-level archives? Geo-Mar Lett 40:699–711. https://doi.org/10.1007/s00367-020-00666-4

Gomes MP, Vital H, Nascimento Silva LL et al (2020b) Nature and condition of outer shelf habitats on the drowned Açu Reef, Northeast Brazil. In: Harris P, Baker E (eds) Seafloor Geomorphology as Benthic Habitat, 2nd edn. Elsevier, pp 571–585

Harborne AR (2013) The Ecology, behaviour and physiology of fishes on coral reef flats, and the potential impacts of climate change. J Fish Biol 83(3):417–447. https://doi.org/10.1111/jfb.12203

Hartt CF (1870) Geology and physical geography of Brazil. Fields, Osgood, Massachusetts

Hijmans RJ, Phillips S, Leathwick J et al (2017) dismo: species distribution modeling. R package version 1.1–4. https://CRAN.R-project.org/package=dismo

Hoeksema BW, Cairns S (2022) World List of Scleractinia. *Stephanocoenia intersepta* (Esper, 1795). Accessed through: World Register of Marine Species at: https://www.marinespecies.org/aphia.php?p=taxdetails&id=291119 on 2022-05-27

Jones GP, Munday PL, Caley MJ (2002) Rarity in coral reef fish communities. In: Sale P (ed) Coral reef fishes dynamics and diversity in a complex ecosystem. Academic Press, London, pp 81–101. https://doi.org/10.1016/B978-012615185-5/50006-2

Kemp DW, Cook CB, LaJeunesse TC et al (2006) A comparison of the thermal bleaching responses of the zoanthids *Palythoa caribaeorum* from three geographically different regions in south Florida. J Exp Mar Biol Ecol 335:266–276. https://doi.org/10.1016/j.jembe.2006.03.017

Kikuchi RKP, Leão ZMAN (1997) Rocas (Southwestern Equatorial Atlantic, Brazil): an atoll built primarily by coralline algae. In: Lessios HA, Macintyre IG (eds) Proceedings of the International coral reef symposium ISRS, Panamá, pp 731–736

Kikuchi RKP, Leão ZMAN (1998) The effects of Holocene sea level fluctuation on reef development and coral community structure, Northern Bahia, Brazil. An Acad Bras Cienc 70:159–171

Kikuchi RKP, Oliveira MDM, Leão ZMAN et al (2008) Os recifes de Tinharé-Boipeba-Camamu, Bahia. In: Proceedings Rio Oil&Gas Conference. Instituto Brasileiro do Petróleo, Rio de Janeiro, 15–18 Setembro 2008

Kikuchi RKP, Leão ZMAN, Oliveira MDM (2010) Conservation status and spatial patterns of AGRRA vitality indices in Southwestern Atlantic reefs. Rev Biol Trop 58:1–31. https://doi.org/10.15517/rbt.v58i1.20021

Kikuchi RKP, Leão ZMAN, Oliveira MDM (2023) On the morphology and distribution of reefs and carbonate buildups in the Western South Atlantic. In: Kikuchi RKP, Leão ZMAN, De Araújo ME et al (eds) Brazilian coral reefs. Springer Science—Business Media (in print)

Kinzie RA III, Buddemeier RW (1996) Reefs Happen. Glob Change Biol 2:479–494. https://doi.org/10.1111/j.1365-2486.1996.tb00062.x

Kleypas JA, McManus JW, Meñez LAB (1999) Environmental limits to coral reef development: where do we draw the line? Amer Zool 39:146–159

Laborel J (1969) Les peuplements de madréporaires des côtes tropicales du Brésil. University of Abidjan, Ivory Coast

Laborel-Deguem F, Castro CB, Nunes FLD et al (eds) (2019) Recifes brasileiros: o legado de Laborel. Museu Nacional, Rio de Janeiro, Brazil

Leão ZMAN (1996) The coral reefs of Bahia: morphology, distribution and the major environmental impacts. An Acad Bras Cienc 68:439–452

Leão ZMAN, Ginsburg RN (1997) Living reefs surrounded by siliciclastic sediments: the Abrolhos Coastal reefs, Bahia, Brazil. In: Lessios HA, Macintyre IG (eds) Proceedings of the international coral reef symposium. ISRS, Panamá, pp 1767–1772

Leão ZMAN, Dutra LXC, Spano S (2006) The characteristics of bottom sediments. In: Allen G, Dutra GF, Werner TB et al (eds) A biological assessment of Abrolhos Bank, Brazil, vol 38. RAP Bull Biol Assess, pp 75–81

Leão ZMAN, Kikuchi RKP, Testa V (2003) Corals and coral reefs of Brazil. In: Cortés J (ed) Latin American coral reefs. Elsevier Science, Amsterdam, pp 9–52. https://doi.org/10.1016/B978-044451388-5/50003-5

Leão ZMAN, Kikuchi RKP, Ferreira BP et al (2016) Brazilian coral reefs in a period of global change: a synthesis. Braz J Oceanogr 64:97–116. https://doi.org/10.1590/S1679-875920160916064sp2

Lesser MP, Slattery M, Leichter JJ (2009) Ecology of mesophotic coral reefs. J Exp Mar Biol Ecol 375:1–8. https://doi.org/10.1016/j.jembe.2009.05.009

Lewis A (1997) Effects of experimental coral disturbance on the structure of fish communities on large patch reefs. Mar Ecol Prog Ser 161:37–50. https://doi.org/10.3354/meps161037

Lindner A, Sarti AP, Batista AA et al (2017) A biodiversidade marinha das ilhas da Rebio Arvoredo e entorno. In: Segal B, Freire AS, Lindner A et al. (eds) MAArE—Monitoramento Ambiental da Reserva Biológica Marinha do Arvoredo e Entorno. 1st edn. UFSC/MAArE, Florianópolis, Brazil, pp 211–249

Loiola M, Cruz ICS, Lisboa DS et al (2019) Structure of marginal coral reef assemblages under different turbidity regime. Mar Environ Res 147:138–148. https://doi.org/10.1016/j.marenvres.2019.03.013

Luiz OJ, Santos WCR, Marceniuk AP et al (2021) Multiple lionfish (*Pterois* spp.) new occurrences along the Brazilian coast confirm the invasion pathway into the Southwestern Atlantic. Biol Invasions 23:3013–3019. https://doi.org/10.1007/s10530-021-02575-8

Maida M, Ferreira BP (1997) Coral reefs of Brazil: an overwiew. In: Lessios HA, Macintyre IG (eds) Proceedings of the international coral reef symposium ISRS, Panamá, pp 263–274

Magris RA, Grech A, Pressey RL (2018) Cumulative Human Impacts on Coral Reefs: Assessing Risk and Management Implications for Brazilian Coral Reefs. Diversity 10. https://doi.org/10.3390/d10020026

Magris RA, Costa MDP, Ferreira CEL et al (2020) A blueprint for securing Brazil's marine biodiversity and supporting the achievement of global conservation goals. Divers Distrib 27:198–215. https://doi.org/10.1111/ddi.13183

Manel S, Williams HC, Ormerod SJ (2001) Evaluating presence-absence models in ecology: the need to account for prevalence. J Appl Ecol 38:921–931. https://doi.org/10.1046/j.1365-2664.2001.00647.x

Mantelatto MC, Creed JC, Mourao GG et al (2011) Range expansion of the invasive corals *Tubastraea coccinea* and *Tubastraea tagusensis* in the Southwest Atlantic. Coral Reefs 30:397–397. https://doi.org/10.1007/s00338-011-0720-z

Mazzei EF, Bertoncini AA, Pinheiro HT et al (2017) Newly discovered reefs in the southern Abrolhos Bank, Brazil: anthropogenic impacts and urgent conservation needs. Mar Pollut Bull 114:123–133. https://doi.org/10.1016/j.marpolbul.2016.08.059

Mies M, Güth AZ, Tenório AA et al (2018) In situ shifts of predominance between autotrophic and heterotrophic feeding in the reef-building coral *Mussismilia hispida*: an approach using fatty acid trophic markers. Coral Reefs 37:677–689. https://doi.org/10.1007/s00338-018-1692-z

Milliman JD, Barretto HT (1975) Part I. Background. In: Milliman JD, Summerhayes CP (eds) Upper continental margin sedimentation off Brazil. E. Schweizerbartsche Verlarsbuchhandlung, Sttutgart, pp 1–10

Morais RA, Ferreira CEL, Floeter SR (2017) Spatial patterns of fish standing biomass across Brazilian reefs, Southwestern Atlantic. J Fish Biol 91:1642–1667. https://doi.org/10.1111/jfb.13482

Moss RH, Edmonds AJ, Hibbard KA et al (2010) The next generation of scenarios for climate change research and assessment. Nature 463:747–756. https://doi.org/10.1038/nature08823

Moura RL, Secchin NA, Amado-Filho GM et al (2013) Spatial patterns of benthic megahabitats and conservation planning in the Abrolhos Bank. Cont Shelf Res 70:109–117. https://doi.org/10.1016/j.csr.2013.04.036

Mies M, Francini-Filho RB, Zilberberg C et al (2020) South Atlantic Coral Reefs Are Major Global Warming Refugia and Less Susceptible to Bleaching. Front Mar Sci 7. https://doi.org/10.3389/fmars.2020.00514

Mora C, Sale P (2002) Are populations of coral reef fish open or closed? Trends Ecol Evol 17:422–428. https://doi.org/10.1016/S0169-5347(02)02584-3

Muehe D (ed) (2018) Panorama da erosão costeira no Brasil. Ministério do Meio Ambiente, Brazil

Munday PL, Leis JM, Lough JM et al (2009) Climate change and coral reef connectivity. Coral Reefs 28:379–395. https://doi.org/10.1007/s00338-008-0461-9

Munday PL, McCormick MI, Nilsson GE (2012) Impact of global warming and rising CO_2 levels on coral reef fishes: what hope for the future? J Exp Biol 215:3865–3873. https://doi.org/10.1242/jeb.074765

Naimi B, Hamm NA, Groen TA et al (2014) Where is positional uncertainty a problem for species distribution modelling. Ecography 37:191–203. https://doi.org/10.1111/j.1600-0587.2013.00205.x

Neves EG, Andrade SCS, Da Silveira FL et al (2008) Genetic variation and population structuring in two brooding coral species (*Siderastrea stellata and Siderastrea radians*) from Brazil. Genetica 132:243–254. https://doi.org/10.1007/s10709-007-9168-z

Neves EG, Johnsson R, Sampaio C et al (2006) The occurrence of Scolymia cubensis in Brazil: revising the problem of the Caribbean solitary mussids. Zootaxa 1366:45–54. https://doi.org/10.11646/zootaxa.1366.1.2

Neves EG, Silveira FL, Pichon M et al (2010) Cnidaria, Scleractinia, Siderastreidae, *Siderastrea siderea* (Ellis and Solander, 1786): hartt expedition and the first record of a caribbean siderastreid in tropical Southwestern Atlantic. Check List 6:505–510. https://doi.org/10.15560/6.4.505

Nunes F, Fukami H, Vollmer SV et al (2008) Re-evaluation of the systematics of the endemic corals of Brazil by molecular data. Coral Reefs 27:423–432. https://doi.org/10.1007/s00338-007-0349-0

Nunes FLD, Lami TG, Castro CB et al (2019) Atualizações taxonômicas e notas explicativas sobre os táxons citados. In: Laborel-Deguem F, Castro CB, Nunes FLD et al (eds) Recifes brasileiros: o legado de Laborel. Museu Nacional, Rio de Janeiro, pp 345–359

Pereira PHC, Ternes MLF, Nunes JACC et al (2021) Overexploitation and behavioral changes of the largest South Atlantic parrotfish (Scarus trispinosus): Evidence from fishers' knowledge. Biol Conserv 254. https://doi.org/10.1016/j.biocon.2020.108940

Pereira PHC, Lima GV, Pontes AVF, Côrtes LGF, Gomes E, Sampaio CLS, Pinto TK, Miranda RJ, Cardoso ATC, Araujo JC, Seoane JCS (2022) Unprecedented coral mortality on Southwestern Atlantic coral reefs following major thermal stress. Front Mar Sci 9725778. https://doi.org/10.3389/fmars.2022.725778

Pereira Filho GH, Amado-Filho GM, Guimarães SMPB et al (2011) Reef fish and benthic assemblages of the trindade and Martin Vaz island group, SouthWestern Atlantic. Braz J Oceanogr 59:201–212. https://doi.org/10.1590/s1679-87592011000300001

Pereira-Filho GH, Shintate GS, Kitahara MV et al (2019) The southernmost Atlantic coral reef is off the subtropical island of Queimada Grande (24°S), Brazil. Bull Mar Sci 95:277–287. https://doi.org/10.5343/bms.2018.0056

Phillips SJ, Dudík M (2008) Modeling of species distribution with Maxent: new extensions and a comprehensive evalutation. Ecograpy 31:161–175. https://doi.org/10.1111/j.0906-7590.2008.5203.x

Phillips SJ, Anderson RP, Schapire RE (2006) Maximum entropy modeling of species geographic distributions. Ecol Modell 190:231–259. https://doi.org/10.1016/j.ecolmodel.2005.03.026

Pinheiro HT, Bender MG, Longo GO et al (2023) Origins, biogeography, and macroecology of the Southwestern Atlantic reef biodiversity. In: Kikuchi RKP, Leão ZMAN, De Araújo ME et al (eds) Brazilian coral reefs. Springer Science—Business Media (in print)

Pinheiro HT, Rocha LA, Macieira RM et al (2018) South-western Atlantic reef fishes: zoogeographical patterns and ecological drivers reveal a secondary biodiversity centre in the Atlantic Ocean. Diversity Distrib 24:951–965. https://doi.org/10.1111/ddi.12729

Principe SC, Acosta AL, Andrade JE et al (2021) Predicted shifts in the distributions of Atlantic reef-building corals in the face of climate change. Front Mar Sci 8. https://doi.org/10.3389/fmars.2021.673086

R Core Team (2018) R: a language and environment for statistical computing. R foundation for statistical computing, Vienna, Austria. Available online at https://www.R-project.org/

Riahi K, Rao S, Krey V et al (2011) RCP 8.5—a scenario of comparatively high greenhouse gas emissions. Clim Change 109:33–57. https://doi.org/10.1007/s10584-011-0149-y

Sampaio CLS, Miranda RJ, Maia-Nogueira R et al JCC (2012) New occurrences of the nonindigenous orange cup corals *Tubastraea coccinea* and *T. tagusensis* (Scleractinia: Dendrophylliidae) in southwestern Atlantic. Check List 8:528–530. https://doi.org/10.15560/8.3.528

Santos CLA, Vital H, Amaro VE et al (2007) Mapping of the submerged reefs in the coast of the Rio Grande do Norte, near Brazil: Macau to Maracajau. Rev Bras Geofis 25:27–36. https://doi.org/10.1590/s0102-261x2007000500004

Seoane JCS, Arantes RCM, Castro CB (2012) Benthic habitat mapping at Recife de Fora, Brazil: Imagery and GIS. In: Proceedings of the 12th international coral reef symposium, Cairns, Australia, 9–13 July 2012

Soares MO, Lotufo TMC, Vieira LM et al (2016) Brazilian Marine animal forests: a new world to discover in the Southwestern Atlantic. In: Rossi S, Bramanti L, Gori A et al (eds) Marine Animal Forests. Springer International Publishing, Cham, pp 1–38

Soares MO, Teixeira CEP, Ferreira SMC et al (2019) Thermal stress and tropical reefs: mass coral bleaching in a stable temperature environment? Mar Biodivers 49:2921–2929. https://doi.org/10.1007/s12526-019-00994-4

Soares MO (2020) Marginal reef paradox: a possible refuge from environmental changes? Ocean Coast Manag 185. https://doi.org/10.1016/j.ocecoaman.2019.105063

Soares MO, Teixeira CEP, Bezerrra LEA et al (2020) Oil spill in South Atlantic (Brazil): environmental and governmental disaster. Marine Policy 115. https://doi.org/10.1016/j.marpol.2020.103879

Soares MO, Rossi S, Gurgel AR et al (2021) Impacts of a changing environment on marginal coral reefs in the Tropical Southwestern Atlantic. Ocean Coast Manag 210. https://doi.org/10.1016/j.ocecoaman.2021.105692

Soares MO, Matos E, Diniz B et al (2022). Climate change and local human impacts threaten Brazilian coral reefs In: Kikuchi RKP, Leão ZMAN, De Araújo ME et al (eds) Brazilian coral reefs. Springer Science—Business Media (in print)

Spalding MD, Fox HE, Allen GR et al (2007) Marine Ecoregions of the world: a bioregionalization of coastal and shelf areas. Bioscience 57:573–583. https://doi.org/10.1641/B570707

Suggett D, Kikuchi RKP, Oliveira MDM et al (2012) Photobiology of corals from Brazil's nearshore marginal reefs of Abrolhos. Mar Biol 159:1461–1473. https://doi.org/10.1007/s00227-012-1925-6

Testa V (1997) Calcareous algae and corals in the inner shelf of Rio Grande do Norte, NE Brazil. In: Lessios HA, Macintyre IG (eds) Proceedings of the international coral reef symposium ISRS, Panamá, pp 737–742

Tittensor DP, Baco AR, Brewin PE et al (2009) Predicting global habitat suitability for stony corals on seamounts. J Biogeogr 36:1111–1128. https://doi.org/10.1111/j.1365-2699.2008.02062.x

Tittensor DP, Micheli F, Nyström M et al (2007) Human impacts on the species-area relationship in reef fish assemblages. Ecol Lett 10(9):760–772. https://doi.org/10.1111/j.1461-0248.2007.01076.x

Tyberghein L, Verbruggen H, Pauly K et al (2012) Bio-ORACLE: a global environmental dataset for marine species distribution modelling. Glob Ecol Biogeogr 21:272–281. https://doi.org/10.1111/j.1466-8238.2011.00656.x

Zuur AF, Ieno EN, Elphick CS (2010) A protocol for data exploration to avoid common statistical problems. Methods Ecol Evol 1:3–14. https://doi.org/10.1111/j.2041-210X.2009.00001.x

Chapter 6
Geodiversity and Biodiversity of the Tropical Shelf of Northeastern Brazil

Helenice Vital, José Maria Landim Dominguez, Alex Cardoso Bastos, and Tereza Cristina Medeiros de Araujo

Abstract Continental shelves are areas of high heterogeneity that are poorly understood and increasingly influenced by anthropic activities. The tropical passive continental shelf of northeastern Brazil is the narrowest in the country and in some stretches is among the world's narrowest. Narrow continental shelves are unusual features on passive continental margins. This is clearly reflected in the small number of studies in the international literature. In this chapter, we present a synthesis of the current knowledge about seafloor morphology, its associated benthic ecosystems, and the role of eustatic variations in the evolution of this tropical shelf. Major human uses are also discussed.

Keywords Continental shelf · Shelf sedimentation · Siliciclastic–carbonate sedimentation

6.1 A General Overview of the Brazilian Tropical Shelf

The Brazilian tropical shelf extends for almost 3000 km from the Cape Orange (4° N) to the Abrolhos bank (19° S) (Fig. 6.1). Located in a passive margin, the shelf varies considerably in shape and width. It is rather wide near the Amazon River mouth (330 km) and narrower in front of Salvador (8 km). It can be divided into two parts: the Northern Brazil Shelf (NBS), from Cape Orange to the Parnaiba Delta, and the Northeastern Brazil Shelf (NEBS) from the Parnaiba Delta to Abrolhos Bank (Vital et al. 2010a; Vital 2014).

H. Vital (✉)
Federal University of Rio Grande do Norte (UFRN), Natal, Rio Grande do Norte, Brazil
e-mail: helenice.vital@ufrn.br

J. M. L. Dominguez
Federal University of Bahia (UFBA), Salvador, Bahia, Brazil

A. C. Bastos
Federal University of Espírito Santo (UFES), Vitória, Espírito Santo, Brazil

T. C. M. de Araujo
Federal University of Pernambuco (UFPE), Recife, Pernambuco, Brazil

J. M. Landim Dominguez et al. (eds.), *Tropical Marine Environments of Brazil*,
The Latin American Studies Book Series, https://doi.org/10.1007/978-3-031-21329-8_6

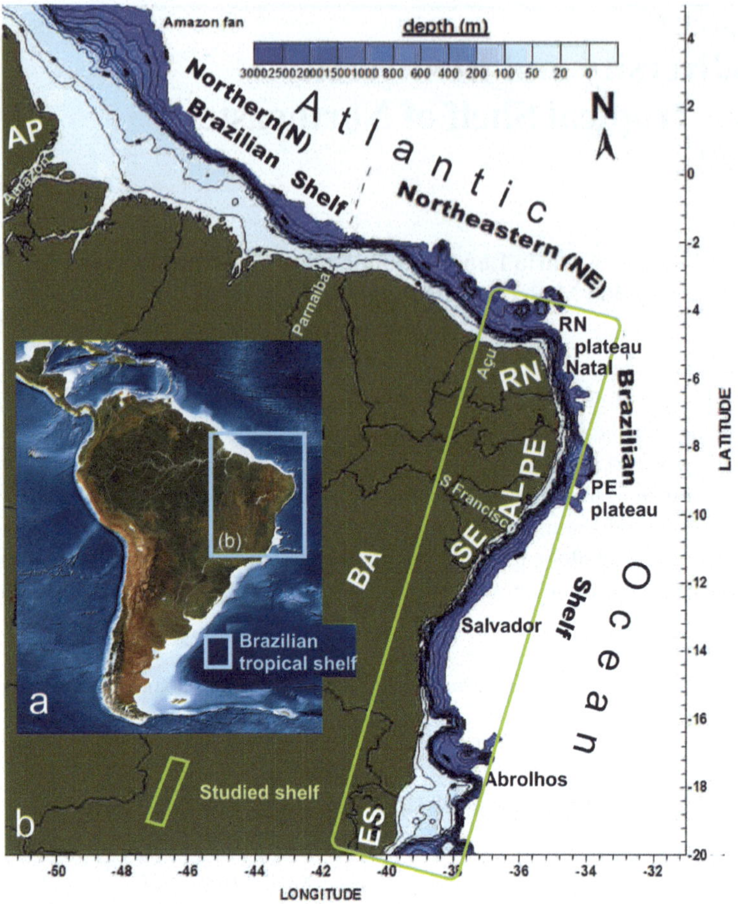

Fig. 6.1 **a** Location map (*Source* Planetary vision) and **b** bathymetric map of the Brazilian tropical shelf (modified from Vital 2014). State abbreviations: RN—Rio Grande do Norte, PE—Pernambuco, SE—Sergipe, AL—Alagoas, BA—Bahia, ES—Espirito Santo

This continental margin originated from the separation between South America and Africa, and its evolution followed the classical model for passive margins, including rift, transitional, and open marine stages (Asmus 1981).

The NBS is subjected to a very energetic forcing, including near-resonant semidiurnal macro tides, large buoyancy flux from the Amazon River discharge, wind stress from the northeasterly trade winds, and strong along-shelf flow associated with the North Brazil Current (Beardsley et al. 1995; Geyer and Kineke 1995; Nogueira Neto and Silva 2014; Prestes et al. 2018). Because of the large riverine inflows, most of the shelf is covered with siliciclastic muds and sands (Kuehl et al. 1986, 1996; Nittrouer et al. 1986, 1996; Figueiredo et al. 2008).

Table 6.1 The Brazilian tropical shelf (modified from Vital 2014)

	Northern Brazil Shelf (NBS)	Northeastern Brazil Shelf (NEBS)
Length	~1000 km	2000 km
Average width	+300 km	40 km
Tide, waves, and currents	>5 m (macrotidal), 1–2 m wave height and 13 s period, up to 200 cm s^{-1}	Mostly 2–3.5 m tidal range (mesotidal), 1–2 m wave height and 7.5 s period, currents up to 100 cm s^{-1}
Dominant process (wave/current/tides)	Tides, currents	Waves, mixed tidal and waves, currents
Av. depth of the shelf break	~100 m	60 m
Siliciclastic/carbonate/autigenic (%)	Siliciclastic (90%)	Middle to outer shelf carbonate Inner shelf siliciclastic to mixed
Modern/relict/palimpsest (%)	Inner shelf modern Outer shelf relict	Middle to outer shelf most modern Inner shelf modern and palimpsest
Tectonic trend over the last glacial cycle	Subsiding	Uplift and subsidence

The NEBS is subjected to the full strength of the westerly flowing South Equatorial Current, combined with eastern-southeastern trade winds, meso to micro tides, and moderate wave energy (Knoppers et al. 1999; Hazin et al. 2008; Domingues et al. 2017; Ribeiro et al. 2018). It is a narrow and open shelf almost entirely covered by carbonate sediments due to reduced riverine input (Coutinho and Morais 1970; Testa and Bosence 1999; Vital et al. 2008; Dominguez 2009) (Table 6.1).

Despite its extension and recognized importance, little is known about its marine habitats and resources. In this sense, since its creation, the Brazilian National Institute on Science and Technology in Tropical Marine Environments (inctAmbTropic) has made it possible to advance the evaluation of the spatial heterogeneity of shelf substrates in the north–northeastern region of Brazil.

This chapter presents a review of the existing knowledge of the NEBS, more specifically along the sector extending from the states of Rio Grande do Norte (RN) to Espirito Santo (ES) (Fig. 6.1b).

This review emphasizes seafloor mapping, which can then be integrated with the concept of habitat mapping, defined as the geological, sedimentological, and geomorphological characteristics of the seafloor that can define benthic habitats, together with other environmental drivers (temperature, water chemistry, depth, etc.) (Nunes 2009; Moura et al. 2013; Goes et al. 2019; Lucatelli et al. 2020; Fontes et al. 2020; Gomes et al. 2020a; Quaresma et al. 2020; Rebouças et al. 2020; Vital et al. 2020).

Greene et al. (1999) pointed out that habitat does not have a fixed scale and could be recognized as physiographic provinces (e.g., continental shelf), bottom shapes (e.g., submerged dunes), or types of sediment (e.g., muddy deposits). Lundblad et al. (2006) emphasize the need for an information baseline. According to these authors, based on what the seafloor looks like, a biologist, geologist, ecologist, geophysicist, or other interested parties will supplement that framework with specific datasets. A biologist may add a layer of information about the amount of relief or the thickness of sediments. A geologist may add data revealing sediment size or rock type. Depending on the interest of the research, different layers of information are needed. However, there is a consensus in the literature that mapping the seafloor is the basis for knowing the distribution of geodiversity and its relationship with marine biodiversity (Lundblad et al. 2006; Brown et al. 2011, 2012; Diesing et al. 2016; Lecours et al. 2016; Baker and Harris 2020; Harris and Baker 2020). It is, therefore, necessary to link the mapping of the seabed to the concept of habitat, that is, to map the nature, distribution, and extent of physical environments as a way of seeking to predict or have a forecast basis for the occurrence of associated biological communities (Quaresma et al. 2020).

6.2 The Northeastern Brazil Shelf

On the NEBS, the climate varies from tropical dry semiarid to tropical humid with few rivers emptying in this region, with the exception of the São Francisco, Jequitinhonha, and Doce rivers.

From a morphodynamic point of view, this is a wave-dominated to tide-modified coast with active sea cliffs carved into Miocene tablelands alternating with reefs or dune-barrier sections and beach-ridge terraces (Pereira et al. 2016; Short and Klein 2016; Vital et al. 2016).

It is characterized by reduced width (on average 40 km) and shallow depths (shelf break commonly starts at an average depth of 60 m) (Fig. 6.2a) when compared with other parts of the Brazilian continental shelf (Araujo et al. 2004; Vital et al. 2010a) and other passive margin shelves of the world (Kennett 1982; Harris et al. 2014). The mixed character of sedimentation is typical of this shelf, with coexisting siliciclastic and carbonate systems (e.g. Pessoa Neto 2003; Vital et al. 2005; Bastos et al. 2013; Dominguez et al. 2013; Puga-Bernabéu 2017), the former being dominant in the proximal nearshore regions and the latter in the distal portion or shelf edge (Fig. 6.2b). Although the narrower portions of the shelf are almost completely covered by carbonate (essentially bioclastic) sediments (Dominguez et al. 2013), the presence of shelf-incised valleys can produce sharp lateral shifts in sediment distribution, with siliciclastics confined to these valleys and carbonate sediments occurring in their margins (Vital et al. 2008; Gomes et al. 2015; Vieira et al. 2019).

The NEBS is mostly 20 km wide (Fig. 6.2), with its narrowest portion (8 km) located off the coast of Salvador city, whose physiography is attributed to a strong structural control by Halla et al. (2020). The edge of the continental shelf generally

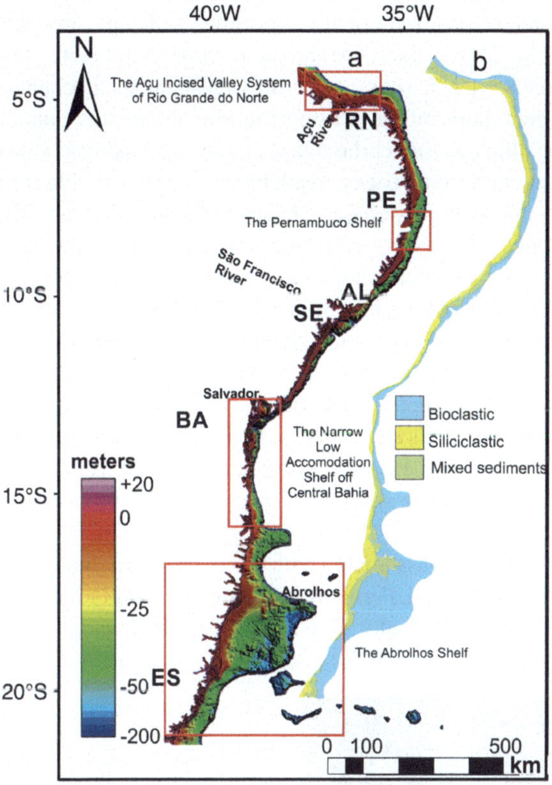

Fig. 6.2 Main features of the Northeastern Brazil Shelf (NEBS). **a** Relief image of NEBS (*Source* GEBCO database). **b** Surface sediment distribution. Red rectangles indicate sectors detailed in the text

runs parallel to subparallel to the coast. At the Abrolhos bank, the shelf reaches 200 km in width as a result of volcanic activity that took place between the Palaeocene and the Eocene (Mohriak 2006). The shelf break at the Abrolhos bank and around the RN plateau starts at a depth of approximately 80 m.

The shelf can be subdivided using depth as a criterion into inner (up to 15–20 m), middle (15–25 m), and outer shelf from 25 to 30 m until the shelf break (Coutinho 1996; Vital 2014). A diversity of bedforms, such as hydraulic dunes, reefs, and incised valley systems, are present on this shelf and can be observed in satellite images, even in regions close to the shelf break (Vianna et al. 1991; Tabosa et al. 2007; Vital et al. 2005, 2008, 2018; Silveira et al. 2020) and by using acoustic data (e.g., Schwarzer et al. 2006; Bastos et al. 2015; Cetto et al. 2021).

Carbonate build-ups and reefs are a dominant feature along the Brazilian tropical shelf (Figs. 6.3 and 6.4) (see also Chap. 5). These features can be classified according to their origin in (i) sandstone reef banks—cemented beach and shoreface deposits known as sandstone reefs or beachrocks, usually parallel to the shoreline, and recording ancient positions of the coastline, with the most continuous chain occurring along the 25 m isobath (Araujo et al. 2004; Vital et al. 2008; Fontes et al. 2017) and (ii) reefs—rigid, rocky structures resistant to mechanical wave action and marine currents, built by the action of marine organisms (animals and plants), capable of segregating calcium carbonate skeletons, and using as substrate, sedimentary, igneous, or metamorphic rocks, available at a depth suitable for their establishment and development. In this chapter, beachrocks and reefs are discussed only as geomorphic features or types of geohabitats present in the study area.

Both shallow water and mesophotic reefs are present on the NEBS (Fig. 6.4). The shallowest ones are known for a long time (Branner 1905; Laborel 1969) and at least in part are located in marine protected areas such as the Abrolhos Marine National Park (BA e ES states), the Costa dos Corais (AL and PE states), and Recifes de Corais (RN state) and Marine Protected areas (Ferreira and Maida 2006; Laborel-Deguem et al. 2019). Mesophotic reefs have been described more recently across the mid and outer shelves of the ES, BA, PE, and RN states at depths from 25 to the shelf edge (e.g., Camargo 2016; Silva et al. 2018; Fontes et al. 2020; Gomes et al. 2020a, b; Bastos et al. 2021).

Other important features associated with this continental shelf include the plateaus of Rio Grande do Norte and Pernambuco states (Fig. 6.1) as well as a series of submarine canyons (e.g., Salvador, São Francisco, Natal, and Açu).

Fig. 6.3 Underwater photos of beachrock chain along 25–30 m depth in the NEBS: **a** Rio Grande do Norte shelf; **b** dense carpet of macroalgae cover on the Pernambuco shelf; **c** reef flank region surrounded by sandy sediments on the Pernambuco shelf; **d** gorgonians and black corals cover on the Sergipe shelf. *Source* a: Vital et al. (2008); b, c: Fontes et al. (2020); d: Fontes et al. (2017)

Fig. 6.4 Coral reefs from NEBS. **a** Multibeam 3D model of California reef on the Abrolhos shelf; **b** aerial view of Costa dos Corais, shallow water reefs from Alagoas-Pernambuco; **c** aerial view of Maracajau, shallow water reefs from the Rio Grande do Norte; **d** and **e** underwater photos of shelf edge reefs from Pernambuco; **f** multibeam 3D model of Açu reef field on outer Rio Grande do Norte shelf; **g** and **h** application of benthic terrain modeller (BTM) analysis for the Açu reef field. *Source* a: Bastos et al. (2021); b: ICMBIO; c: IDEMA; d, e: Camargo (2016); f, g, h: Gomes et al. (2020a)

Below, we provide a detailed description of four sectors of the NEBS that illustrate the diversity of existing geomorphic features. For each sector, the focus will be on the geological setting, relief, sediment supply and sediment distribution, and the main features present, such as incised valleys and carbonate build-ups.

6.2.1 The Açu Incised Valley System of Rio Grande do Norte

The Rio Grande do Norte shelf is a narrow (40 km), shallow (70 m at the shelf break), and mixed carbonate–siliciclastic shelf. This part of the shelf is the submerged portion of the Potiguar Sedimentary Basin (Vital et al. 2008, 2020; Gomes et al. 2016, 2020a, b). The mixed sedimentation in the Potiguar basin began in the Neocampanian and extends to the present day (Pessoa Neto 2003). This basin has been subjected during its history to E–W compression, which has been released along preexisting fault lines lying approximately NE–SW (Milani and Thomaz-Filho 2000). As a result, the shelf configuration is strongly affected by vertical movements along these faults.

The distribution of sediments, seabed features, and general physiographic breaks allow us to divide the shelf into the inner, middle, and outer shelves, showing a clear seaward trend of increasing carbonate content (Vital et al. 2008, 2010a; Gomes et al. 2014) (Fig. 6.5).

The boundaries between the inner, middle, and outer shelves have been proposed by different authors using water depth, geomorphology, and sediment type as criteria (Gomes and Vital 2010; Vital et al. 2008, 2010a; Vital 2014; Almeida et al. 2015; Gomes et al. 2014, 2015, 2016): the inner shelf extends to a water depth of 15 m and is characterized by the presence of very large longitudinal subaqueous dunes composed mainly of quartz sands. The middle shelf, between 15 and 25 m depth, is characterized by a very large subaqueous transversal dune field composed of fine- to medium-grained siliciclastic sand on the crests, while the troughs are composed of coarse bioclastic gravel stabilized by living coralline algal maerl and Halimeda meadows (Testa and Bosence 1998, 1999; Tabosa 2006; Santos et al. 2007). The transition from the middle to the outer shelf, around the 25 m isobath, is defined by aligned submerged beachrock outcrops, composed of sandstones and grainstones cemented by carbonate, commonly colonized by seaweeds, rhodoliths, and sponges (Santos et al. 2007; Vital et al. 2008; Cabral Neto et al. 2013).

Incised valleys associated with the Açu River are infilled with muddy sediments (Costa Neto 1997; Vital et al. 2008, 2010a, b). Mesophotic knoll reefs are found on both sides of these incised valleys (Gomes et al. 2016; Silva et al. 2018). The outer shelf (25 m to shelf break) exhibits a terraced morphology inherited from the bedrock, underfilled incised valleys, and heads of submarine canyons indenting the shelf break (Fig. 6.5a, b). It is a hotspot of biodiversity with respect to soft sediment communities and the presence of the Açu Reef (Gomes et al. 2020a).

The study by Vital et al. (2020) indicates that the seabed features and associated grain size are responsible for the structure and composition of the benthic macroinfaunal and epifaunal communities. Fine-grained sediments inside the Açu incised

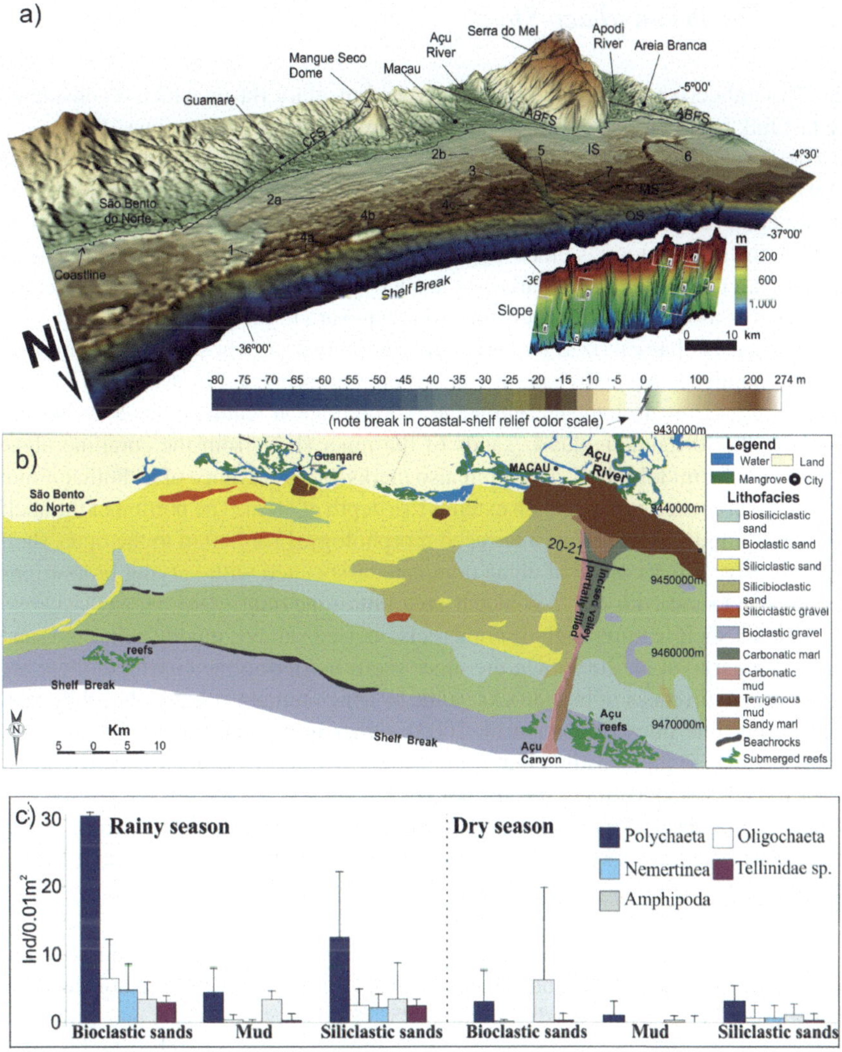

Fig. 6.5 Açu incised valley shelf system. **a** Digital terrain model and multibeam bathymetry to the upper slope; **b** sedimentary facies; **c** average density of the most abundant taxon during rainy and dry seasons. *Source* a: modified from Gomes et al. (2014) and Almeida et al. (2015); b: modified from Vital et al. (2008); c: modified from Vital et al. (2020)

shelf valley exhibit less diverse and less abundant macroinfauna assemblages than adjacent sandy bottoms, both bioclastic and siliciclastic, which harbour faunas of greater abundance and diversity (Fig. 6.5c).

6.2.2 The Pernambuco Shelf

The Pernambuco shelf is in the Pernambuco Sedimentary Basin, which is considered the last link between Africa and South America (Rand and Mabesoone 1982).

The Pernambuco shelf is narrow (average width of 35 km), shallow, and has a shelf break located between −50 and −60 m (Manso et al. 2003). According to Araujo et al. (2004), the PE shelf exhibits a smooth relief from 0 to 20 m. From 20 to 30 m, the morphology is more rugged, characterized by the alternation of valleys and crests, cutting perpendicularly to the shelf, mainly north of the city of Recife. This relief is associated with shelf subaerial exposure in the past.

The shelf is characterized by carbonate sedimentation whose major component is coralline algae. A subdivision for this shelf was proposed by Coutinho (1996), considering sedimentological criteria and morphological features: the 20 m isobath separates the terrigenous quartz sands of the inner shelf from the coralline algae deposits on the middle shelf. This limit also marks the appearance of Lithothamnium in free and branched forms, which extends to a depth of 40 m. The portion of the shelf between 23 and 40 m exhibits a terraced morphology. From 40 m to the outer shelf edge, massive blocks made of algae dominate, associated with varying proportions of biodetrital sand with 10–15% bluish-grey limestone mud.

Regional shelf features include coral reefs, incised valleys, muddy belts, and sand banks (Camargo et al. 2007). On the inner shelf, bordering the coastline, there are numerous occurrences of beachrocks, some of which remain completely submerged even during low tides (Manso et al. 2003). Other important features include reef ridges parallel to the coastline that developed on the top of beachrock substrata or Pleistocene bedrock (Camargo et al. 2015).

The Pernambuco Plateau is an extension of thinned continental crust of the Pernambuco Sedimentary Basin, extending from the shelf break to water depths of 4000 m (Fig. 6.1).

Seven substrate/habitat types with distinct acoustic characteristics were identified in the portion of the shelf offshore Tamandaré (Fig. 6.6b). These are comprised of three coral reef types I, II, and III (respectively P1, P2, and P3), carbonate gravelly bottoms (P4), gravel-sandy bottoms (P5), sandy bottoms (P6), and carbonate muddy bottoms (P7) (Fontes et al. 2020). These authors broadly classified benthic communities in the region into two groups governed by substrate type: hard and soft substrates.

Hard substrate areas consist of coral reefs, submerged beachrocks, and Pleistocene limestone outcrops. Coral reefs are complex environments that support algae, massive corals, zoanthids, fishes, crustose coralline algae, and barnacles, among others. It is observed that the more three-dimensional reefs, with the presence of cavities and caves, support less algae and more corals and herbivorous fishes. In less three-dimensional reefs, two main types of macroalgae, *Sargassum* sp. and *Dictyopteris* sp., are abundant. Algae may be displaced from some hard substrate areas by current and wave activity.

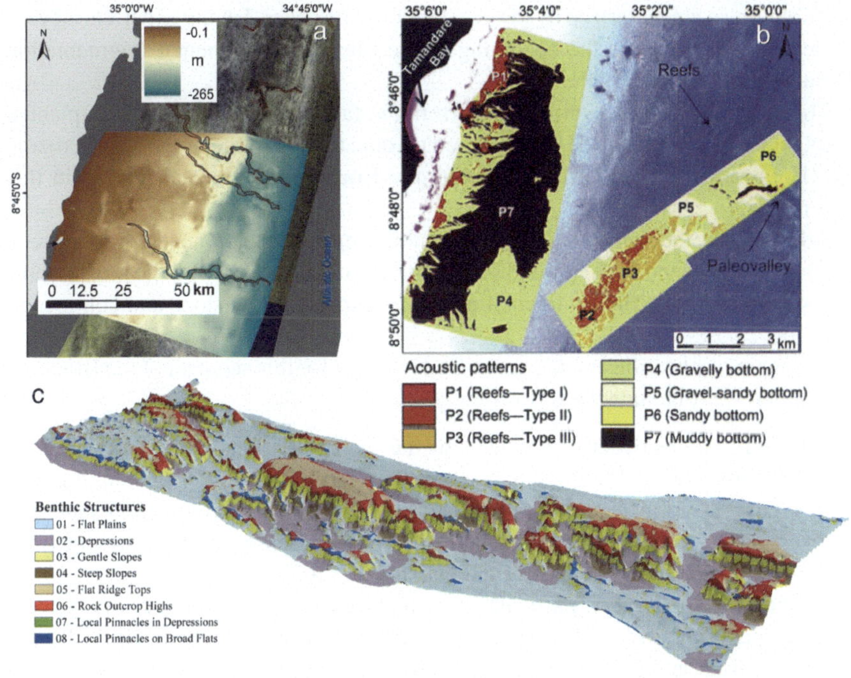

Fig. 6.6 Pernambuco shelf. **a** Incised valleys of the Tamandaré region; **b** map of acoustic backscatter patterns (P1 to P7) of the Tamandare's inner shelf, with Landsat 8 image on background; **c** benthic structures. *Source* a: Camargo (2016); b: Fontes et al. (2020); c: Goes et al. (2019)

Soft substrate areas (such as gravelly and muddy bottoms) are inhabited mostly by foraminifera, ostracodes, molluscs, echinoderms, and bryozoans. Patches of algae are also found in sediment-covered areas but are less widespread than on hard substrate areas.

The results from the BTM analysis revealed eight types of benthic structures (Goes et al. 2019): Flat Plains, Depressions, Gentle Slopes, Steep Slopes, Flat Ridge Tops, Rock Outcrop Highs, Local Pinnacles in Depressions, and Local Pinnacles on Broad Flats (Fig. 6.6c).

6.2.3 The Narrow, Low-Accommodation Shelf of Central Bahia

The state of Bahia has the most extensive stretch of a narrow continental shelf in Brazil, with widths varying between 8 and 37 km (Fig. 6.7). This shelf, like the rest of the continental shelf in northeastern Brazil, is also very shallow (break < 60 m). This platform is also characterized by low accommodation space, having experienced very limited subsidence since the beginning of the Mesozoic continental fragmentation (Dominguez et al. 2011, 2013).

These characteristics seem to result from the fact that this platform is implanted directly on the São Francisco craton, a geotectonic unit of Archean-Paleoproterozoic age (Heilbron et al. 2017) that intersects the Brazilian continental margin in this region.

Additionally, the stability of the craton, inhibiting the development of large reliefs, may have contributed to the limited sediment flux to this region, with the most important sedimentation of the continental margin occurring on the slope and continental elevation.

The main implications of this geological heritage for this segment of the Brazilian continental shelf include the following:

1. The limited subsidence experienced by the shelf means that features of the structural framework of the marginal sedimentary basins (Fig. 6.7) have, until today, been expressed in the morphology and bathymetry of the continental shelf and exert control on sedimentation and on the shelf seascape as documented by Dominguez et al. (2011, 2013) and Halla et al. (2020).
2. During the Quaternary, particularly in the last 1 million years, the average sea-level position was −62 m (Lea et al. 2002; Waelbroeck et al. 2002; Berger 2008; Blum and Hattier-Womack 2009), implying that the shelf was exposed subaerially almost continuously during most of the Quaternary period. As a result of this extended subaerial exposure, the submarine geomorphology of the shelf was determined by long-term erosional processes as opposed to sedimentation (Dominguez et al. 2013; Halla et al. 2020). The platform was only completely flooded during brief interglacial periods.
3. The stability conferred by the presence of the craton also favoured the preservation, along the emerged region of the coastal zone, of sedimentary rocks that accumulated in the marginal sedimentary basins of Almada, Camamu, and Recôncavo during the initial phases of Mesozoic continental fragmentation (Fig. 6.7), a situation practically absent in the rest of the Brazilian continental margin. The differential erosion between the more resistant rocks of the crystalline basement and the less resistant sedimentary rocks produced a series of topographically lowered regions along the coastal zone that were flooded during the highstands, giving rise to shallow bays (Dominguez et al. 2013; Dominguez 2015) (Fig. 6.7). Smaller bays such as Almada were completely filled during the current highstand (Dominguez et al. 2009), while larger bays such as Camamu and Todos os Santos bays are still underfilled. These bays work as natural traps for sediments with muds accumulating in their innermost parts, while large volumes of sand have accumulated in their entrances in association with ebb tidal deltas, thus influencing sedimentation on the continental shelf in their vicinity (Dominguez et al. 2012, 2013; Coni e Mello et al. 2020) (Fig. 6.7a, b).
4. The combination of a narrow platform, a shelf break starting at −60 m, and prolonged subaerial exposure favoured the development of incised valleys of

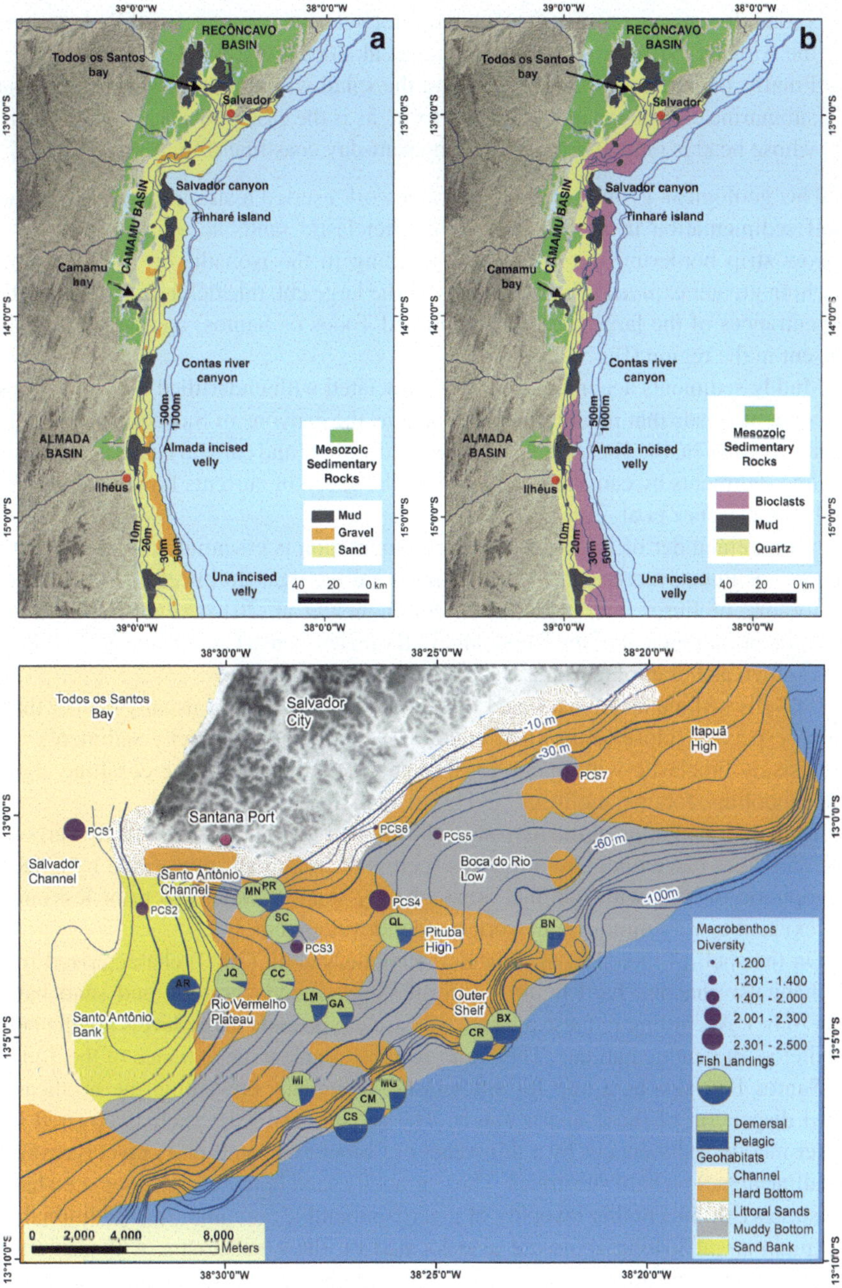

Fig. 6.7 Narrow shelf of central Bahia. **a** Bathymetric map with surficial sediment textural facies; **b** bathymetric map with surficial sediment types; **c** geohabitats of Salvador city. *Source* c: Modified from Rebouças et al. (2020)

varied dimensions, some of which have not yet been completely infilled, such as the Almada incised valley (Dominguez et al. 2009) (Fig. 6.7).

5. Finally, the reduced sediment flux and the small width of the shelf favoured the submarine canyons to indent the platform, as is the case of the Salvador canyon, whose head is only 10 km from the present-day coastline (Fig. 6.7).

This geological–geomorphological framework exerted a strong control over the shelf sedimentation in the region. Thus, siliciclastic sands accumulate only in a narrow strip bordering the coastline, extending to the isobaths of 15–20 m and reaching greater expressiveness precisely in the large ebb tide deltas associated with the entrances of the large bays (Camamu and Todos os Santos) and tidal channels present in the region (Fig. 6.7a).

Muddy sediments accumulate mainly associated with underfilled incised valleys and canyon heads that indent the shelf, such as the canyons of Salvador and Rio de Contas (Fig. 6.7a, b). These underfilled incised valleys and canyon heads act as traps for fine sediments by causing a reduction in the speed of currents flowing along the shelf (Dominguez et al. 2012, 2013).

In the remainder of the shelf, sandy sedimentation is essentially bioclastic with major constituents represented by incrusting coralline algae, followed, in importance, by forams, molluscs, and bryozoans (Dominguez et al. 2013) (Fig. 6.7b). These constituents are typical of the Rhodoalgal lithofacies (sensu Carannante et al. 1988), also known as Foramol (sensu Lees and Buller 1972).

Gravelly sediments are restricted to the outer shelf in waters usually deeper than 30 m, especially close to the shelf edge (Fig. 6.7a). These gravelly sediments are exclusively bioclastic with major constituents including incrusting coralline algae and rhodoliths, forams, molluscs, and bryozoans (Fig. 6.7b).

Thus, sedimentation at the narrow, shallow shelf of Bahia is dominantly carbonatic and takes place under humid tropical climate conditions, dominated by a deep oligophotic interior with minimal areas of very shallow water (10 m or less) and nonexistent to very minor framework builder development.

On the shelf of the coast of Salvador, Rebouças et al. (2020) characterized five geohabitats: channel, hard bottom, littoral sands, muddy bottom, and sand bank (Fig. 6.6c). According to these authors, substrate type and energy levels are the best predictors of benthic habitats. The Santo Antonio Bank at the entrance of the Todos os Santos Bay acts as a trap for siliciclastic sediments, preventing the northeastward dispersion of these quartz sands. The remainder of the shelf is covered by either muddy sediments or by a thin veneer of bioclastic sands and gravel (crustose coralline algae) on hard substrates. These gravelly sand bottoms and associated hard substrates exhibit a higher diversity of macrozoobenthic communities. Additionally, the majority of fishing spots are also located on these substrates, with demersal species comprising approximately 75% of the catches. Notwithstanding coastal and migratory pelagic species comprise most of the biomass.

6.2.4 The Abrolhos Shelf

The Abrolhos shelf has a total area of 46,000 km^2 and a width that can reach approximately 220 km (Figs. 6.1, 6.2, and 6.8). In its northern portion, the Abrolhos is marked by an extensive inner shelf (depth < 30 m) that can reach 100 km in width, marked by a low gradient and a very irregular morphology characterized by the presence of emergent and submerged reefs forming pinnacles and reef banks with varying relief heights (Ferreira et al. 2020) and incised valleys in the transition to the outer shelf (Moura et al. 2013; Bastos et al. 2015, 2021; D'Agostini et al. 2019). Moreover, the transition from the inner to the outer shelf also marks the presence of mesophotic reefs and rhodolite beds (Ferreira et al. 2020). Figure 6.8 combines multibeam echo sounder and side scan sonar images, showing the distinct morphological features in the Abrolhos shelf that reflect its high geodiversity, including reef banks, submerged pinnacles, sinkholes ("Buracas"), and incised valleys. The "Buracas", interpreted as sinkholes (Bastos et al. 2013), formed as a result of the dissolution of carbonate deposits associated with karst relief during periods of shelf exposure (Bastos et al. 2016). These sinkholes were mapped between depths of 30 and 65 m and may have a negative relief of more than 30 m.

The southern part of the Abrolhos Bank is marked by the presence of the Abrolhos Depression (Fig. 6.8a). Morphologically, this feature could be associated with an embayment (Bastos et al. 2021) or a paleolagoon (Vicalvi et al. 1978). It is interesting to note that striking features such as incised valleys flow towards the depression. According to D'Agostini (2017), these channels would represent a continental paleodrainage during the last glacial maximum when the sea level reached ~ −120 m. The Abrolhos Depression would have been a lagoon during this period of low sea level, and its origin would be associated with an erosion of up to 35 m of the former shelf surface (Vicalvi et al. 1978). The presence of isolated drowned reefs along the Abrolhos Depression, as well as on its northern and western margins, has been documented in seismic and sonographic surveys (Moura et al. 2013; D'Agostini 2017; Cetto et al. 2021).

The Abrolhos shelf presents mostly a mixed sedimentation regime, with terrigenous sediments in the inner shelf and carbonate sedimentation in the outer shelf (Fig. 6.9a). To the south, mixed sedimentation, interspersed with rhodolith bottoms and reef structures, marks the inner shelf, with the outer shelf defined by the Abrolhos Depression and its high topographic edge (Fig. 6.9a), marked by carbonate mud in the deeper parts and an extensive bank of rhodoliths in the shallower portion. The distribution of rhodoliths occurs predominantly along the middle and outer shelf, that is, mainly in areas with depths greater than 30–40 m.

The benthic terrain model (BTM) map (Fig. 6.9b) mainly highlights the areas that have reliefs above and below what can be considered flat areas. Basically, what has been defined as ridges, channel flanks, channel bottoms, and depressions present reliefs distinct from the flatter trend that characterizes much of this shelf. This map clearly shows a relief variability to the south marked by shelf-incised valleys (Fig. 6.9c) and hard bottoms defined by BTM classes that present changes

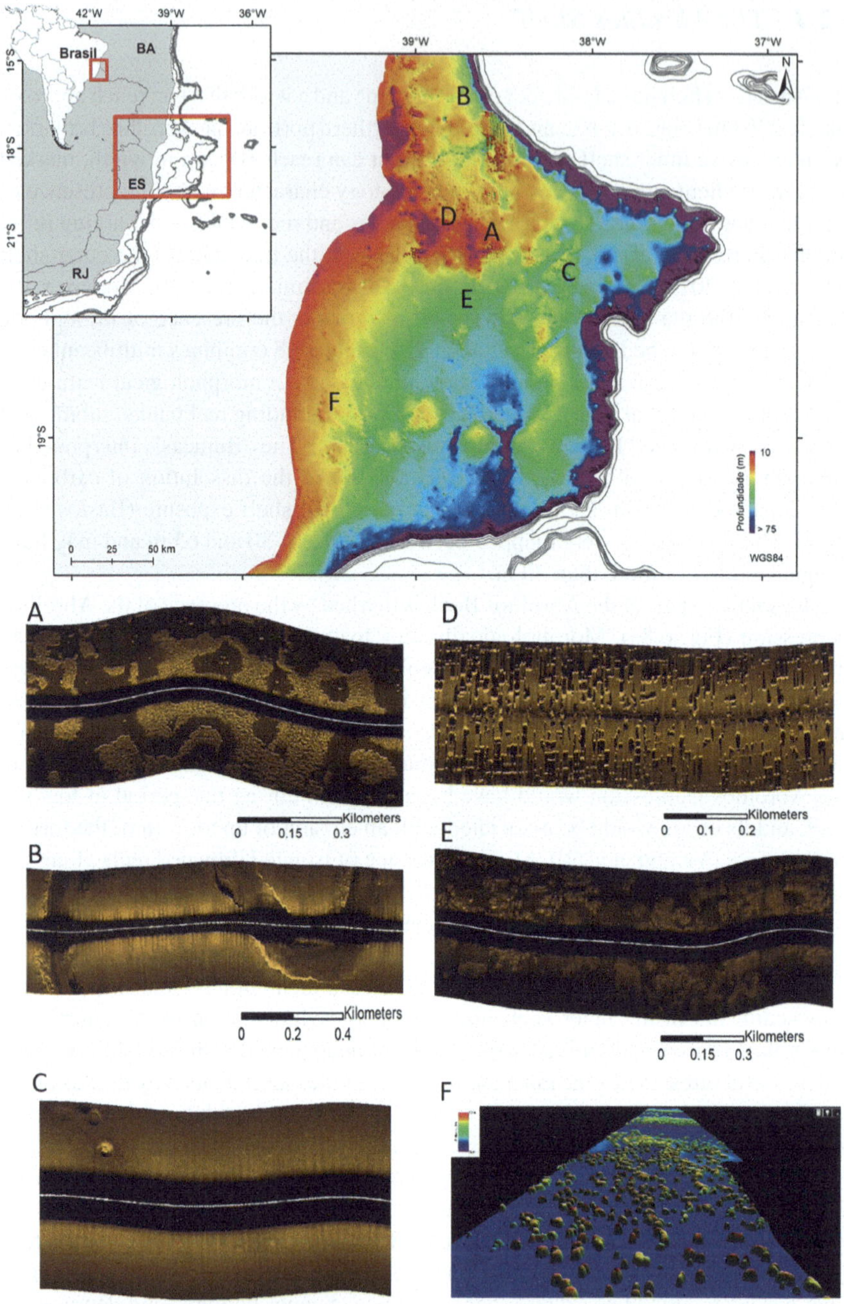

Fig. 6.8 Acoustic images of the main morphological features mapped along the Abrolhos shelf: **a** reef banks; **b** incised valley with bioconstructions along the margin; **c** sinkhole ("Buracas"); **d** reef pinnacles; **e** low-relief reef banks; **f** 3D multibeam image of patches of reef pinnacles in the southern Abrolhos shelf

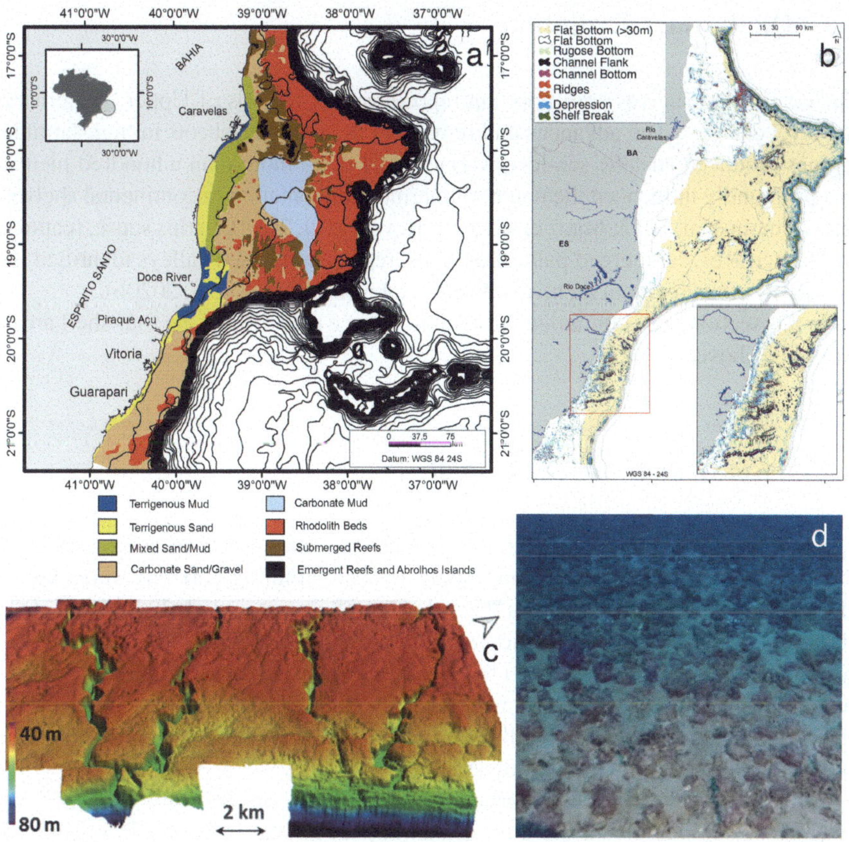

Fig. 6.9 Espirito Santo and Abrolhos shelf. **a** Sedimentary facies map; **b** relief attribute map as part of the benthic terrain model (BTM) analysis; **c** incised valleys from Guarapari in southern Espirito Santo state; **d** image of rhodolith bottom. Bastos et al. (2021)

in slope and depth (Fig. 6.9b). This remarkable relief, heterogeneous and somewhat rougher, can be interpreted as a potential indicator of greater associated biodiversity due to the three-dimensionality of the bottom. This is the sector of the shelf with low sedimentary input and dominance of rhodolith bottoms (Fig. 6.9d).

Abrolhos has a relief with a smoother and flatter tendency, with regions of ridges and valleys, mainly to the south of the Abrolhos Depression and in the area of a valley to the north. The inner shelf in southern Bahia state already indicates the irregularities that would be associated with the reef bottoms, showing that the pattern of ridges and depressions can indicate an important roughness, which also produces three-dimensionality in the bottom. Therefore, it induces an analysis of potential habitats or indicates areas that should be better studied and known.

6.3 Quaternary Evolution

The continental shelves are important repositories of records of Upper Quaternary sea-level changes. The advances and retreats of the great ice sheets in high-latitude regions produced eustatic sea-level variations of up to more than a hundred metres that profoundly influenced the nature of marine sedimentation in continental shelves (e.g. Camoin et al. 2012; Mauz et al. 2015; Rovere et al. 2018). In this sense, features such as terraces, submerged paleoshorelines, reefs, and incised valleys identified on the NEBS act as regional archives of past sea levels (Gomes et al. 2020b).

In this section, we summarize the evolution since the LGM of the four shelf areas described above.

6.3.1 Rio Grande do Norte Shelf

The Quaternary evolution of the Rio Grande do Norte shelf is summarized in Fig. 6.10. At least four seismostratigraphic units and limiting horizons were clearly identified in the RN shelf and Açu incised valley (Schwarzer et al. 2006; Vital et al. 2008, 2010a, b; Gomes 2009; Gomes et al. 2016). They were numbered from I (youngest) to IV (oldest). Unit I is characterized by parallel layers, Unit II comprises sigmoidal-shaped reflectors better recognized in valley-parallel profiles, and Unit III represents incised valley basal infill. Unit IV corresponds to the Pleistocene substrate. Horizon I on top of Unit IV is a subaerial unconformity.

Units I, II, and III mark important depositional phases interrupted by periods of erosion, culminating in the current architecture of the incised valley. Unit III was deposited in a period with a higher flow regime, interspersed with periods of drought, observed in the incision of medium-sized valleys. Unit II was deposited when the valley was completely inundated and is characterized by a uniform infill by finer sediments, possibly in an estuarine or deltaic environment, during the marine transgression. Unit I, being the youngest deposition, is marked by horizontal and prograding plane-parallel reflectors, suggestive of marine fluvial deposition when sea level reached the transition between the inner and middle shelf.

During the LGM, when the sea level fell 120 m below the present sea level (Peltier and Fairbanks 2006), incisions of the Late Pleistocene valleys took place on the exposed shelf (Fig. 6.10b). Coarse sand fluvial sediments were deposited in incised valley areas or in the form of clastic wedges on the outer shelf and slope.

The Açu and Apodi-Mossoró rivers (Fig. 6.5a, b) flow in their current position at least since the Mid Miocene, acting as incised valley systems during the multiple lowerings of the sea level since that time (Pessoa Neto 2003). The incised valleys of these two systems are controlled by tectonic troughs connected to NE–SW faults. These faults caused instability in the slope, initiating the formation of canyons on the shelf edge, some of which are active to this day (Pessoa Neto 2003; Almeida et al.

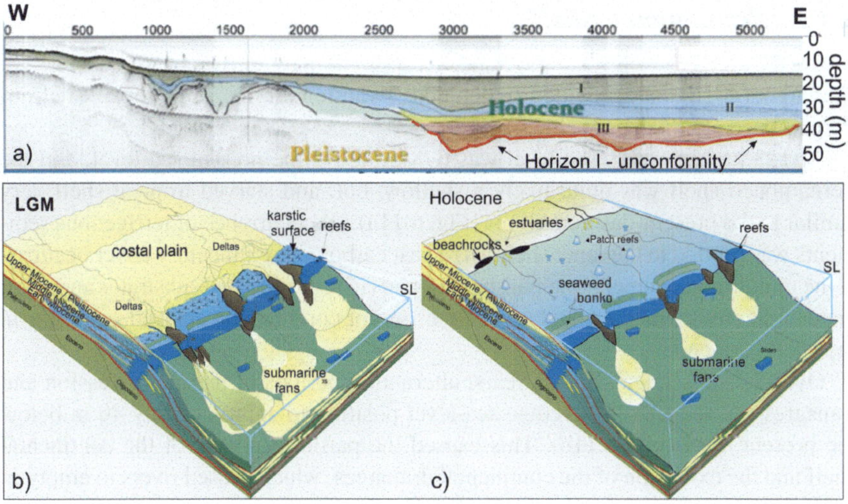

Fig. 6.10 Schematic evolution of the Açu incised valley shelf system. **a** Seismic cross section on the Rio Grande Norte inner shelf, the red line indicates horizon I, a subaerial unconformity separating Pleistocene strata below from Holocene units I, II, III above; **b** Last Glacial Maximum (LGM) characterized by a greater expression of siliciclastic sedimentation; **c** Holocene highstand characterized by carbonate sedimentation. Modified from Gomes (2009) (a) and Pessoa Neto (2003) (b, c)

2015). Clastic sedimentation, conditioned by depositional troughs, was captured by canyons on the outer shelf.

Outside these depositional troughs, there are adjacent elevated zones, where terrigenous sediment influx was lower, allowing the proliferation of the organisms that make up the carbonate factory. The subaerial exposure of these carbonate sediments generated karst relief in these shelf areas. Valley infill took place during the transgressive and highstand stages of the last sea-level cycle (Fig. 6.10).

During lowstands, carbonate sedimentation remained active on the outer shelf, while prograding clastic wedges advanced from the Açu and Mossoró rivers in the depositional troughs (Fig. 6.10b). These wedges seem to have been restricted to the inner shelf at that time. They rarely reached the outer shelf or were simply not preserved in that portion of the basin. Sedimentation on the outer shelf was further dominated by carbonate sedimentation in the form of algal beds and reefs, which appear to have been the only deposits that resisted subsequent erosion during full exposure to the shelf (Pessoa Neto 2003).

During highstands, with the shelf completely flooded, carbonate sedimentation reached its greatest expansion, with the development of reefs and algal beds and macro foraminifera on the outer shelf and the inundation of the inner shelf during peaks of maximum transgression (Fig. 6.10c). The influence of clastic sedimentation was minor and restricted to the portions of the shelf closer to the shoreline.

6.3.2 *Pernambuco Shelf*

Camargo (2016) has provided a detailed evolution of the continental shelf of Pernambuco summarized below.

At 123 ka highstand, sea level was 6–8 m above the present sea level, and the Pernambuco shelf was quite likely a shallow, hot, and starved tropical shelf very similar to the present-day conditions (Fig. 6.11a): (i) occurrence of terrigenous sediments was limited to the inner shelf; (ii) a vast carbonate sedimentary cover occurred at the middle and outer continental shelves; and (iii) consolidated substrates and associated reef environments (algal, sandstone, and/or coral reefs) occurred scattered on the shelf surface.

Over the subsequent 40,000 years, alternating periods of marine regression and transgression reduced the average sea-level position to approximately 40 m below the present level (Fig. 6.11b). This caused the partial exposure of the continental shelf and the extension of the continental drainages, which caused rivers to empty in the external portion of the shelf (40–60 m depth). The erosive action of continental drainage resulted in the incision of river valleys in this wide coastal plain. Sediments transported by these rivers were deposited in the available accommodation space between the −40 m shoreline and the shelf edge. Additionally, shallow water reef environments were also present in this still flooded stretch of the continental shelf.

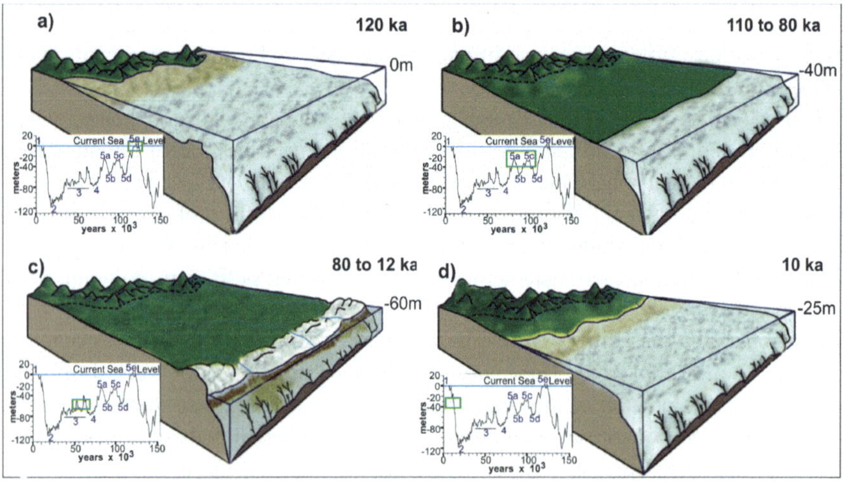

Fig. 6.11 Schematic evolution of the Pernambuco shelf since the LGM. **a** Interglacial highstand at 120 ka. Paleogeographies at approximately 110–80 ka (**b**); between 80 and 12 ka (**c**); and at 10 ka (**d**). Small insets depict the eustatic sea-level curve for the last 120 ka. The red line indicates the average position of the sea level. Camargo (2016)

At approximately 65 ka, the mean sea level was positioned at a depth of −60 m, and the shelf was completely exposed to subaerial erosion (Fig. 6.11c). This resulted in an even greater incision of the river valleys on the outer shelf and slope, the probable karstification of carbonate sediments, and the sedimentary transport directly to the slope.

After the LGM, the Holocene transgression resulted in the rapid drowning of the Pernambuco shelf, which apparently took place during two faster pulses, separated by a sea-level stabilization period at approximately −40 m. Most likely, at the end of the second pulse, the sea level reached a depth of approximately −25 m, and from this point on, the transgression was probably slower and more gradual, giving rise to shelf steps observed at depths of 20–23 and 16–20 m, and the beachrock lines were located on the inner shelf (Fig. 6.11d).

6.3.3 *Salvador Shelf*

The evolution of the Salvador shelf, located in the northern portion of the Central Bahia shelf, is summarized below by Dominguez et al. (2011) and Halla et al. (2020).

The Salvador Shelf was completely exposed to subaerial conditions during the LGM (Fig. 6.12a). This exposure, along with previous glacial maximum events, originated as an erosive surface on which upper Quaternary units were deposited. The sea-level rise after the LGM (20 ka BP) flooded the incised valleys. The infilling of these valleys, represented by stratigraphic unit Su03, is thought to have been favoured by low rates of eustatic sea-level rise after MWP-1A, which culminated in the Younger Dryas (12.9–11.7 ka BP) (Fig. 6.12b) when sea level was mostly stabilized. The complete infill of these valleys is believed to have occurred at the end or soon after the Younger Dryas. The infill of the inner bay near the shoreline also possibly dates back to this period. In this inner region, the infill may have been favoured by the presence of a bathymetric high along the middle shelf, which could have acted as a sill, helping to trap sediments from the continent (Fig. 6.12b).

Faster flooding of the shelf occurred after the Younger Dryas as a result of an increase in sea-level rise rates (MWP-1B), drowning the middle shelf sill. This drowning is recorded in shelf deposits associated with the maximum transgression surface, characterized by a thin stratigraphic unit (Su04) that occurs across the entire Salvador shelf and marks the end of the transgressive system tract (Fig. 6.12c).

After 7–8 ka BP, as the eustatic sea level stabilized, deposition of the highstand deposits, represented by stratigraphic unit 05 (Su05), began (Fig. 6.12d).

The only reported dates for the central Bahia shelf are from Dominguez et al. (2013), which were recorded from vibracores collected from the shelf break. These data show that carbonate sedimentation in the shelf break region began 10,000 cal yrs BP, coincident with the shelf inundation after the Younger Dryas (Dominguez et al. 2013). These cores were composed of packstone/wackstone facies dominated by encrusting coralline algae, rhodoliths, molluscs and forams. Sediment accumulation rates of 0.3–0.4 mm/year were estimated for these carbonate sediments.

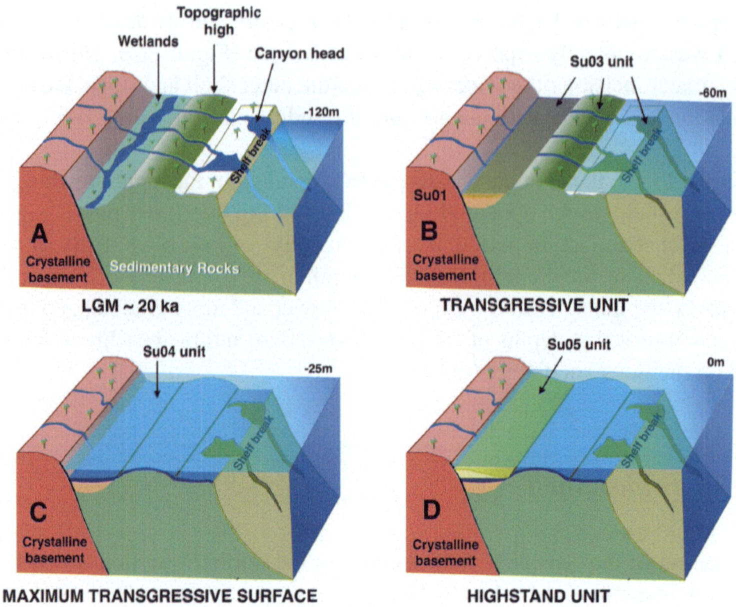

Fig. 6.12 Schematic evolution of the Salvador shelf since the LGM. **a** During the LGM, the sea level was 120 m below the current position, exposing the entire Salvador shelf to subaerial conditions. **b** As a consequence of the decrease in sea-level rise rates after MWP1A, incised valleys began infilling (unit Su03). They were probably completely infilled when the sea level was 60 below the current position (Younger Dryas). **c** When the sea level reached the −25 m isobath, the continental shelf was already completely flooded (unit Su04). **d** After sea-level rise stabilized at approximately 7 ka, the highstand deposits began accumulating (unit Su05). Modified from Halla et al. (2020)

6.3.4 Abrolhos Shelf

With the end of the LGM, the Abrolhos shelf became progressively flooded and was initially restricted to the channels and incised valleys (Bastos et al. 2021; Cetto et al. 2021). The high rates of sea-level rise at the beginning of the transgression possibly did not allow the development of shelf edge reefs, although carbonate build-ups have been observed at a depth of approximately 100–150 m (D'Agostini et al. 2019), both on the northern and southern slopes of the Abrolhos bank. The rapid shelf drowning and low sedimentary input could be the main reason for the development of rhodolith bottoms, which are known to be transgressive facies (Nalin et al. 2008). Figure 6.13 shows paleogeographic reconstructions of the shelf region after the three major melt-water pulses during deglaciation. After MWP1A, the entire shelf remained exposed (Bastos et al. 2021). Bastos et al. (2016) have shown evidence that portions of the shelf deeper than 70 m were already drowned by 12,780 cal yrs BP, based on radiocarbon dating from the walls of the sinkholes. After MWP1B, more distal portions of the shelf were drowned, in particular, the Abrolhos depression, a paleolagoon/estuary first described by Vicalvi et al. (1978) (Fig. 6.13b). At approximately 11,500 cal yrs

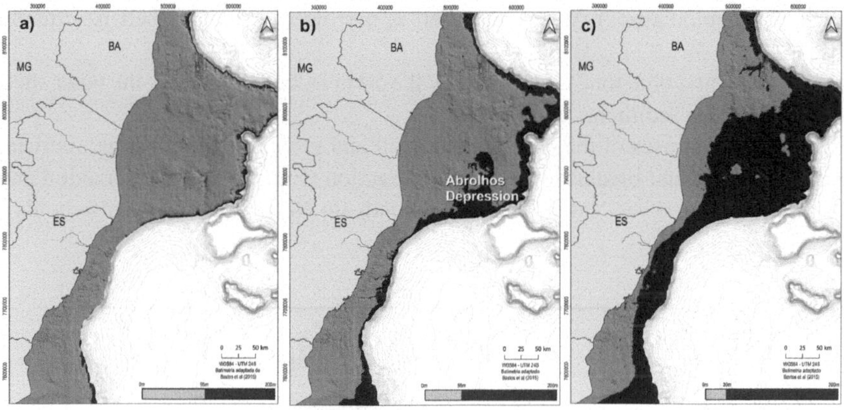

Fig. 6.13 Paleogeography of the Abrolhos shelf after meltwater pulses during the last deglaciation: **a** after MWP1A—sea level = −95 m; **b** after MWP1B—sea level = −55 m; and **c** after MWP1C—sea level = −30 m. Bastos et al. (2021)

BP, sedimentary records from this region show that at depths between 55 and 58 m, the environmental conditions changed from typically estuarine to open marine conditions (Vicalvi et al. 1978; D'Agostini 2017; Silva 2017). The sinkholes on the outer shelf were completely drowned at approximately 10,300 cal yrs BP (Bastos et al. 2016). Even after MWP1C, large portions of the inner shelf were exposed subaerially (Fig. 6.13c).

6.4 Shelf Environments and Human Uses

There is still no single geohabitat map available for the NEBS. The habitat maps that have already been generated and published have used different methodologies (Goes et al. 2019; Ferreira et al. 2020; Gomes et al. 2020a; Lucatelli et al. 2020; Quaresma et al. 2020; Rebouças et al. 2020; Vital et al. 2020). It is also worth mentioning that there is no detailed geomorphological map for these regions. Additionally, few places have coverage with multibeam bathymetry (Almeida et al. 2015; Bastos et al. 2021). In fact, the number of multibeam echo sounders available to the scientific community in the country is restricted to those installed on three Brazilian Navy ships, and only five universities have their own multibeam systems, of which only one has its own vessel.

Figure 6.14 presents an effort to synthesize the major characteristics and human uses of the NEBS using the GEBCO database. Depths are set from 0 to −200 m, where warmer colours represent shallower water and cooler colours represent deeper water. The red colour represents the land–ocean interface (emphasizing the connection with rivers on the continent). The shades of orange correspond to the inner shelf (consisting predominantly of siliciclastics), and yellow corresponds to the middle shelf (mixed siliciclastics–bioclastics). Where the shelf is narrower, a middle shelf is virtually absent.

The green and blue tones, separated by the 50 m isobath, represent the outer shelf where bioclastic sediments dominate.

Major human uses of the NEBS include oil and gas extraction, fishing, tourism, and siliciclastic and bioclastic aggregate extraction (Fig. 6.14). Salt extraction and shrimp farming are very intensive in Rio Grande do Norte state, which also holds a

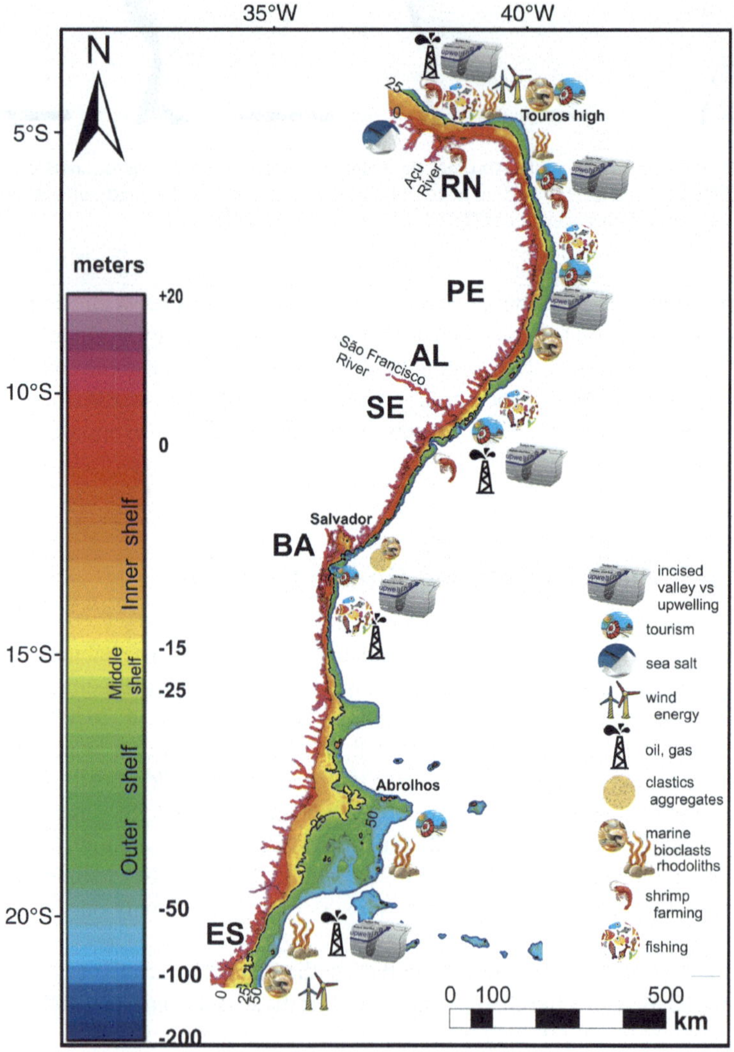

Fig. 6.14 The Northeastern Brazil Shelf (NEBS) and major human uses in the area. Depths from the GEBCO database

very high potential for offshore wind energy generation. The Abrolhos shelf, in turn, presents the largest extension of rhodolith beds of the entire NEBS (Fig. 6.14).

Upwelling has been documented at a number of canyon heads in the region (e.g. Nogueira 2014; Eichler et al. 2019; Thevenin et al. 2019; Vieira et al. 2019; Silva et al. 2022). Eichler et al. (2019) showed that local upwelling associated with canyon heads might be responsible for nourishing living coral–algal systems on the mesophotic outer shelf of the Rio Grande do Norte shelf, such as the Açu Reef. They have also suggested that foraminifer assemblages associated with high organic matter content at the sediment–water interface could serve as a potential tool to reconstruct paleoenvironmental and climate change in these regions. Silva et al. (2022) observed the seasonal fertilization of the waters of the northeast coast of Brazil, breaking the paradigm of predominantly oligotrophic waters in this region, thus pointing to the development of future research in the NEBS.

6.5 Concluding Remarks

Changes in sea level resulting from tectonic–eustasy interactions are assumed to be the main controlling factor in creating accommodation space for mixed shelf sedimentation in the NEBS. Climate exerts an important control on sediment flux to the coastal ocean and carbonate sedimentation.

Upwelling associated with canyon heads linked to incised valleys on the NEBS can produce distinctive foraminifer assemblages whose signatures can be used as a potential tool to reconstruct paleoenvironmental and climate changes.

One of the major challenges facing the marine economy today is to map the morphology and sediment cover of the seabed using recent advances in the acquisition, processing, analysis, and dissemination of data. Technological innovations based on big data and artificial intelligence, including intensification in the use of remote sensors, will be used extensively and on a large scale in the mapping and identification of materials on the ocean floor. However, to use these technologies, it is necessary to acquire real data for calibration.

The absence of robust scientific data for the NEBS, associated with the strong pressure for its use, points to research opportunities to carry out mapping and characterization of the seabed, essential to promote the adequate management and sustainable exploitation of mineral and energy resources, contributing to reducing impacts and associated risks and uncertainties related to its potential uses.

Acknowledgements The following research grants have made this study possible: InctAmbTropic (CNPq/FAPESB No. 565054/2010-4, 465634/2014-1 and inc004/2019).

References

Almeida NM, Vital H, Gomes MP (2015) Morphology of submarine canyons along the continental margin of the Potiguar Basin, NE Brazil. Mar Pet Geol 68:307–324. https://doi.org/10.1016/j.marpetgeo.2015.08.035

Araujo TCM, Seoane JCS, Coutinho PN (2004) Geomorfologia da plataforma continental de Pernambuco. In: Eskinazi-Leça E, Neumann-Leitao S, Costa MF (eds) Oceanografia: um cenário tropical. Edições Bagaço, Recife, Brazil, pp 39–57

Asmus HE (1981) Estruturas e tectonismo da margem continental brasileira, e suas implicações nos processos sedimentares e na avaliação do potencial de recursos minerais: relatório final. PETROBRAS, Rio de Janeiro, Projeto REMAC Series, vol 9

Baker EK, Harris PT (2020) Chapter 2—habitat mapping and marine management. In: Harris PT, Baker E (eds) Seafloor geomorphology as benthic habitat, 2nd edn. Elsevier, pp 17–33. http://doi.org/10.1016/B978-0-12-814960-7.00002-6

Bastos AC, Moura RL, Filho A et al (2013) Buracas: novel and unusual sinkhole-like features in the Abrolhos Bank. Cont Shelf Res 70:118–125. https://doi.org/10.1016/j.csr.2013.04.035

Bastos AC, Quaresma VS, Marangoni MB et al (2015) Shelf morphology as an indicator of sedimentary regimes: a synthesis from a mixed siliciclastic-carbonate shelf on the eastern Brazilian margin. J S Am Earth Sci 63:125–136. https://doi.org/10.1016/j.jsames.2015.07.003

Bastos AC, Amado Filho GM, Moura RL et al (2016) Origin and sedimentary evolution of sinkholes (Buracas) in the Abrolhos continental shelf, Brazil. Palaeogeogr Palaeoclimatol Palaeoecol 462:101–111. https://doi.org/10.1016/j.palaeo.2016.09.009

Bastos AB, Quaresma V, Vieira F et al (2021) Plataforma Continental do Espírito Santo e Abrolhos. In: Vital H, Dias MS, Bastos AC (eds) Plataforma Continental Brasileira: Estados do Espirito Santo e Ceará, 1st edn. PGGM Projetos e Produções 2, pp 7–70

Beardsley RC, Candela J, Limeburner R et al (1995) The M2 tide on the Amazon shelf. J Geophys Res 100(C2):2283–2319. https://doi.org/10.1029/94JC01688

Berger W (2008) Sea level in the late Quaternary: patterns of variation and implications. Int J Earth Sci (Geol Rundsch) 97:1143–1150. https://doi.org/10.1007/s00531-008-0343-y

Blum MD, Hattier-Womack J (2009) Climate change, sea-level change, and fluvial sediment supply to deepwater depositional systems. In: Kneller B, Martinsen OJ, McCaffrey B (eds) External controls on deep-water depositional systems. SEPM special publication, 92. SEPM, Tulsa, Oklahoma, pp 15–40

Branner JC (1905) Stone reefs on the NE coast of Brazil. Bull Geol Soc Am 16:1–12. https://doi.org/10.1130/GSAB-16-1

Brown CJ, Smith SJ, Lawton P et al (2011) Benthic habitat mapping: a review of progress towards improved understanding of the spatial ecology of the seafloor using acoustic techniques. Estuar Coast Shelf Sci 92:502–520. https://doi.org/10.1016/j.ecss.2011.02.007

Brown CJ, Sameoto JA, Smith SJ (2012) Multiple methods, maps, and management applications: purpose made seafloor maps in support of ocean management. J Sea Res 72:1–13. https://doi.org/10.1016/j.seares.2012.04.009

Cabral Neto I, Córdoba VC, Vital H (2013) Morfologia, microfaciologia e diagênese de beachrocks costa-afora adjacentes à costa norte do Rio Grande do Norte, Brasil. Geociências 32:471–490

Camargo JMR (2016) Geodiversidade e Paisagens Submersas de uma Plataforma Continental Tropical no Nordeste do Brasil. Ph.D. dissertation, Universidade Federal de Pernambuco, Brazil

Camargo JMR, Araújo TCM, Maida M et al (2007) Morfologia da plataforma continental interna adjacente o município de Tamandaré, sul de Pernambuco—Brasil. Rev Bras Geofís 25(Supl. 1):79–89

Camargo JMR, Araujo TCM, Ferreira BP et al (2015) Topographic features related to recent sea level history in a sediment-starved tropical shelf: linking past, present and future. Reg Stud Mar Sci 2:203–211. https://doi.org/10.1016/j.rsma.2015.10.009

Camoin GF, Seard C, Deschamps P et al (2012) Reef response to sea-level and environmental changes during the last deglaciation: integrated ocean drilling program expedition 310, Tahiti sea level. Geology 40:643–646. https://doi.org/10.1130/G32057.1

Carannante C, Esteban M, Milliman JD et al (1988) Carbonate lithofacies as paleolatitude indicators: problems and limitations. Sediment Geol 60:333–346. https://doi.org/10.1016/0037-0738(88)90128-5

Cetto PH, Bastos AC, Lanniruberto M (2021) Morphological evidences of eustatic events in the last 14,000 years in a far-field site, East-Southeast Brazilian continental shelf. Mar Geol 442. https://doi.org/10.1016/j.margeo.2021.106659

Coni e Mello AC, Dominguez ML, Souza LAP (2020) The Santo Antônio bank: a high-resolution seismic study of a deflected ebb-tidal delta located at the entrance of a large tropical bay, eastern Brazil. Geo-Mar Lett 40:965–975. https://doi.org/10.1007/s00367-019-00624-9

Costa Neto LX (1997) Evolução geológica-geomorfológica recente da plataforma continental interna ao largo do Delta do rio Açu, Macau – RN. MS thesis, Universidade Federal Fluminense, Brazil

Coutinho PN (1996) Levantamento do Estado da Arte da Pesquisa dos Recursos Vivos Marinhos do Brasil-Oceanografia Geológica. Região Nordeste. Programa REVIZEE. Ministério do Meio Ambiente, dos Recursos Hídricos e da Amazônia Legal, Brasília, Brazil

Coutinho PN, Morais JO (1970) Distribution de los sedimentos en la plataforma continental norte-nordeste do Brasil. Arq Cien Mar 10(1):79–90

D'Agostini DP (2017) A Plataforma Continental de Abrolhos: contexto paleoambiental, sismoestratigrafia e domínios sedimentares. Ph.D. thesis, Universidade Federal do Espírito Santo, Brazil

D'Agostini DP, Bastos AC, Amado Filho GM et al (2019) Morphology and sedimentology of the shelf-upper slope transition in the Abrolhos continental shelf (East Brazilian margin). Geo-Mar Lett 39:117–134. https://doi.org/10.1007/s00367-019-00562-6

Diesing M, Mitchell P, Stephens D (2016) Image-based seabed classification: what can we learn from terrestrial remote sensing? ICES J Mar Sci 73:2425–2441. https://doi.org/10.1093/icesjms/fsw118

Domingues EC, Schettini CAF, Truccolo EC et al (2017) Hydrography and currents on the Pernambuco continental shelf. Rev Bras Rec Hidricos 22. https://doi.org/10.1590/2318-0331.0217170027

Dominguez JML (2009) The coastal zone of Brazil. In: Dillemburg S, Hesp P (eds) Geology and geomorphology of Brazilian Holocene coastal barriers. Springer, Heidelberg, pp 17–51. http://doi.org/10.1007/978-3-540-44771-9_2

Dominguez JML (2015) The Todos os Santos Bay—an ephemeral high-stand feature incised into an aborted cretaceous rift. In: Vieira BC, Salgado AAR, Santos LJC (eds) Landscapes and landforms of Brazil. Springer, Netherlands, pp 55–63. http://doi.org/10.1007/978-94-017-8023-0_6

Dominguez JMLD, Andrade ACS, Almeida AB et al (2009) The Holocene barrier strandplains of the state of Bahia. In: Dillemburg S, Hesp P (eds) Geology and geomorphology of Brazilian Holocene coastal barriers. Springer, Heidelberg, pp 253–288. http://doi.org/10.1007/978-3-540-44771-9_8

Dominguez JML, Ramos JMF, Rebouças RC et al (2011) A Plataforma Continental do Município de Salvador: geologia, usos múltiplos e recursos minerais. Série Arquivos Abertos 37, CBPM, Salvador, Brazil

Dominguez JML, Nunes AS, Rebouças RC et al (2012) A Plataforma Continental do Estado da Bahia. In: Barbosa JSF, Mascarenhas JF, Corrêa Gomes LC et al (eds) Geologia da Bahia, vol 2. CPRM, Salvador, Bahia, Brazil, pp 427–496

Dominguez JML, Silva RP, Nunes AS et al (2013) The narrow, shallow, low-accommodation shelf of central Brazil: sedimentology, evolution and human uses. Geomorphology 203:46–59. https://doi.org/10.1016/j.geomorph.2013.07.004

Eichler PPB, Silva LLN, Andrade AU et al (2019) Organically enriched sediments and foraminiferal species from the Açu Reef, indicators of upwelling in NE Brazil? Mar Geol 19. https://doi.org/10.1016/j.margeo.2019.106016

Ferreira BP, Maida M (2006) Monitoramento dos Recifes de Coral do Brasil. Ministério do Meio Ambiente, Brasilia, Brazil

Ferreira LC, Bastos AC, Amado Filho GM et al (2020) Chapter 30—submerged reefs in the Abrolhos shelf: morphology and habitat distribution In: Harris PT, Baker E (eds) Seafloor geomorphology as benthic habitat, 2nd edn. Elsevier, pp 519–532. http://doi.org/10.1016/B978-0-12-814960-7.00030-0

Figueiredo Jr AG, Silveira OFM, Ayres Neto A et al (2008) A geologia e Geomorfologia da Margem Equatorial Brasileira. Relatório Técnico do Projeto Piatam OCEANO, Petrobras, Rio de Janeiro, Brazil

Fontes LCS, Santos JR, Santos LA et al (2017) Sedimentos Superficiais da Plataforma Continental de Sergipe-Alagoas. In: Fontes LCS, Kowsmann RO, Puga-Bernabeú A (eds) Geologia e Geomorfologia da Bacia Sergipe-Alagoas. PETROBRAS, Editora UFS, Brazil, pp 62–96

Fontes VC, Gomes MP, Vital H et al (2020) Chapter 33—reefs distribution and inter reef sedimentation on Tamandaré continental shelf, NE-Brazil. In: In: Harris PT, Baker E (eds) Seafloor geomorphology as benthic habitat, 2nd edn. Elsevier, pp 561–569. http://doi.org/10.1016/B978-0-12-814960-7.00033-6

Geyer WR, Kineke GC (1995) Observations of currents and water properties in the Amazon frontal zone. J Geophys Res 100:2321–2339. https://doi.org/10.1029/94JC02657

Goes ER, Brown C, Araujo TCM (2019) Geomorphological classification of the benthic structures on a tropical continental shelf. Front Mar Sci. https://doi.org/10.3389/fmars.2019.00047

Gomes MP (2009) Aquisição, processamento e análise de dados de sísmica de alta resolução na plataforma continental norte do Rio Grande do Norte: Vale inciso do rio Açu. M.S. thesis, Universidade Federal do Rio Grande do Norte, Brazil

Gomes MP, Vital H (2010) Revisão da compartimentação geomorfológica da Plataforma Continental Norte do Rio Grande do Norte, Brasil. Rev Bras Geoc 40:321–329

Gomes MP, Vital H, Bezerra FHR et al (2014) The interplay between structural inheritance and morphology in the equatorial continental shelf of Brazil. Mar Geol 355:150–161. https://doi.org/10.1016/j.margeo.2014.06.002

Gomes MP, Vital H, Eichler PPB et al (2015) The investigation of a mixed carbonate-siliciclastic shelf, NE Brazil: side-scan sonar imagery, underwater photography, and surface-sediment data. Ital J Geosci 134:9–22. https://doi.org/10.3301/IJG.2014.08

Gomes MP, Vital H, Stattegger K et al (2016) Bedrock control on the Assu incised valley morphology and sedimentation in the Brazilian equatorial shelf. Int J Sediment Res 31:181–193. https://doi.org/10.1016/j.ijsrc.2015.04.002

Gomes MP, Vital H, Nascimento Silva LL et al (2020a) Chapter 34—nature and condition of outer shelf habitats on the drowned Açu Reef, NE Brazil. In: Harris PT, Baker E (eds) Seafloor geomorphology as benthic habitat, 2nd edn. Elsevier, pp 571–585. http://doi.org/10.1016/B978-0-12-814960-7.00034-8

Gomes MP, Vital H, Droxler AW (2020b) Terraces, reefs, and valleys along the Brazil northeast outer shelf: deglacial sea-level archives? Geo-Mar Lett 40:699–711. https://doi.org/10.1007/s00367-020-00666-4

Greene HG, Yoklavich MM, Starr RM et al (1999) A classification scheme for deep seafloor habitats. Oceanol Acta 22:663–678

Halla MMS, Dominguez JML, Gomes LCC (2020) Structural controls on the morphology of an extremely narrow, low-accommodation, passive margin shelf (Eastern Brazil). Geo-Mar Lett 40:865–878. https://doi.org/10.1007/s00367-019-00605-y

Harris PT, Baker EK (eds) (2020) Seafloor geomorphology as benthic habitat, 2nd edn. Elsevier

Harris PT, Macmillan-Lawler M, Rupp J et al (2014) Geomorphology of the oceans. Mar Geol 352:4–24. https://doi.org/10.1016/j.margeo.2014.01.011

Hazin FHV, Wor C, Oliveira JEL et al (2008) Resultados obtidos por meio do fundeio de um correntógrafo na plataforma continental do Estado do Rio Grande do Norte, Brasil. Arq Ciênc Mar 41:30–35

Heilbron M, Cordani UG, Alkmim FF (eds) (2017) São Francisco Craton, Eastern Brazil. Regional geology reviews: São Francisco Craton, Eastern Brazil: tectonic genealogy of a miniature continent. Springer International Publishing, pp 3–13. http://doi.org/10.1007/978-3-319-01715-0

Kennett JP (1982) Marine geology. Prentice-Hall, Englewood Cliffs, New Jersey

Knoppers B, Ekau W, Figueiredo AG (1999) The coast and shelf of east and northeast Brazil and material transport. Geo-Mar Lett 19:171–178. https://doi.org/10.1007/s003670050106

Kuehl SA, DeMaster DJ, Nittrouer CA (1986) Nature of sediment accumulation on the Amazon continental shelf. Cont Shelf Res 6:209–225. https://doi.org/10.1016/0278-4343(86)90061-0

Kuehl SA, Nittrouer CA, Allison MA et al (1996) Sediment deposition, accumulation and seabed dynamics in an energetic fine-grained coastal environment. Cont Shelf Res 16:787–815. https://doi.org/10.1016/0278-4343(95)00047-X

Laborel J (1969) Les peuplements de madréporaires des côtes tropicales du Brésil. University of Abidjan, Ivory Coast

Laborel-Deguem F, Castro CB, Nunes FLD et al (eds) (2019) Recifes brasileiros: o legado de Laborel. Museu Nacional, Rio de Janeiro, Brazil

Lea DW, Martin PA, Pak DK (2002) Reconstructing a 350 ky history of sea level using planktonic Mg/Ca and oxygen isotope records from a Cocos Ridge core. Quat Sci Rev 21:283–293. https://doi.org/10.1016/S0277-3791(01)00081-6

Lecours V, Brown CJ, Devillers R et al (2016) Comparing selections of environmental variables for ecological studies: a focus on terrain attributes. PLoS One 11. https://doi.org/10.1371/journal.pone.0167128

Lees A, Buller AT (1972) Modern temperate-water and warm-water shelf carbonate sediments contrasted. Mar Geol 13:67–73. https://doi.org/10.1016/0025-3227(72)90011-4

Lucatelli D, Goes ER, Brown CJ et al (2020) Geodiversity as an indicator to benthic habitat distribution: an integrative approach in a tropical continental shelf. Geo-Mar Lett 40:911–923. https://doi.org/10.1007/s00367-019-00614-x

Lundblad E, Wright D, Miller J et al (2006) A benthic terrain classification scheme for American Samoa. Mar Geodesy 29:89–111. https://doi.org/10.1080/01490410600738021

Manso VAV, Corrêa ICS, Guerra NC (2003) Morfologia e Sedimentologia da Plataforma Continental Interna entre as Praias Porto de Galinhas e Campos - Litoral Sul de Pernambuco, Brasil. Pesq Geociênc 30:17–25. http://doi.org/10.22456/1807-9806.19587

Mauz B, Vacchi M, Green A et al (2015) Beachrock a tool for reconstructing relative sea level in the far-field. Mar Geol 362:1–16. https://doi.org/10.1016/j.margeo.2015.01.009

Milani EJ, Thomaz Filho A (2000) Sedimentary basins of South America. In: Cordani UG, Milani EJ, Thomaz Filho A et al (eds) Tectonic evolution of South America. Proceedings of the 31st international geological congress, Rio de Janeiro

Mohriak WU (2006) Interpretação geológica e geofísica da Bacia do Espírito Santo e da região de Abrolhos: petrografia, datação radiométrica e visualização sísmica das rochas vulcânicas. Bol Geociênc Petrobrás 14:133–142

Moura RL, Secchin NA, Amado-Filho GM et al (2013) Spatial patterns of benthic megahabitats and conservation planning in the Abrolhos Bank. Cont Shelf Res 70:109–117. https://doi.org/10.1016/j.csr.2013.04.036

Nalin R, Campbell SN, Basso D et al (2008) Rhodolith-bearing limestones as transgressive marker beds: fossil and modern examples from North Island, New Zealand. Sedimentology 55:249–274. https://doi.org/10.1111/j.1365-3091.2007.00898.x

Nittrouer CA, Kuehl SA, DeMaster DJ et al (1986) The deltaic nature of Amazon shelf sedimentation. Geol Soc Am Bull 97:444–458. https://doi.org/10.1130/0016-7606(1986)97%3c444:TDNOAS%3e2.0.CO;2

Nittrouer CA, Kuehl SA, Figueiredo AG et al (1996) The geological record preserved by Amazon shelf sedimentation. Cont Shelf Res 16:817–841. https://doi.org/10.1016/0278-4343(95)00053-4

Nogueira MLS (2014) Caracterização Morfológica-Sedimentar do Vale Inciso Apodi-Mossoro e Plataforma Continental Adjacente - Bacia Potiguar Offshore. Ph.D. dissertation, Universidade Federal do Rio Grande do Norte, Brazil

Nogueira Neto AV, Silva AC (2014) Seawater temperature changes associated with the North Brazil current dynamics. Ocean Dyn 64:13–27. https://doi.org/10.1007/s10236-013-0667-4

Nunes AS (2009) A Utilização da geologia na identificação dos habitats mais adequados para o estabelecimento de áreas marinhas protegidas na Costa do Dendê, Bahia, Brasil. Ph.D. dissertation, Universidade Federal da Bahia, Brazil

Peltier WR, Fairbanks RG (2006) Global glacial ice volume and last glacial maximum duration from an extended Barbados sea level record. Quat Sci Rev 25:3322–3337. https://doi.org/10.1016/j.quascirev.2006.04.010

Pereira P, Araujo TCM, Manso VAV (2016) Chapter 10—tropical sandy beaches of Pernambuco state. In: Short A, Klein A (eds) Brazilian beach systems. Coastal research library, vol 17. Springer, Cham, pp 251–279. http://doi.org/10.1007/978-3-319-30394-9_10

Pessoa Neto OC (2003) Estratigrafia de sequencias da plataforma mista Neogênica na Bacia Potiguar, Margem Equatorial Brasileira. Rev Bras Geociênc 33:263–278

Prestes YO, Silva AC, Jeandel C (2018) Amazon water lenses and the influence of the North Brazil current on the continental shelf. Cont Shelf Res 160:36–48. https://doi.org/10.1016/j.csr.2018.04.002

Puga-Bernabéu A (2017) Contextualização global da margem continental de SE-AL, NE do Brasil: Morfologia e dinâmica sedimentar durante o Pleistoceno tardio e Holoceno. In: Fontes LCS, Kowsmann RO, Puga-Bernabeú A (eds) Geologia e Geomorfologia da Bacia Sergipe-Alagoas. PETROBRAS, Editora UFS, Brazil, pp 232–264

Quaresma VS, Bastos AC, Menandro PS et al (2020) Mapeamento geológico de habitat marinhos: abordagem e estudo de caso. In: Muehe D, Lins-de-Barros FM, Pinheiro L (eds) Geografia Marinha: oceanos e costas na perspectiva de geógrafos. PGGM, Rio de Janeiro, Brazil, pp 267–291

Rand HM, Mabesoone JM (1982) Northeastern Brazil and the final separation of South America and Africa. Palaeogeogr Palaeoclimatol Palaeoecol 38:163–183. https://doi.org/10.1016/0031-0182(82)90002-5

Rebouças RC, Dominguez, JML, Avena PP et al (2020) Chapter 27—continental shelf habitats off a large South American metropolis: Salvador City, Eastern Brazil. In: Harris PT, Baker E (eds) Seafloor geomorphology as benthic habitat, 2nd edn. Elsevier, pp 473–485. http://doi.org/10.1016/B978-0-12-814960-7.00027-0

Ribeiro F, Pimenta FM, Vital H (2018) Inner shelf currents off Ponta Negra Beach, Natal, RN, Brazil. Rev Bras Geofís 36:43–58. http://doi.org/10.22564/.v36i1.841

Rovere A, Khanna P, Bianchi CN et al (2018) Submerged reef terraces in the Maldivian Archipelago (Indian Ocean). Geomorphology 317:218–232. https://doi.org/10.1016/j.geomorph.2018.05.026

Santos CLA, Vital H, Amaro VE et al (2007) Mapeamento de Recifes Submersos na Costa do Rio Grande do Norte, NE Brasil: Macau a Maracajau. Rev Bras Geofís 25:27–36. https://doi.org/10.1590/S0102-261X2007000500004

Schwarzer K, Stattegger K, Vital H et al (2006) Holocene coastal evolution of the Rio Açu area (Rio Grande do Norte, Brazil). J Coast Res SI 39:140–144

Short A, Klein AHF (eds) (2016) Brazilian beach systems. Coastal research library, vol 17. Springer, Berlin

Silva AED (2017) Ecolução Sedimentar Recente ao Longo de uma Plataforma Continental com Estilo Contrastante de Sedimentação. Ph.D. dissertation, Universidade Federal do Espírito Santo, Brazil

Silva LLN, Gomes MP, Vital H (2018) The Açu reef morphology, distribution, and inter reef sedimentation on the outer shelf of the NE Brazil equatorial margin. Cont Shelf Res 160:10–22. https://doi.org/10.1016/j.csr.2018.03.011

Silva MVB, Ferreira BP, Maida M et al (2022) Flow-topography interactions in the western tropical Atlantic boundary off Northeast Brazil. J Mar Sys 227. https://doi.org/10.1016/j.jmarsys.2021.103690

Silveira CBL, Strenzel GMR, Maida M et al (2020) Pushing satellite imagery to new depths: seascape feature mapping in a tropical shelf. Remote Sens Appl Soc Environ 19. https://doi.org/10.1016/j.rsase.2020.100345

Tabosa WF (2006) Caracterização Geologica e Sedimentologica da plataforma continental adjacente a São Bento e Caiçara do Norte. Ph.D. dissertation. Universidade Federal do Rio Grande do Norte, Brazil

Tabosa WF, Amaro VE, Vital H (2007) Análise do Ambiente Costeiro e Marinho, a partir de Produtos de Sensoriamento Remoto na Região de São Bento do Norte, NE Brasil. Rev Bras Geofis 25:37–48. https://doi.org/10.1590/S0102-261X2007000500005

Testa V, Bosence DWJ (1998) Carbonate-siliciclastic sedimentation on high-energy, ocean-facing, tropical ramp, NE Brazil. In: Wright VP, Burchette TP (eds) Carbonate ramps. Geological Society London Special Publications, vol 149, pp 55–71. http://doi.org/10.1144/GSL.SP.1999.149.01.05

Testa V, Bosence DWJ (1999) Physical and biological controls on the formation of carbonate and siliciclastic bedforms on the North-East Brazilian shelf. Sedimentology 46:279–301. https://doi.org/10.1046/j.1365-3091.1999.00213.x

Thevenin MR, Pereira J, Lessa GC (2019) Shelf-break upwelling on a very narrow continental shelf adjacent to a western boundary current formation zone. J Mar Sys 194:52–65. https://doi.org/10.1016/j.jmarsys.2019.02.008

Vianna ML, Solewicz R, Cabral A et al (1991) Sandstream on the Northeast Brazilian shelf. Cont Shelf Res 2:509–524. https://doi.org/10.1016/0278-4343(91)90008-T

Vicalvi MA, Costa MPA, Kowsmann RO (1978) Depressão de Abrolhos: uma paleolaguna holocênica na plataforma continental leste brasileira. Bol Téc Petrobrás 21:279–286

Vieira FV, Bastos AC, Quaresma VS et al (2019) Along-shelf changes in mixed carbonate-siliciclastic sedimentation patterns. Cont Shelf Res 187. https://doi.org/10.1016/j.csr.2019.103964

Vital H (2014) Chapter 4—the north and northeast Brazilian tropical shelves. In: Chiocci FL, Chivas AR (eds) Continental shelves of the world: their evolution during the last Glacio-Eustatic cycle. Geological Society London Memoirs, vol 4, pp 35–46. http://doi.org/10.1144/M41.12

Vital H, Silveira IM, Amaro VE (2005) Carta Sedimentólogica da Plataforma Continental Brasileira - Área Guamaré a Macau (NE Brasil), Utilizando Integração de Dados Geológicos e Sensoriamento Remoto. Rev Bras Geofís 23:233–241

Vital H, Stattegger K, Amaro VE et al (2008) A modern high-energy siliciclastic–carbonate platform: continental shelf adjacent to northern Rio Grande do Norte state, NE Brazil. In: Hampson G, Steel RJ, Burgess PM et al (eds) Recent advances in models of siliciclastic shallow-marine stratigraphy. SEPM Special Publications, Tulsa, vol 90, pp 177–190. http://doi.org/10.2110/pec.08.90.0177

Vital H, Gomes MP, Tabosa WF et al (2010a) Characterization of the Brazilian continental shelf adjacent to Rio Grande do Norte state, NE Brazil. Braz J Oceanogr 58:43–54

Vital H, Furtado SFL, Gomes MP (2010b) Response of Apodi Mossoró estuary-incised valley system (NE Brazil) to sea-level fluctuations. Braz J Oceanogr 58:13–24

Vital H, Silveira IM, Tabosa WF et al (2016) Chapter 8—beaches of Rio Grande do Norte. In: Short A, Klein A (eds) Brazilian beach systems. Coastal research library, vol 17. Springer, Cham, pp 201–229. http://doi.org/10.1007/978-3-319-30394-9_8

Vital H, Silveira IM, Lima ZMC et al (2018) Rio Grande do Norte. In: Muehe D (ed) Panorama da erosão costeira no Brasil. Ministério do Meio Smbiente, Brasília, Brazil, pp 289–326

Vital H, Leite TS, Viana MG et al (2020) Chapter 35—seabed character and associated habitats of an equatorial tropical shelf: The Rio Grande do Norte shelf, Northeast Brazil. In: Harris PT, Baker E (eds) Seafloor geomorphology as benthic habitat, 2nd edn. Elsevier, pp 587–603. http://doi.org/10.1016/B978-0-12-814960-7.00035-X

Waelbroeck C, Labeyrie L, Michel E et al (2002) Sea-level and deep water temperature changes derived from benthic foraminifera isotopic records. Quat Sci Rev 21:295–305. https://doi.org/10.1016/S0277-3791(01)00101-9

Chapter 7
Zoo- and Ichthyoplankton Communities of Pelagic Ecosystems in the Western Tropical Atlantic

Ralf Schwamborn, Sigrid Neumann-Leitão, Simone Maria de Albuquerque Lira, Claudeilton S. Santana, Nathália Lins-Silva, Gabriela Guerra A. A. Figueiredo, Patrícia Silva Basílio Dantas, Denise Fabiana de Moraes Costa Schwamborn, Catarina da Rocha Marcolin, Christiane Sampaio de Souza, Laura Rodrigues da Conceição, Paulo de Oliveira Mafalda Jr, Paula Cilene Alves da Silveira, Delzenira Silva do Nascimento da Costa, Rayssa de Lima Cardoso, Anna Evelin Coimbra Libório, Juliana Franco Lima, Elisabeth Cabral Silva Falcão, Paula Nepomuceno Campos, Xiomara Franchesca García Díaz, Leiliane Souza da Silva, Eduardo Tavares Paes, Nuno Filipe Alves Correia de Melo, Jéssica dos Santos Lima Pantoja, Elton Alex Correa da Silva, Glauber David Almeida Palheta, Ana Carolina Melo Rodrigues-Inoue, Francielly Alcântara de Lima, and Jussara Moretto Martinelli-Lemos

Abstract Zoo- and ichthyoplankton are key components of pelagic ecosystems. This study investigates zoo- and ichthyoplankton communities in seven Brazilian tropical marine environments that differ considerably in their abiotic and biological settings. The two study areas off Pará and Maranhão (Northern Brazil) present extremely wide continental shelves lined by mangroves. Two narrow oligotrophic shelf areas are located in northeastern Brazil (Pernambuco and Bahia). Three oceanic areas

R. Schwamborn (✉) · S. Neumann-Leitão · S. M. de A. Lira · C. S. Santana · N. Lins-Silva · G. G. A. A. Figueiredo · P. S. B. Dantas · D. F. de M. C. Schwamborn
Federal University of Pernambuco (UFPE), Recife, Pernambuco, Brazil
e-mail: rs@ufpe.br

C. da R. Marcolin
Federal University of Sul da Bahia (UFSB), Porto Seguro, Bahia, Brazil

C. S. de Souza · P. de O. Mafalda Jr
Federal University of Bahia (UFBA), Salvador, Bahia, Brazil

L. R. da Conceição
University of São Paulo (USP) São Paulo, São Paulo, Brazil

P. C. A. da Silveira · D. S. do N. da Costa · R. de L. Cardoso · A. E. C. Libório · J. F. Lima
Federal University of Maranhão (UFMA), São Luís, Maranhão, Brazil

E. C. S. Falcão
Federal University of Rio Grande do Sul (UFRGS), Imbé, Rio Grande Do Sul, Brazil

J. M. Landim Dominguez et al. (eds.), *Tropical Marine Environments of Brazil*, The Latin American Studies Book Series, https://doi.org/10.1007/978-3-031-21329-8_7

make up the unique Brazilian island systems (Rocas Atoll, Fernando de Noronha and St. Peter and St. Paul's Archipelagos). The waters off Salvador (Bahia) are influenced by oceanic water masses, resulting in the occurrence of tropical oceanic species. Higher temperatures and salinities that fluctuate seasonally lead to a lower density and diversity of copepods. Off Tamandaré (Pernambuco), ichthyoplankton, copepods, and decapods abounded inshore, especially during the dry season. In northern shelf systems (Pará and Maranhão), fish larvae were more abundant during the rainy season. Off Pará, caridean shrimp larvae were found closer to the coast, while penaeids abounded offshore. Off oceanic islands, some groups, such as fish eggs, were significantly more abundant downstream (island biomass effect). In general, strong inshore-offshore gradients in plankton communities were observed, and for some taxa, seasonal variations in their density were observed.

Keywords Continental shelf · Oceanic islands · Diversity · Ecology

7.1 Introduction

In the pelagic environment, zooplankton are the main component responsible for the consumption, processing, and transfer of synthesized or assimilated material by phytoplankton and bacteria; therefore, zooplankton metabolic activity has important ecological implications in fisheries in regional and global biogeochemical cycles, and it influences many other pelagic processes (Levinton 1995; Paranaguá et al. 2004). The structure and distribution of zooplankton communities depend on numerous abiotic (physical, chemical, and geological) and biotic factors inherent to the community (Bernal 1990).

Most animal phyla have representatives in marine zooplankton, including the holoplankton that spend their entire lives in the plankton and the meroplankton, that as adults are nektonic or benthic, but the early stages of their life cycle are spent as eggs or larvae in the plankton (Wickstead 1979; Boltovskoy 1981; Pereira and Soares-Gomes 2002). Among benthic organisms, approximately 80% of those living in shallow tropical waters have planktonic larvae (Gross and Gross 1996).

Fish eggs, larvae and small juveniles are referred to as ichthyoplankton (Lalli and Parsons 1997). Many marine fish release free-floating eggs and have planktonic larvae (Franco-Gordo et al. 2002). Larval fish assemblages play an essential role in the recruitment success and temporal and spatial variations of fish populations, as well as in marine food webs (Moser and Smith 1993; Franco-Gordo et al. 2002). An understanding of the recruitment processes requires knowledge of their small-, medium- and large-scale distributions in relation to abiotic and biotic parameters

P. N. Campos · X. F. G. Díaz · E. T. Paes · N. F. A. C. de Melo · J. dos S. L. Pantoja · E. A. C. da Silva · G. D. A. Palheta
Federal Rural University of Amazônia (UFRA), Belém, Pará, Brazil

L. S. da Silva · A. C. M. Rodrigues-Inoue · F. A. de Lima · J. M. Martinelli-Lemos
Federal Universidade of Pará (UFPA), Belém, Pará, Brazil

(Röpke 1993). The distribution patterns of the fish larvae in any region of the ocean are related to the reproductive activity of the adult population and to the topographic and hydrographic features that affect larval dispersal (Nonaka et al. 2000).

According to the horizontal distribution, zooplankton can be neritic (occurring from the coast to the shelf break, being particularly rich in meroplankton, due to the proximity of the substrate) or oceanic (located away from the coastal regions, it is essentially composed of holoplankton forms).

The relationship between some zooplankton species and the aquatic environment is so strong that it can be used to characterize water masses (Wickstead 1979). These communities respond very strongly to even mild changes in the environment (stenopotent) and can be used as biological indicators (Chuecas 1998). In oceanic areas, which cover the shelf break and the oceanic basins, plankton densities are extremely low, characterized by oligotrophic waters, which can be locally enriched by the phenomenon called "topographic upwelling", contributing to higher productivity and fishing potential (Heywood et al. 1990; Signorini et al. 1999).

Zooplankton sampling is usually performed using nets of various mesh sizes. Thus, depending on the equipment used, sampling will capture a zooplankton community of a certain size range and taxonomic composition. In addition, the vertical distribution of the various fractions of plankton size in relation to environmental conditions is an important aspect of the structure and function of marine communities (Böttger-Schnack 1996). Among others, an important effect of this distribution and the migration movements associated with it is the substantial transport of biomass and energy from the upper to lower layers, directly influencing the higher trophic levels (Angel 1985, 1993). Because it presents a significant proportion of planktonic biomass, the epipelagic layer plays an important role in these processes in tropical/subtropical regions (Longhurst and Williams 1979; Gallienne and Robins 2001). However, the zooplankton community of tropical oceanic regions is generally rarely investigated (Schnack-Schiel et al. 2010).

Pugh and Boxshall (1984) highlight that the vertical distribution of oceanic epiplankton is related to the physical structure of the water column. Thus, the dynamics of the physical environment regulate interactions within the community, determining its structure and function (Haury et al. 1990). Cummings (1984) points out that the vertical distribution pattern of pelagic copepod species is differentiated mainly according to depth, which is considered the primary dimension of the habitat. In tropical regions, because it is the factor with the greatest variation as a function of depth, the thermocline plays a fundamental role in forming the structure of communities, creating an important diversity of niches (Longhurst 1985), in addition to often acting as a barrier to several species of plankton (Haney 1988). For most plankton, circumglobal distributions are more common than endemism, and a high proportion of tropical marine zooplankton species have been described, mainly for the surface layer (Boltovskoy et al. 2002).

7.1.1 Zooplankton and Ichthyoplankton Research in Northern Brazil

The northern coast of Brazil (from Maranhão to Amapá State) (Fig. 7.1) is a complex and unique large-scale system. It includes the largest source of freshwater and sediments running off into the oceans on our planet, the Amazon River (Kineke et al. 1991). This region sustains one of the largest mangrove areas in the world (Beys-da-Silva et al. 2014) and a unique and extensive reef system (Moura et al. 2016). This entire productive zone exports an important quantity of particles to the Atlantic Ocean, controlled by extremely high rainfall and high energy, due to the strong macrotidal currents, one of the strongest western boundary currents (the North Brazil Current), and strong trade winds (Gibbs 1970).

The zooplankton community is fundamental in this system's food web because it provides food for many predators (Smith et al. 2016). At the same time, zooplankton predators such as cnidarian medusae control the abundance of phytoplankton and the initial phases of the fish populations in the region (Krumme et al. 2015). Despite its important ecosystem functions, there are many gaps in the understanding of the ecology of this community (Leite et al. 2020). The scarce information that exists in the published literature indicates that zooplankton are abundant and diverse in this region, with a mixture of estuarine and oceanic species typical of the North Brazil Current and influenced by the Amazon plume (Neumann-Leitão et al. 2018).

Copepods and appendicularians are the main holoplanktonic groups (Neumann-Leitão et al. 2018). Copepods are the most abundant and diverse group with different

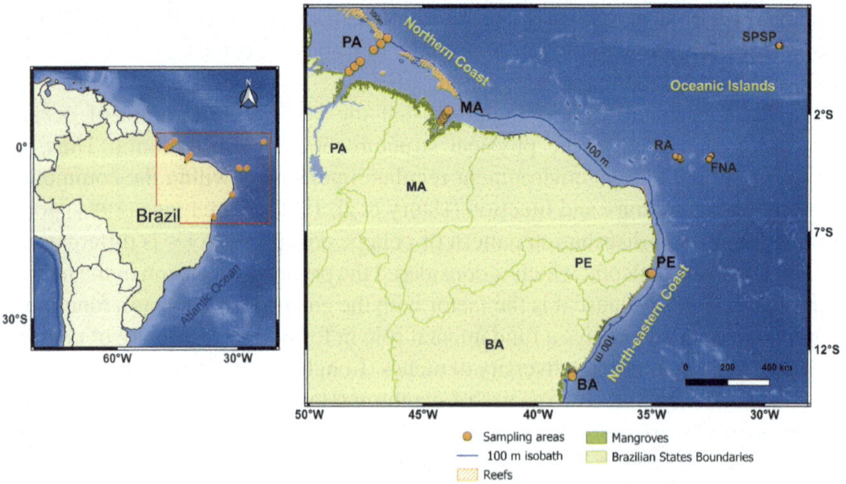

Fig. 7.1 Location of study areas discussed in the text. SPSP: St. Peter and St. Paul Archipelago, FNA: Fernando de Noronha Archipelago, RA: Rocas Atoll, BA: narrow shelf off Salvador (Bahia State), PE: narrow shelf off Tamandaré (Pernambuco State), MA: wide continental shelf off Maranhão State, PA: wide continental shelf off Pará State. Reefs and mangroves are also indicated

trophic strategies, with a dominance of omnivore–herbivore groups in estuarine waters and carnivorous and detritivorous groups in oceanic waters (Neumann-Leitão et al. 2018). Filtering organisms such as Appendicularia, which efficiently retain large amounts of various types of particles (organic and inorganic) in their gelatinous houses, are found in high densities in this region (Neumann-Leitão et al. 2018). The material that is not consumed on the surface is exported to deeper layers as marine snow, probably representing a significant contribution to the region's carbon pump and fertilizing benthic ecosystems, such as the unique reef system that characterizes the outer shelf. In relation to meroplankton, decapod larvae are generally the most abundant group, showing a coastal-oceanic gradient in abundance and composition; they are better represented in estuarine and coastal waters (Melo et al. 2014a, b). Future studies are essential to elucidate the ecological role of zooplanktonic biodiversity of the Amazon Coastal System and adjacent oceanic waters.

Despite the great importance of fish resources in the region, fish eggs and larvae in northern Brazil are poorly studied, and most studies concentrate on freshwater (Araújo-Lima et al. 2001; Lima and Araújo-Lima 2004; Leite et al. 2006, 2007; Silveira 2003, 2008, 2011; Cardoso 2014; Zacardi et al. 2016, 2017, 2020; Costa 2017; Paes et al. 2020) and estuarine regions (Barletta-Bergan et al. 2002a, b; Barletta et al. 2005; Bonecker et al. 2007; Barletta and Barletta-Bergan 2009; Costa et al. 2011; Costa 2013; Mangas et al. 2013; Sarpedonti et al. 2008, 2013; Soares et al. 2014a, b; Zacardi et al. 2016; Conceição 2016; Cardoso 2016; Zacardi and Bittencourt 2017; Silva et al. 2011a, b, 2017; Lima 2019; Soares 2019; Soares et al. 2020; Soares 2021; Cardoso et al. 2021; Rodrigues 2021). In the marine ecosystems of the northern Brazilian continental shelf, ichthyoplankton studies are rare, such as those conducted by Bittencourt et al. (2007), Zacardi et al. (2008), Ferreira et al. (2014), and Bonecker and Castro (2018), which included coastal and oceanic waters.

7.1.2 Zooplankton and Ichthyoplankton Research in Northeastern Brazil

In northeastern Brazil, most zooplankton studies were performed through collections with nets and generally in neritic areas, which cover estuaries and the continental shelf. In general, few studies have been conducted in this region. The tropical Southwest Atlantic is one of the least known marine environments, mainly in relation to many holoplankton groups. In this region, a low diversity is expected when compared to the species-rich benthic community because the distribution barriers for marine plankton are few and diffuse (Angel 1993). New research carried out within the context of INCTAmbTropic since 2010, whose results are presented in this chapter, has been filling many gaps and contributing greatly to a better knowledge of zooplankton in the area.

Previous research on zooplankton in northeastern Brazil (from Bahia to Maranhão State) (Fig. 7.1) is restricted to some groups, such as Cnidaria (Vannucci 1958; Tosetto et al. 2021), Copepoda (Kanaeva 1960; Björnberg 1963; Araújo 2006; Melo et al. 2014a, b), Chaetognatha (Vannucci and Hosoe 1956; Hosoe 1956; Gusmão 1986; Melo et al. 2020); Larvacea (Björnberg and Forneris 1956), Thecosomata and Gymnosomata (Larrazábal and Oliveira 2003), Decapoda (Silva-Falcão et al. 2007; Santana et al. 2018), protists (Costa et al. 2018), Radiolaria (Costa et al. 2019), and ichthyoneuston (Andrade 2000; Santana et al. 2020a, b).

Boltovskoy (1981, 1999) published an atlas on zooplankton for the South Atlantic, with a clear emphasis on temperate and subtropical waters off southern Brazil, Uruguay and Argentina. Thus, information on northeastern Brazilian zooplankton is still scarce, and most of the published studies are primarily descriptive (Vannucci and Queiroz 1963; Paranaguá 1966, 1967/1969; Klein and Moreira 1977; Medeiros and Björnberg 1978; Sankarankutty et al. 1990, 1995; Nascimento-Vieira et al. 1990; Neumann-Leitão et al. 1991/1993, 1999, 2008a, b, 2009, 2019; Ekau and Knoppers 1996, 1999; Gusmão et al. 1997; Schwamborn 1997; Lessa et al. 1999; Gusmão 2000; Cavalcanti and Larrazábal 2004; Garcia-Diaz 2007; Garcia-Diaz et al. 2009; Melo 2009; Melo et al. 2012; Campos et al. 2017; Campelo et al. 2018, 2019; Conceição et al. 2021). Lira et al. (2014) described zooneuston, and Souza et al. (2020) conducted size spectra modeling of mesozooplankton.

In northeastern Brazil, the main highlights were ichthyoplankton studies carried out by Mafalda Jr et al. (2006) for the North Coast of Bahia State and Mafalda Jr et al. (2008) in Todos os Santos Bay, Bahia (Fig. 7.1). These studies are more expressive in estuarine and coastal areas. Some describe new records of specific taxa in estuarine waters, such as Exocoetidae and Microdesmidae (Castro et al. 2008; Severi et al. 2008), but the majority are contributions to the community structure and spatiotemporal distribution of fish larvae in estuarine ecosystems (Marques et al. 2015; Santos and Severi 2019).

Although few studies have been conducted on the continental shelf of Northeast Brazil, it is known that its fish larvae assemblage is diverse, with more than 70 species, and the participation of meso- and bathypelagic species contributes to the increase in planktonic diversity and biomass (Santana et al. 2020a, b).

7.1.3 Zooplankton and Ichthyoplankton Research off the Oceanic Islands of Northeastern Brazil

A high abundance of zooplankton often occurs in waters off oceanic islands compared to surrounding oligotrophic oceanic areas (Hernández-León 1991). The "island mass effect" (IME) (i.e., a topographically forced upwelling of nutrients near oceanic islands, Doty and Oguri 1956) is well known and described worldwide, although the influencing forces are variable and not all are clear (Palacios 2002). Among the IME processes concerning zooplankton structure are tidal currents, frontal dynamics, boundary layers, topographically produced eddies, flow variations, geostrophic currents and Ekman drift (Palacios 2002; Gove et al. 2016). On the other hand, biological processes (life cycles, reproduction, growth, physiology, competition, and predation) also affect the food web (Longhurst and Pauly 1987; Sigman and Hain 2012).

The presence of the isolated Fernando de Noronha Archipelago (FNA), Rocas Atoll (RA) and Saint Peter and Saint Paul Archipelago (SPSPA) (Fig. 7.1) in the equatorial Atlantic has long been known to increase the biological productivity nearby (Souza et al. 2013). This means that the presence of these archipelagos alters the oligotrophy of the typical tropical structure, which includes a strong, permanent thermocline. This alteration involves upwelling by local processes such as surface current divergence, winds, and interactions between the currents and the submarine relief (Travassos et al. 1999; Araujo and Cintra 2009), providing nutrients to the photic zone (Neumann-Leitão et al. 1999; Lessa et al. 2009). As a result, an increase in productivity occurs at numerous trophic levels (Doty and Ogury 1956; González-Quirós et al. 2003), ensuring maintenance (Boehlert and Genin 1987; Rogers 1994; Genin 2004) and increasing the diversity of planktonic communities (Boltovskoy 1981, 1999; Melo et al. 2012).

FNA, RA and SPSPA (Fig. 7.1) are ideal areas for the development of studies on oceanic zooplankton (Quevedo et al. 2003; Campelo et al. 2019; Figueiredo et al. 2020). Several studies were conducted in SPSPA focused on the temporal and spatial distribution of pico-, nano, meso- and macrozooplankton (Advincula 1999; Galvão 2000; Cavalcanti 2002; Garcia-Díaz 2007; Melo et al. 2012; Correia 2014). Studies in RA were limited to some specific groups, such as Tintinnina (Nogueira et al. 2005, 2008; Nogueira and Sassi 2011), Chaetognaths (Souza et al. 2014) and Foraminifera (Tinoco 1965), or related to macrozooplankton (Pinto et al. 1997). Recently, Dantas (2019) described the microzooplankton in this area. FNA studies emphasized zooneuston (Lira et al. 2014), chaetognaths (Souza et al. 2014) and Decapoda larvae (Santana et al. 2018).

In the oceanic islands and seamounts off northeastern Brazil, ichthyoplankton are dominated by epipelagic species, such as flying fish, and mesopelagic species, such as myctophids (Lessa et al. 1999; Lima et al. 2016; Santana et al. 2020a, b). These islands represent a true oasis within the oligotrophic waters of the western Tropical Atlantic, where the density of fish eggs and larvae is significantly higher than that found in the adjacent oceanic areas (Lessa et al. 1999; Souza and Mafalda Jr 2019). It is curious that these assemblages have little representation of reef-associated fish species (Lima et al. 2016; Santana et al. 2020a, b) since the adult reef ichthyofauna of these islands are quite diverse, with approximately 113 species of shallow and deep reef habitats (Pinheiro et al. 2020). Despite being planktonic, benthic habitats can significantly influence the structure of the fish larvae assemblage; therefore, proximity to the islands influences the composition of the samples, as observed in a study in the SPSPA where larvae sampling in the inshore bay area revealed a high abundance and richness of reef fish species (Macedo-Soares et al. 2012).

7.1.4 Sampling Design and Abiotic Conditions

This chapter describes and compares zoo- and ichthyoplankton communities from seven tropical marine pelagic ecosystems that differ considerably in their abiotic and biological settings (Fig. 7.1): two areas off Pará and Maranhão (Northern Brazil) are characterized by extremely wide mangrove-fringed continental shelves, two narrow oligotrophic shelf areas located in northeastern Brazil (Pernambuco and Bahia) and three oceanic areas encompassing the FNA, RA and SPSPA.

These sites encompass several gradients regarding climate (humid in the north, semiarid in the northeast), seasonal cycles (timing and seasonal amplitude of rainfall and winds, e.g., with peak rainy seasons from June to August off Pernambuco State and from January to June off Pará State), shelf width (more than 300 km off Pará, less than 10 km off Salvador, Bahia), tidal amplitude (from 2 m in Bahia to 8 m in Maranhão), mean temperature (increasing equatorwards, Fig. 7.2), salinity (lower salinities are found only on the northern shelf, especially in the plume of the Pará River (Fig. 7.2) and seasonality of temperature (seasonal temperature amplitude decreasing towards the equator) (Fig. 7.2). These sites are generally characterized by eusaline marine waters (Fig. 7.2), except for the northernmost transect, across the Pará shelf and Pará River plume, which receives considerable freshwater input from numerous rivers (especially from the vast Tocantins-Araguaia River system).

7.2 Seasonal and Interannual Variability of the Zooplankton and Ichthyoplankton in Neritic and Offshore Waters off Salvador, Bahia, Brazil.

The continental shelf off Salvador is one of the narrowest shelves in the world, with widths varying from 9 to 17 km, and the shelf break is located at 70–80 m (Ekau and Knoopers 1999; Pereira and Lessa 2009) (Fig. 7.3).

The Salvador continental shelf near the entrance to Todos os Santos Bay is vital for the reproductive behavior of fish. Many marine species of commercially important fishes that use Todos os Santos Bay as nurseries for their young breed on the continental shelf. The shelf is under the influence of the Brazil Current (BC), which is responsible for transporting tropical water (AT) with salinities between 36 and 38 and temperatures between 24–26 °C (Cirano et al. 2006). Salinity and temperature vary little in the surface of the shelf waters, presenting a homogeneous character (Castro Filho and Miranda 1998). Furthermore, in this area, fluvial input is negligible, making this shelf typical of the Northeast continental shelf, which is characterized by warm and oligotrophic waters (Ekau and Knoppers 1999).

In this area, plankton samples were collected between April 2013 and October 2014, totaling 10 campaigns (Fig. 7.3).

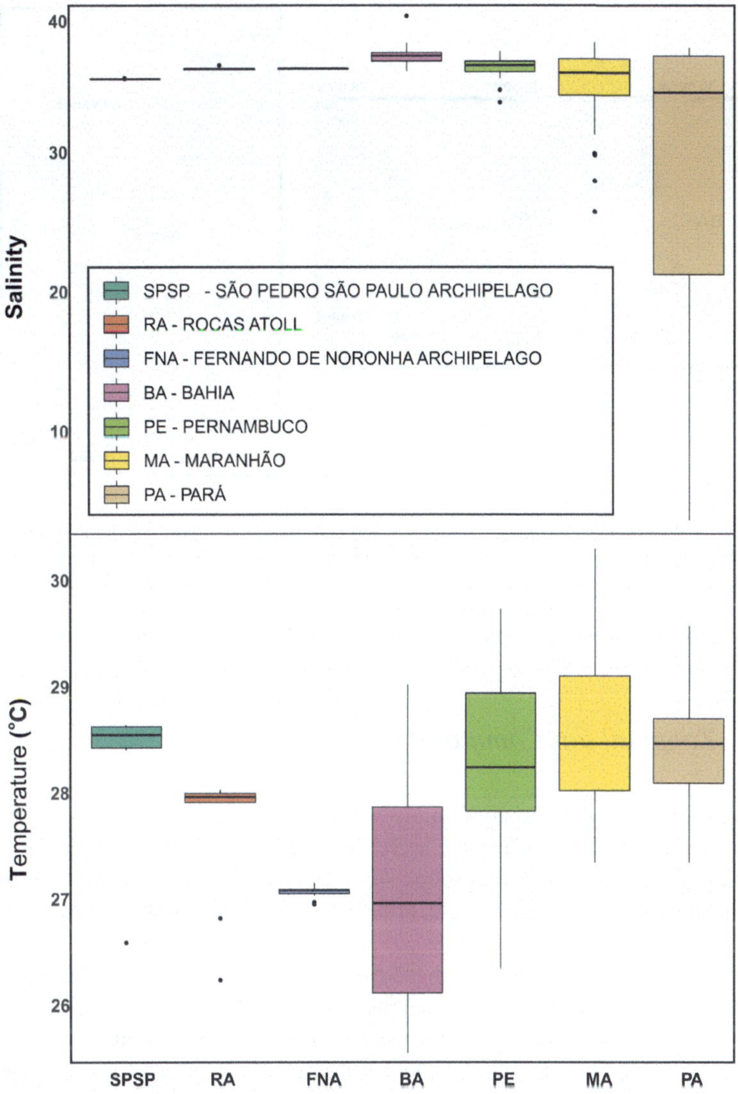

Fig. 7.2 Boxplots of surface (1 m) salinity and temperature measured at sampling stations off oceanic islands and on the continental shelf. SPSP: St. Peter and St. Paul Archipelago, FNA: Fernando de Noronha Archipelago, RA: Off Rocas Atoll, BA: narrow shelf off Salvador (Bahia State), PE: narrow shelf off Tamandaré (Pernambuco State), MA: wide shelf off Maranhão State, PA: wide shelf off Pará State (northern Brazil)

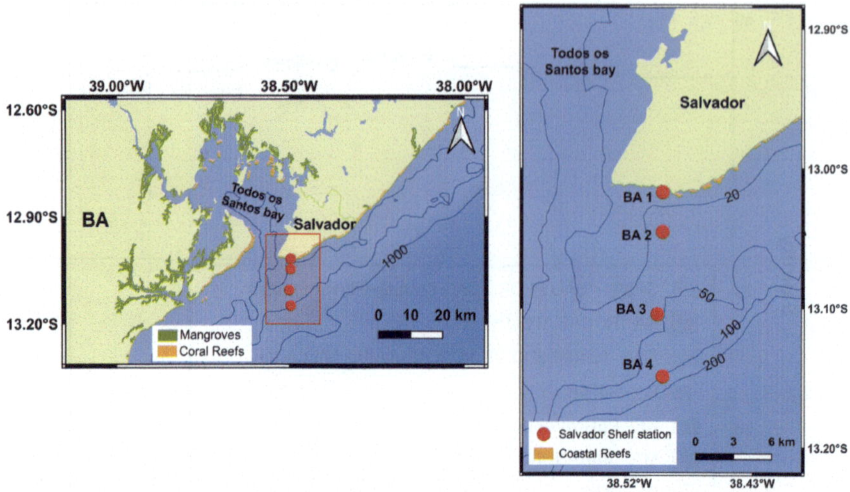

Fig. 7.3 Location map of sampling stations at the continental shelf off Salvador (Bahia State). All plankton samples were collected at four fixed stations, BA1, BA2, BA3 and BA4, located approximately 0.5, 4, 8 and 17 km from the coast, respectively. Zooplankton and ichthyoplankton sampling were performed through subsurface horizontal hauls with a conical plankton net (300 μm mesh size and 60 cm mouth diameter) for 10 min at a speed of 2–3 knots with a flowmeter coupled to the mouth of the net to obtain the volume of water filtered

7.2.1 Zooplankton Community

In general, the zooplankton community was represented by organisms that play a key role within the marine trophic web; the Copepoda, Chaetognatha, Larvacea, Molluska and other Crustacea groups stood out, and these results are similar to those found by Neumann-Leitão et al. (1999), Cavalcanti and Larrazábal (2004), and Neumann-Leitão et al. (2008a, b).

Density values of zooplankton ranged from 11.07 to 1023.13 ind./m^3. The highest values were registered in June and August 2013 and March 2014, all in the rainy season (Fig. 7.4). The lowest values were registered in the months of April and October 2013 (Fig. 7.4). In general, the density presented a coast-to-ocean gradient, with a tendency of the highest values occurring at stations closer to the coast and the smallest values occurring at more oceanic stations. This behavior is common to tropical marine regions and has already been reported for other regions of the Brazilian coast (Lopes et al. 1999; Neumann-Leitão et al. 2008a, b; Dias and Bonecker 2009).

A total of 131 taxa were identified, with holoplankton representing 94% of the community and meroplankton representing 6%, varying the importance and contribution of different taxa between campaign and sampling stations. Overall, Copepoda was the most dominant group, representing approximately 68% of the total community. The relative abundance of dominant or characteristic copepod species showed spatial and temporal variability.

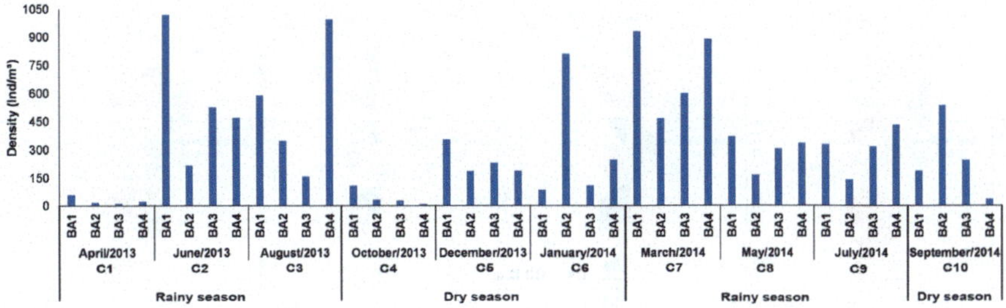

Fig. 7.4 Density (ind./m^3) of the zooplankton community in neritic and offshore waters off Salvador (Bahia State) from April 2013 to October 2014. BA1, BA2, BA3 and BA4 indicate the sampling stations. C1 through C10 indicate campaign number

Farranula gracilis was among the dominant species during campaigns C1, C3, C4 and C5. This species showed high representation (above 50%) at stations B3 and B4 during campaigns C4 and C5. *Temora turbinata* dominated (more than 70%) at station BA1 during campaigns C2 and C4. Campaign 6 was marked by the high representation of *Onchocorycaeus latus*, mainly in the more oceanic stations (BA3 and BA4). During campaign 3, *Clausocalanus furcatus* and Pontellidae (juvenile) dominated; however, they showed an inverse spatial pattern. *C. furcatus* showed a gradient of ocean-to-coast increase, while Pontellidae increased from coast to ocean. *Undinula vulgaris* and *Paracalanus parvus* dominated at station BA4 during campaign C2. Campaigns C9 and C10 were dominated by *Paracalanus* sp. (Fig. 7.5).

The ordination diagram elaborated with the first two canonical axes of the redundancy analysis explained 81% of the accumulated percentage variation in the relationship between the composition of the copepod assemblage and the oceanographic structure (Fig. 7.6). Through the ordination diagram, it was possible to verify a clear separation between the sampling stations as a function of the oceanographic gradient and the composition of the copepod assemblage (Fig. 7.6). The correlations of environmental variables with the first two canonical axes were used to assess the relative importance of each variable, where absolute values of correlation coefficients >0.4 are significant and can be considered biologically important.

During campaigns C2, C3, C7 and C10, the stations located on the left side of the ordination diagram were linked to the highest values of nitrate and phytoplankton sedimented volume and to the highest densities of *C. furcatus*, *T. stylifera*, *C. velificatus*, *O. venusta*, *F. gracilis* and *T. turbinata*, indicative of productive water mass. However, they were associated with lower values of temperature, salinity, dissolved oxygen and phosphorus (Fig. 7.6). During campaigns C4 and C5, the stations located on the lower right side of the diagram were linked to the highest values of temperature, salinity, dissolved oxygen and phosphorus, as well as the highest densities of *O. latus*, *Paracalanus* sp., and *Clausocalanus* sp. (Fig. 7.6). However, they were linked to lower values of sedimented phytoplankton volume and nitrate, thus being associated with less productive water mass.

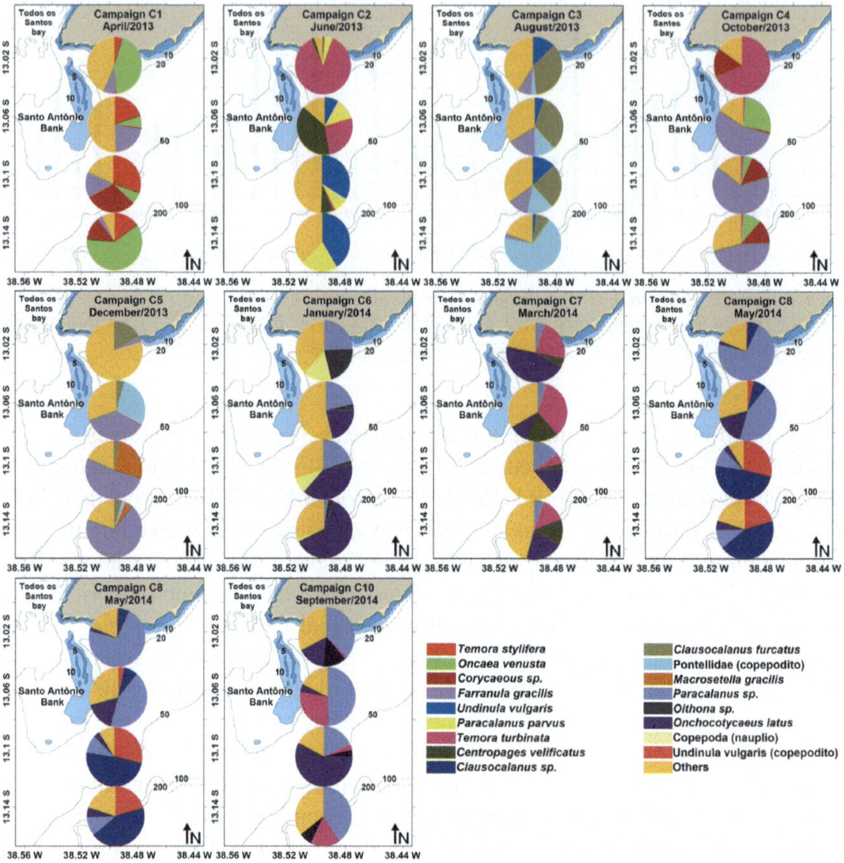

Fig. 7.5 Percent relative abundance of dominant copepod species in neritic and offshore waters off the coast of Salvador (Bahia State) during campaigns C1–C10

The redundancy analysis showed that seasonal and interannual variability of the copepods was influenced by the oceanographic structure of the tropical water mass and that their higher temperature and salinity conditions may reflect low densities and diversity.

The significant occurrence of oceanic species on the continental shelf may be a consequence of its small width (9 km), and thus being strongly influenced by conditions of the tropical water (Neumann-Leitão et al. 2008a, b).

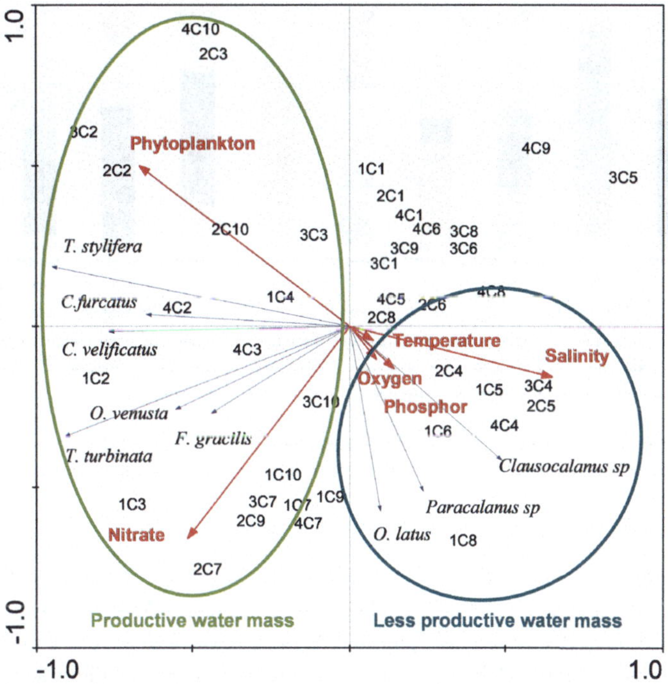

Fig. 7.6 Ordering diagram for redundancy analysis relating the sampling stations, the composition of the copepod assemblage and the oceanographic structure in the neritic and offshore waters off Salvador (Bahia State) from April 2013 to October 2014 (Campaigns C1 through C10)

7.2.2 Ichthyoplankton Community

From the 236 larvae collected during the ten campaigns, 21 families were identified. Seventy-five percent of the total larvae taken represented six characteristic families in increasing order of abundance: Gobiidae (6.44%), Haemulidae (6.87%), Blenniidae (8.15%), Pomacentridae (14.16%), Gerreidae (17.17%), and Sparidae (22.32%). The families Ophidiidae and Holocentridae were collected only in August 2013 (C3), and Tetraodontidae and Cynoglossidae were collected only in December 2013 (C5) and January 2014 (C6), respectively. The total number of larvae collected during December 2013 (C5) and January 2014 (C6) was less than that of the other campaigns. In these months, the lowest number of families was observed (Fig. 7.7). The wide larval distribution, especially Gobiidae, might be a consequence of a high larval drift from the coral reef area where massive spawning occurs (Mafalda Jr and Rubín 2006), such as the coral reefs within Todos os Santos Bay and along the Salvador coastline.

In relation to the habitats of adult fish groups, three dominant fish groups were recognized in the neritic and offshore waters off Salvador: epipelagic, demersal and benthic fishes. The relative abundance of epipelagic fish larvae (Clupeidae) was

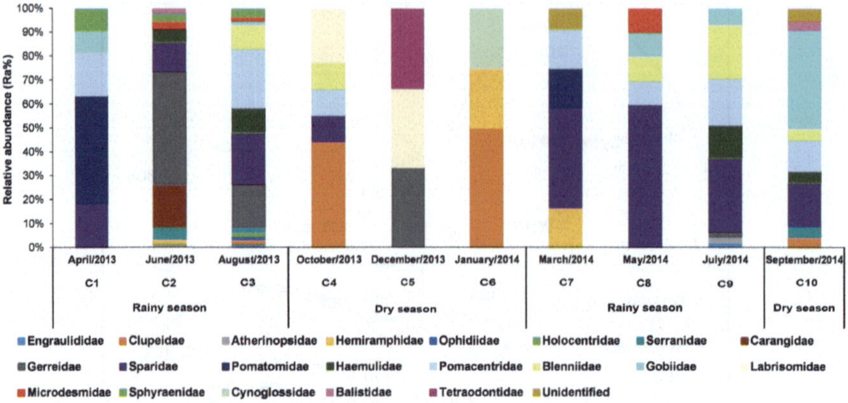

Fig. 7.7 Relative abundance (Ra%) of the ichthyoplankton community in the neritic and offshore waters off Salvador (Bahia State) from April 2013 to October 2014 (Campaigns C1 through though C10)

highest during October 2013 (C4) and January 2014 (C6). For the Sparidae family (Demersal), the levels of relative abundance were highest in March, May and July of 2014 (C7, C8, C9), decreased in September 2014 (C10) and were absent in December 2013 (C5) and January 2014 (C6). The relative abundance of Gobiidae (benthic) peaked in September 2014 (C10) (Fig. 7.7).

The dominance of the demersal association during the rainy season was mainly related to coastal waters (Mafalda Jr et al. 2004), with higher temperature and ammonia concentration and, secondarily, with the highest levels of turbidity and biomass of phytoplankton. The occurrence of the pelagic association during the dry season was related to tropical waters (Mafalda Jr et al. 2004), with higher oxygen and salinity contents. According to Olavo et al. (2005), pelagic fish are mostly captured between spring–summer and demersal fish, mainly in autumn–winter. This seasonality reflects the reproductive population connectivity existing between reef, beach and mangrove environments.

Fish larvae showed a rather heterogeneous distribution, with the density at each station ranging from 0 to 190 larvae per 100 m^3 (Fig. 7.8). A general trend was observed toward lower density at most sampling stations, with abundance generally no more than 25 larvae per 100 m^3. Higher densities were found in June 2013 (C2) and August 2013 (C3) (rainy season) (Fig. 7.8). Regarding fish eggs, the density observed between April 2013 (C1) and September 2014 (C10) ranged from 0 to 5329 per 100 m^3. The highest density value was recorded at station BA1 in March 2014 (C7) (rainy season). The month of January 2014 (C6) (dry season) registered the lowest density of fish eggs (0–17 eggs/100 m^3) (Fig. 7.8).

Studies that interpret ichthyoplankton structure in terms of adult characteristics often find a spatially heterogeneous distribution of larvae that are attributable to adult characteristics (Gaughan et al. 1990; Yoklavich et al. 1992). Knowing the geographical locations of spawning adults can also help explain spatial heterogeneity

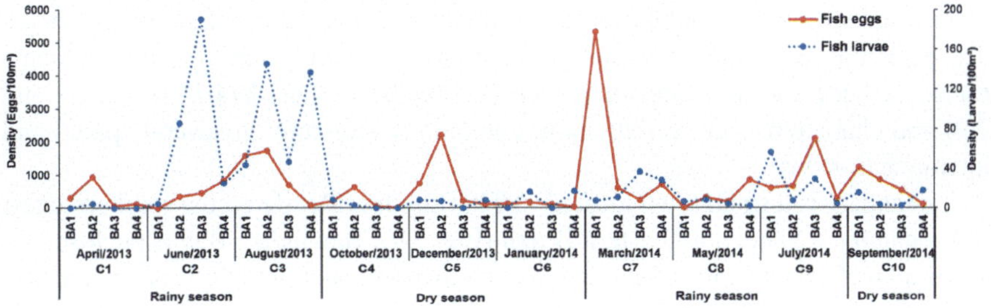

Fig. 7.8 Density (ind./100 m^3) of the ichthyoplankton community in the neritic and offshore waters off Salvador (Bahia State) from April 2013 to October 2014 [samples were collected at the subsurface with a 300 μm mesh net (Campaigns C1 although C10)]

in ichthyoplankton structure (Rakocinski et al. 1995). Coastal regions adjacent to estuaries and bays present favorable conditions for the development of fish species that use these areas as nurseries or to protect their eggs and larvae.

The ichthyoplanktonic community observed in the study area is influenced, among other factors, by the characteristics of the continental shelf of Salvador and by the proximity to the entrance of Todos os Santos Bay. The bathymetric profile of the shelf edge between 50 and 150 m is characterized by the presence of shelf brake reefs, which contribute to population connectivity. The circulation in Todos os Santos Bay forces an exchange of water mass with the continental shelf, where the tides tend to exit the bay in the more surficial layers and to enter the layers below (Lessa et al. 2009). Surface water runoff promotes the transport of larvae from the spawning site (bay) to a nursery area (continental shelf) and vice versa. Thus, both factors contribute to the formation of fish larvae assemblages.

7.3 Spatial, Seasonal and Interannual Dynamics of the Zoo- and Ichthyoplankton at the Continental Shelf off Tamandaré, Pernambuco, Brazil

Coastal and shelf environments in the western tropical Atlantic present a complexity of productive environments (Nixon et al. 1986; Schwamborn 1997; Schwamborn et al. 2001; Neumann-Leitão et al. 2018). The material transported by tropical rivers enriches planktonic production, determining the taxonomic composition of plankton and their distribution patterns (Santana et al. 2020a, b). Copepods generally dominate zooplankton in abundance, but meroplanktonic organisms are also often abundant due to the export of larvae of benthic organisms living in a true mosaic of coastal ecosystems, such as estuaries, reefs, sandbanks, and macroalgal beds (Neumann-Leitão and Matsumura-Tundisi 1998; Schwamborn et al. 2001; Ferreira and Maida 2006; Brito-Lolaia et al. 2020; Lins-Silva et al. 2021). Despite the large ecological importance of this area, the investigation of planktonic communities along the narrow continental shelf off Pernambuco State (Brazil) remains scarce. Here, we provide a description of the composition and distribution of zoo- and ichthyoplankton along a transect adjacent to the Tamandaré shelf (Pernambuco State, Brazil) (Figs. 7.1 and 7.9) and characterize the variations in community structure on several spatial and interannual scales.

Tamandaré is a small town located 110 km south of Recife on the southern coast of Pernambuco State, northeastern Brazil (Fig. 7.9). Tamandaré Bay is enclosed by a coastal reef system and is included in two environmental protection areas, the first created in Brazil to protect coastal reefs.

The climate is "warm and humid" throughout the year, with a well-defined seasonality. Strong wings (SE to E) occur from June to October ("windy season"). The rainy season generally occurs between March and August, with a peak in July, while the dry season is generally from September to February (Ferreira et al. 2003; Grego et al. 2009; Venekey et al. 2011).

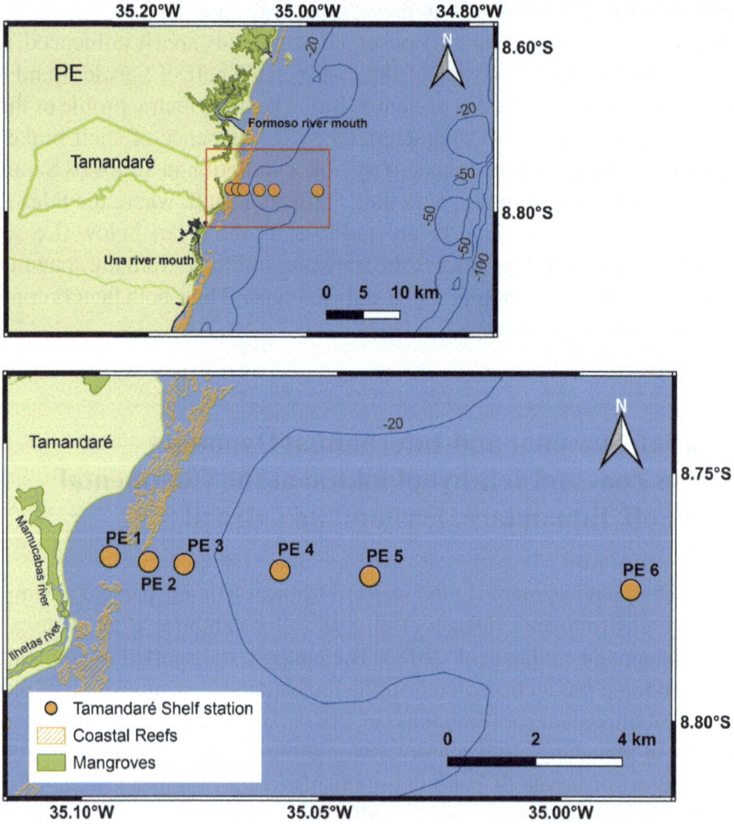

Fig. 7.9 Map showing the location of the sampling stations on the narrow oligotrophic continental shelf off Tamandaré (Pernambuco State). Sampling was performed from April 2013 to May 2015 in bimonthly intervals

The main economic activities along the southern Pernambuco coast are agriculture (mostly sugar cane plantations), tourism and fisheries (Moura and Passavante 1993; Araújo and Costa 2003). The adjacent nearshore continental shelf off Tamandaré is characterized by warm and calm waters. A sandstone reef (beach rock) runs parallel to the coast and functions as a substrate for the development of algae, zoanthids and corals. Coastal drift varies according to wind regime and season but runs dominantly in a south-north direction.

On the southern shore of Tamandaré Bay, there are two small creek inlets, Mamucabas and Ilhetas (Fig. 7.9). Approximately 11 km south of the study area is the inlet of the regionally relevant Una River (Fig. 7.9), which is characterized by significant contributions of water and sediments to the shelf in occasional devastating events of torrential rainfall (Moura et al. 2016).

The Tamandaré shelf has an average width of 35 km with a gentle slope, reaching depths between 50 and 65 m at the shelf break (Manso et al. 2003; Domingues et al. 2017).

Zooplankton samples were collected during 12 campaigns from April 2013 to May 2015 at bimonthly intervals at six fixed sampling stations (PE1–PE6) (Fig. 7.9). Common plankton nets with 300 μm mesh size and 60 cm in mouth (maximum 0.6 m of depth) were used with flowmeters coupled and towed at the subsurface for 5 min with a speed of 2–3 knots. At each station, water transparency was estimated with a Secchi disk (Cialdi and Secchi 1865; Kirby et al. 2021). Other abiotic parameters, such as temperature and salinity, were measured with a CTD probe (YSI CastAway).

In the laboratory, all samples were divided into size fractions and then quartered with a Motada quarterer (Omori and Ikeda 1984), resulting in aliquots (subsamples) of 1000–2000 organisms each. These subsamples were digitized using ZooScan (Gorsky et al. 2010) equipment (Hydroptic model ZSCAN03, resolution: 2400 dpi) and ZooProcess software (http://www.obs-vlfr.fr/LOV/ZooPart/ZooScan). Plankton Identifier (PkID) software was used for a semiautomatic classification of organisms based on the ZooScan vignettes (i.e., after a quick preclassification using the random forest algorithm, the taxonomic assignment of each vignette was verified manually).

Redundancy analysis was applied to identify whether differences in community structure could be explained by environmental and spatial–temporal variables. We included abundance data and a spatiotemporal factor matrix (rainy vs. dry season, sampling stations, years, temperature, salinity, and Secchi depth). Hellinger transformation was applied to abundance data prior to RDA (Legendre and Gallagher 2001). Nonnumeric factors (rainy vs. dry, stations, years) were transformed into dummy variables, represented by 1 and 0. Prior to building the RDA model, all nonsignificant variables ($p > 0.05$) were excluded in a step-by-step procedure using Monte Carlo permutation tests with 9999 permutations, considering collinearities between variables (Ter Braak and Smilauer 2002) and following the recommendations of Legendre and Legendre (1998) and Legendre and Gallagher (2001). All analyses were performed using CANOCO software (version 4.5; Ter Braak and Smilauer 2002).

7.3.1 Zooplankton Composition and Spatial Distribution

Twenty-six zooplankton taxa were registered for the Tamandaré shelf. Of these, four zooplankton groups were the most important in terms of abundance in both seasons: copepods of the orders Calanoida and Cyclopoida, followed by decapods (e.g., larvae of brachyuran crabs in zoea and megalopa stages and caridean shrimp zoeae) and chaetognaths (Fig. 7.10).

Copepods are well known to be the dominant group in marine zooplankton and are present in all marine environments and at all depths (Silva et al. 2003; Costa et al. 2008). A high abundance of decapods was also observed, following the coast-to-shelf gradient with higher values in coastal areas, which could possibly be linked to their life cycle, since larvae are initially released by adults inhabiting shallow coastal waters such as estuaries, mangroves and reefs (Schwamborn and Bonecker 1996; Melo Júnior et al. 2007; Schwamborn et al. 2008; Santana et al. 2020a, b). The phylum Chaetognatha also showed a high abundance, mainly at offshore shelf stations. This group of gelatinous predators is often abundant in marine waters (Harzsch et al. 2015). The other zooplankton were mainly composed of ichthyoplankton (fish eggs and larvae), Thaliacea (salps), gastropods, echinoderm larvae, and cnidarian medusae.

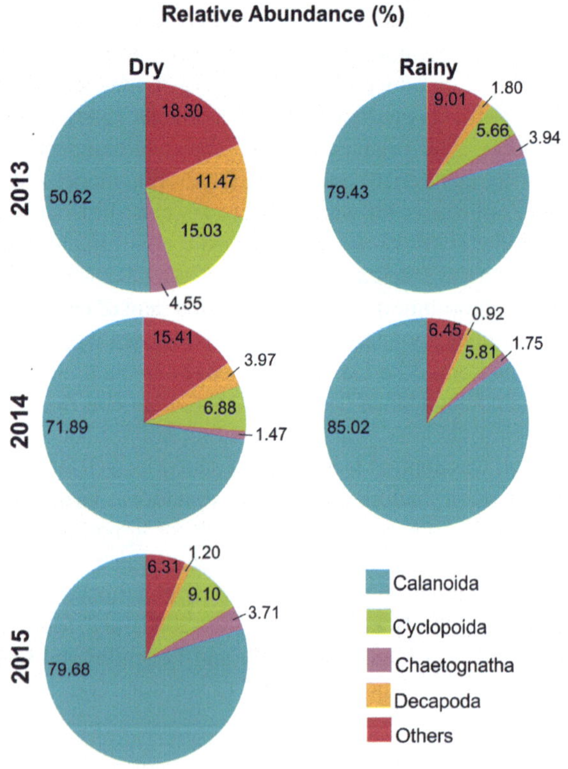

Fig. 7.10 Zooplankton relative abundance for dry and rainy seasons at the Tamandaré shelf (Pernambuco State) in 2013, 2014 and 2015

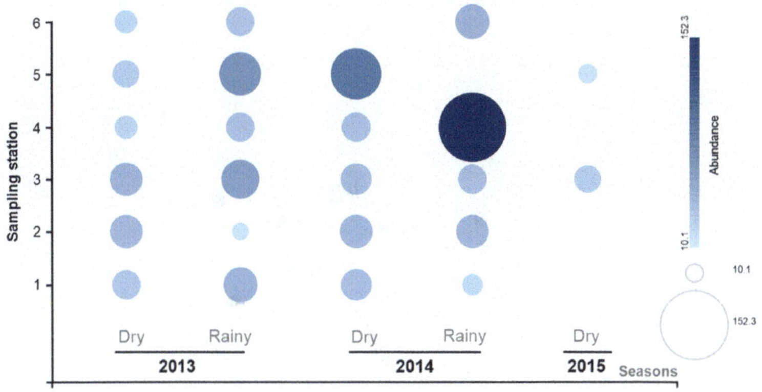

Fig. 7.11 Zooplankton total abundance for dry (from September to February) and rainy (from March to August) seasons at shelf sampling stations in 2013, 2014 and 2015. The diameters and color range (light to dark) of the circles are proportional to the value of each respective abundance interval

No differences were recorded in the zooplankton composition between the years. However, differences between the seasons were recorded, where calanoid copepods showed high values of relative abundance during the rainy season (Fig. 7.10). A higher total zooplankton abundance was found in the rainy season, mainly at stations PE3, PE4 and PE5 (Fig. 7.11). These mid-shelf stations are influenced by estuarine plumes from adjacent rivers, such as the Una River. The existence of a peak in total abundance at the midshelf, with higher abundance than nearshore, is surprising and deviates from the expected nearshore-offshore gradient.

Redundancy analysis showed that the spatial–temporal variables significantly ($p < 0.05$) explained 31% of the variance in zooplankton abundance. The abiotic variables (salinity, temperature, and Secchi depth) were not significant (Fig. 7.12). Two main RDA axes were obtained, separating rainy and dry seasons and coastal and neritic stations (Fig. 7.12).

There was a high contribution of gelatinous organisms, such as Appendicularia and Chaetognatha, invertebrate larvae (e.g., Echinodermata and Polychaeta), and cyclopoid and harpacticoid copepods in the offshore direction. Conversely, in the inshore direction, the most representative groups were calanoids, copepod nauplii, fish larvae and decapods. In relation to the seasons, the dry period was represented mainly by fish larvae and decapods, while the rainy period was represented by salps (Thaliaceae) and bryozoan larvae.

In tropical neritic areas, a coast-to-shelf gradient in plankton communities has been well documented by other studies from Brazil. For example, decapod larvae and fish larvae show high abundance and biomass values in more coastal areas due to their life cycle and the presence of important, highly productive coastal ecosystems, such as mangroves (Schwamborn and Bonecker 1996; Melo Júnior et al. 2007, Schwamborn et al. 2008; Santana et al. 2020a, b).

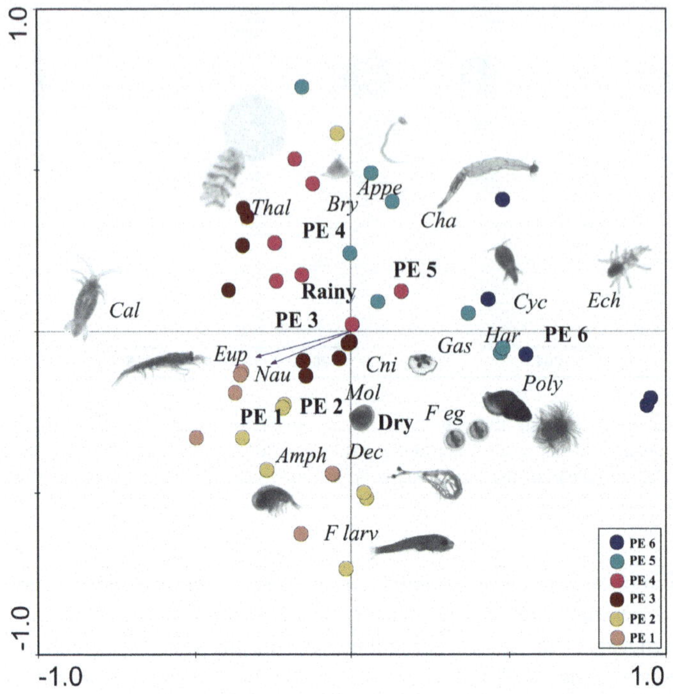

Fig. 7.12 Ordination diagram of the redundancy analysis. Spatiotemporal and abiotic factors and species classification in blue. Zooplankton abundance with a wp2 net (300 μm mesh size) at Tamandaré (April–August 2013–2015). Seasons: rainy and dry cycles: stations PE1, PE2, PE3, PE4, PE5 and PE6): Zooplankton taxa: Cal: Calanoida; Cyc: Cyclopoida; Har: Harpacticoida; Nau: Nauplius of copepods; Appe: Appendicularia; Cha: Chaetognatha; Dec: Decapoda; Eup: Euphausiacea; Amp: Amphipoda; Bry: Bryozoa; Cni: Cnidaria; Ech: Echinodermata; Mol: Molluska; Gas: Gastropoda; Pol: Polychaeta; Thal: Thaliacea; F eg: Fish eggs; F lar: Fish larvae

On the other hand, offshore ecosystems off northeastern Brazil have been characterized by the dominance of large zooplankton (Neumann Leitão et al. 2019a, b). This was also observed in our results, with a high contribution of Chaetognatha and Appendicularia. However, unlike other studies, we also found small organisms, such as cyclopoid and harpacticoid copepods, at more offshore stations (4–6 stations). In general, these organisms are associated with more nearshore coastal stations (Nascimento-Vieira et al. 2010; Melo et al. 2010; Neumann Leitão et al. 2019a, b), as was the case for copepod nauplii and calanoid copepods.

Some groups were strongly associated with the temporal vector. Fish larvae and eggs, decapods and polychaetes were associated with the dry season, while the associated groups were salps (Thaliaceae), and calanoid copepods were associated with the rainy season. The same gradients were recorded from Alagoas to Sergipe, located immediately south of the Tamandaré area, for fish larvae and eggs and decapods (Schwamborn et al. 2019). During the rainy period, there is an increase in nutrient inputs into shelf waters (e.g., from the Una River plume), and these inputs boost primary and secondary production (Platt 1985; Marcolin et al. 2013; Sato et al. 2015; Figueiredo et al. 2020). Tamandaré is one of the few areas where a clear seasonal pattern could be detected.

The total density of fish larvae ranged from 0 to 1440.90 ind. 100 m^3, the highest values were observed in August 2013 (PE 4), March 2014 (PE 4 and PE 5) and May 2014 (PE 4). Fish eggs ranged from 0 to 2498.50 ind. 100 m^3. August 2013 (PE 5) and September 2014 (PE 4) had the highest values (Fig. 7.13).

A total of 90 fish larvae images were analyzed from samples collected on the Tamandaré shelf. Although 57.7% of fish larvae could not be identified due to their small size or scanning position or because they were damaged, it was possible to identify nine different families (Table 7.1). Taxa common from coastal areas, such as Pomacentridae and Carangidae, were identified, the latter being represented by *Oligoplites* and *Chloroscombrus*. Fish larvae from coastal areas near reef ecosystems are mainly composed of resident families, such as Pomacentridae, which contribute to enhancing plankton biomass (Santos et al. 2019). Monacanthidae was represented by the genera *Stephanolepis* and *Monacanthus*, which are quite common in continental shelf waters.

Regarding the stages of larval development, newly hatched larvae dominated, especially in the preflexion stage, which corresponded to 78% of the total larvae analyzed (Fig. 7.14). The dominance of newly hatched larvae in plankton samples is related to the occurrence of reproduction events as well as a complex suite of variables that reduce the abundance of later stages, such as mortality and advection (Silva-Falcão et al. 2013). Additionally, the higher density of fish eggs and larvae in the dry season corroborates other studies carried out in tropical coastal areas (Silva-Falcão et al. 2013; Santos and Severi 2019).

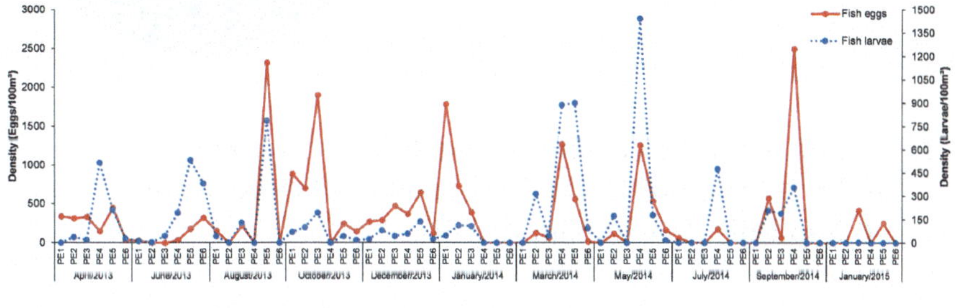

Fig. 7.13 Density (ind./100 m^3) of the ichthyoplankton (teleost eggs and larvae) at the continental shelf off Tamandaré, Pernambuco State

Table 7.1 Taxonomic composition, abundance (N), developmental stages (PF: preflexion; FL: flexion and PO: postflexion) and mean size (mm) of fish larvae collected on the Tamandaré shelf and analyzed through Zooscan images

Taxon	N	Stage	Mean size (mm)
Perciformes	20	PF	2.35
Atherinopsidae	2	PF	
Carangidae	3	PO	3.23
Lutjanidae	1	PF	5.00
Monachantidae	2	PO	5.50
Mugilidae	2	PF, FL	
Pomacentridae	5	PF	1.93
Sparidae	1	FL	3.91
Sphyraenidae	1	FL	4.86
Tetraodontidae	1	PO	3.39

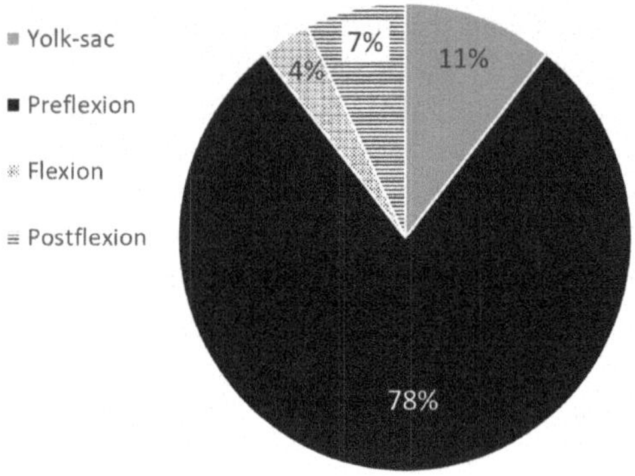

Fig. 7.14 Proportion of developmental stages of the fish larvae collected on the Tamandaré shelf and analyzed through Zooscan images

7.4 Seasonal and Interannual Variability of the Zoo- and Ichthyoplankton in Neritic and Offshore Waters off São Luís, Maranhão, Brazil

The continental shelf off São Luís (Fig. 7.15) is subjected to the influences of the input of significant continental waters, trade winds, the North Brazil Current (NBC) and its subsurface component, the North Brazil Undercurrent (NBUC), which transports a large volume of water to the northwest, parallel to the coast (Johns et al. 1998; Schott et al. 1998; Silva 2006). The region is also characterized by semidiurnal macrotides with tidal ranges varying from 4 to 7 m. Tidal currents can reach speeds of more than 6 m/s, and the climate is tropical humid with well-defined dry (July–December) and rainy (January to June) seasons (PORTOBRAS 1988; Araújo et al. 2009).

All ichthyo- and zooplankton samples were collected at seven fixed stations, MA1, MA2, MA3, MA4, MA5, MA6 and MA7 (Fig. 7.15), separated by approximately

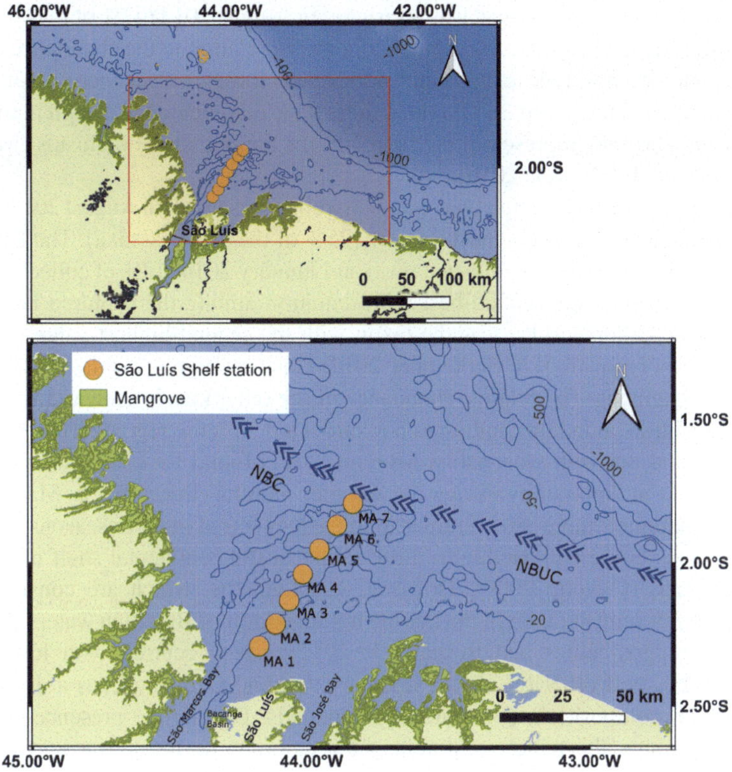

Fig. 7.15 Location map of sampling stations at the continental shelf off São Luís (Maranhão State). Current branches indicated are the North Brazil Current (NBC) and the North Brazil Undercurrent (NBUC). All plankton samples were collected at seven fixed stations, MA1, MA2, MA3, MA4, MA5, MA6 and MA7, located approximately 0.8, 16.9, 32.2, 49.1, 64.1, 81.3 and 96.6 km from the coast, respectively

1 km. Ichthyo and zooplankton sampling was performed through horizontal subsurface hauls with a conical plankton net (300 and 200 μm mesh size and 60 cm mouth diameter, respectively). The hauls lasted 3 to 10 min at a speed of 2 knots, with a flowmeter coupled to the mouth of the net to obtain the volume of filtered water. Ichthyo- and zooplankton samples were collected between April 2013 and September 2014, totaling 9 campaigns.

7.4.1 Zooplankton Community

Copepods (copepodites and adults) were the most abundant group on the continental shelf off São Luís and were recorded throughout the entire sampling period. Their maximum relative abundance (76% of the collected taxa) was recorded in October

2013, and the lowest abundance was registered in August 2013 (47% of the collected taxa) (Fig. 7.16). Therefore, a seasonal influence in relation to the abundance of this taxon cannot be inferred, as both the highest and lowest values were recorded in the dry months. This group, as several authors have commented, has high numerical importance and may represent more than 70% of the plankton organisms (Melo Jr 2005; Lucas 2008; Codina 2010).

The phylum Molluska (Bivalvia) was also well represented. In August 2013, it was the group with the second highest registry (25% of the captured taxa). The Bryozoa in November 2013 (24% of collected taxa) and January 2014 (12% of collected taxa) was the group with the second-best representation among the captured larvae. In March 2014, Oikopleuridae was the taxon with the second highest value (14% of the captured organisms). In September 2014, Gastropoda represented 11% of the captured organisms. The relative abundance of the other taxa is shown in Fig. 7.16. This variation in abundance indicates a relationship with the reproductive period of these taxa and not with seasonality. Moreno (2017) found for southern Brazil that the zooplankton community during the summer months (December to March) had relative abundance values 80% similar to the values found in other seasons.

These results are comparable to those found in the continental shelf off southeastern Brazil (Oliveira 2009). However, chaetognaths, which are considered a numerically dominant taxon in the marine holoplankton of Brazilian waters (Becker 2014; Domingos-Nunes and Resgalla Jr 2012), have occurred with low abundance on the shelf off São Luís. In summary, the São Luís Shelf has a diversified zooplankton composition, dominated by copepods. Despite the presence of other important groups, this pattern of dominance is common for the marine zooplankton community (Boltovskoy 1999).

Density values ranged from 1.2 to 382 ind./m^3. The highest densities were registered in November 2013, followed by September 2014 (203.4 ind./m^3) and May 2014 (196.4 ind./m^3). The lowest density was registered in the month of October

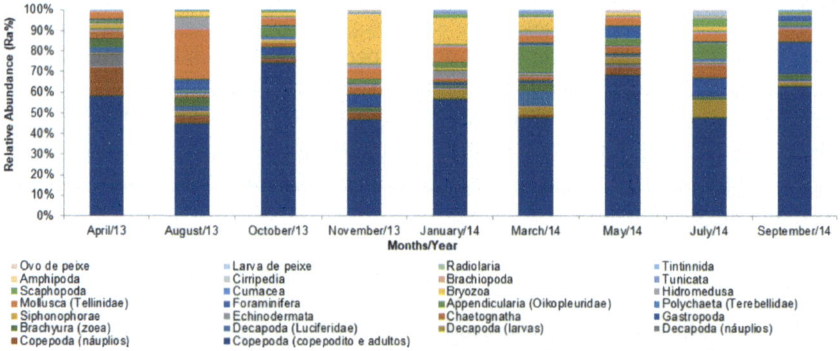

Fig. 7.16 Relative abundance (Ra%) of the zooplankton community from the continental shelf off São Luís (Maranhão State), from April 2013 to September 2014

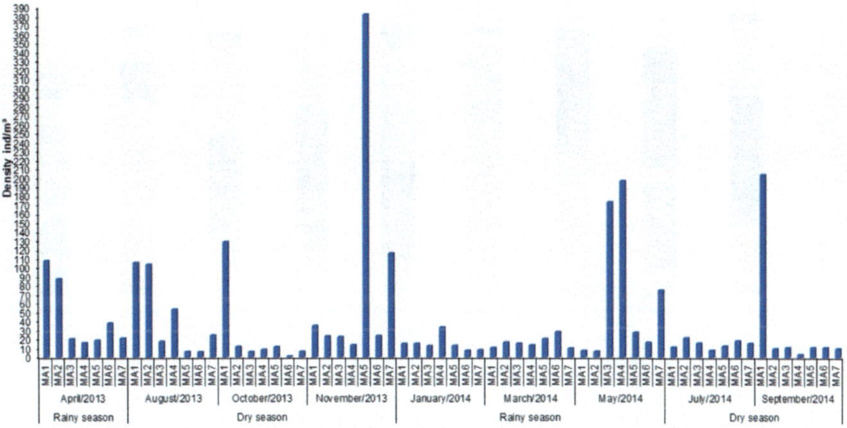

Fig. 7.17 Density (ind./m^3) of the zooplankton community in the continental shelf off São Luís (Maranhão State), from April 2013 to September 2014. This region has well-defined dry (July–December) and rainy (January–June) seasons

2013 (Fig. 7.17). This result enables the observation that seasonality is not an influencing factor on the increase or decrease in the zooplankton density value, as both the highest and the lowest density values were recorded in both seasonal periods, the dry season (July, August, September, October, November) and the rainy season (January, March, April, May). Therefore, the reproductive period of organisms is suggested to be responsible for the variation in density.

7.4.2 Ichthyoplankton Community

Engraulids were dominant throughout almost the entire sampling period. *Anchoviella lepidentostole* (Fowler 1911) stood out as abundant in April 2013 (44% of captured larvae), August 2013 (43% of collected larvae), March 2014 (43% of the captured individuals), January 2014 (35% of the collected individuals), May 2014 (34% of the captured larvae) and July 2014 (11% of the collected larvae) (Fig. 7.18). Thus, *Anchoviella lepidentostole* presented its greatest abundance during the rainy months, a favorable period for its reproductive cycle (Gkanasos et al. 2019). However, *Anchoviella lepidentostole* was also abundant during a dry month (August 2013), which suggests a longer reproductive period for this species, with reproductive peaks throughout the year. The dominance of this species was also observed in the estuarine complex of São Marcos Bay and Bacanga Basin, located onshore of the São Luís shelf (Bonecker et al. 2007; Soares et al. 2014), which reinforces the importance of this well-distributed and commercially exploited species on the North and Northeast coast of Brazil (Piorski et al. 2009).

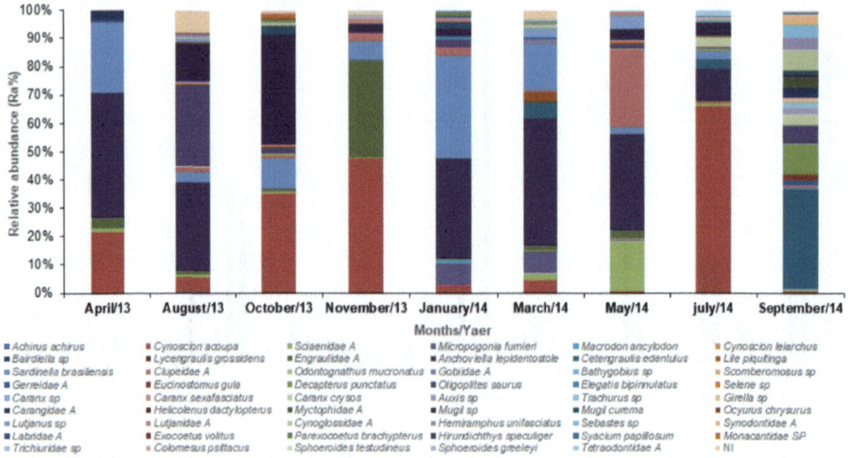

Fig. 7.18 Relative abundance (Ra%) of the ichthyoplankton community at the São Luís shelf from April 2013 to September 2014

Cynoscion acoupa (Lacepède 1801), also called "pescada amarela", a species of great economic importance, was abundant in the months of July 2014 (66% of the captured larvae), November 2013 (47% of individuals collected) and October 2013 (36% of individuals collected) (Fig. 7.18), indicating that both the rainy and the dry months were favorable for its life cycle, with its highest abundance value occurring in the rainy months, thus suggesting this period as its reproductive peak. This species had the highest percentage of abundance, reinforcing its importance as a fishery resource. It is economically important in the North and Northeast regions of Brazil, with high preference in the consumer market of Maranhão (Barletta-Bergan et al. 2002a, b; Matos and Lucena 2006; Zacardi 2015; Almeida et al. 2016). The presence of this species in its larval stage on the São Luis shelf indicates the need for better ecological knowledge of the species from the larval stage to implement adequate management and conservation actions.

Other species that also occurred abundantly were *Mugil* sp. (Linnaeus 1758) in October 2013 (39% of the captured larvae) and August 2013 (18% of the collected larvae); *Oligoplites saurus* (Bloch and Schneider 1801) reached greater abundance in August 2013 (39% of the individuals captured); *Sardinella brasiliensis* in January 2014 (36% of the individuals collected) and in April 2013 (25% of the captured larvae) (Fig. 7.18). It is noteworthy that the species mentioned above are characteristic of this region and important in fish production (Almeida 2008; Piorski et al. 2009; Nunes et al. 2019). Mugilids are euryhaline, and they seek marine waters with greater salinity to spawn. Even though there is divergence from the reproductive period of mugilids, studies of the *Mugil curema* show that their period of reproduction is related to greater food availability, which lasts from August to January (Albieri 2009). These months coincide with the recording of the greatest abundance of *Mugil* sp. in our campaigns. Environmental variability and seasonality can provide these individuals with favorable conditions for their development.

Oligoplites saurus have marine origins and use coastal areas in their early life stage as strategies to maintain populations and fish stocks (Barletta et al. 2016). These strategies of using different habitats possibly favored the higher abundance value of these individuals at more coastal sampling stations (MA1–MA3) in August 2013. These results corroborate Badú (2019), who registered a greater contribution of *Oligoplites saurus* in coastal areas close to the entrance of estuaries.

Sardinella brasiliensis, according to Vazzoler and Rossi-Wongtschowski (1976), presents a long period of reproduction with successive spawning throughout the year, as it has a split spawning strategy; thus, each female releases several oocytes in a single season and a frequency of spawning that takes place between 4 and 11 days (Isaac-Nahum et al. 1983; Dias 1989). This allows us to infer that January is the peak of spawning of this species, in which the highest density value was recorded.

The density of the eggs recorded in 2013 and 2014 ranged from 0.0 to 169.2 eggs/100 m^3. The highest density value was registered in May 2014. The lowest values were recorded in all months during 2014. It is worth noting that the lowest fish egg density values occurred at station MA1 (Fig. 7.19). This may be related to catadromous species, mature individuals who migrate to spawn in the ocean (Ibáñez and Gutierrez 2004).

For fish larvae, the density values recorded during the sampling period ranged from 0.0 to 201.6 larvae/100 m^3. The highest value was registered in July 2014, and the lowest was registered in May 2014. October 2013, November 2013 and September 2014 were the months with the lowest density values (Fig. 7.19).

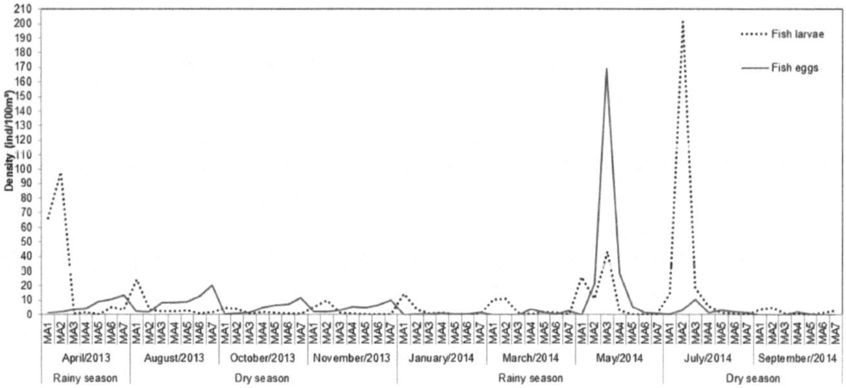

Fig. 7.19 Density (eggs/100 m^3; Larvae/100 m^3) of the ichthyoplankton community from the São Luis shelf from April 2013 to September 2014. The region has well-defined dry (July–December) and rainy (January–June) seasons

The density results reinforce the conclusion about the reproductive peak of *Anchoviella lepidentostole* in April 2013 and May 2014, while July 2014 was favorable for the development of *Cynoscion acoupa*. Both presented higher relative abundances. This reinforces the importance of the rainy season for the reproduction and development of the life cycle of these fish species.

7.5 Spatial, Seasonal and Interannual Dynamics of Zooplankton Biomass and Biodiversity Along a Large-Scale Transect off Pará, Northern Brazil

The Amazon Continental Shelf (ACS) is an area of high contrasts and extraordinary proportions. This region is home to the largest continuous area of mangroves in the world and the largest freshwater discharge system in the world's oceans. It produces a basin-scale, extensive estuarine plume, in which it would be unlikely to find reefs. However, the Great Amazon Reef System (GARS), which was recently discovered on the outer shelf, indicates that there is still much to discover in this area that is so important to humanity. However, we know very little about zooplankton in this region. Of the total 194 adult crab species reported for this area, only 25% of the species have fully described planktonic larval stages, and 62% have no morphological descriptions of any planktonic larval stage (Lima and Martinelli-Lemos 2019), which makes it difficult to identify and, consequently, understand larval ecology. These statistics are no different from those of other taxa. Here, we report new data on larval occurrence and try to elucidate strategies of larval dispersion adopted by key zoo- and ichthyoplankton taxa. This information is essential for understanding life histories at the ACS.

The ACS is an exceptionally energetic coastal region subjected to strong seasonal variability, including the discharges of the Amazon River and transport by the North Brazil Current and trade winds (Silva et al. 2010). The Amazon River flows vary from ~120 × 10^3 m^3 s^{-1} in December to ~300 × 10^3 m^3 s^{-1} in May (mean of ~206 × 10^3 m^3 s^{-1}). For the Amazon flow, we must add the Pará River (i.e., the sum of Pará, Tocantins, and Araguaia flow contributions, with a total mean of ~12 × 10^3 m^3 s^{-1}) that empties south of Marajó Island (Callède et al. 2010; Moura et al. 2016; Fig. 7.20). The ACS is inserted into the so-called tidal embayment of the Amazon (Dominguez 2009) with a shoreline fringed by extensive mangrove forests that extend inland up to 40 km, many deep bays and large-scale estuaries. The outer shelf is characterized by extensive rhodolite beds and biogenic reefs at water depths of up to 100 m (Moura et al. 2016; Lavagnino et al. 2020).

The ACS is influenced by the trade winds almost year round, the macrotidal regime (~8 m), and the discharge of fresh water from rivers in the region, mainly from the Amazon River, which generates a large-scale estuarine plume on the shelf, with seasonally varying vertical and horizontal extension (Castro Filho and Miranda 1998). The ACS is a region of low atmospheric pressure, high rainfall, and high

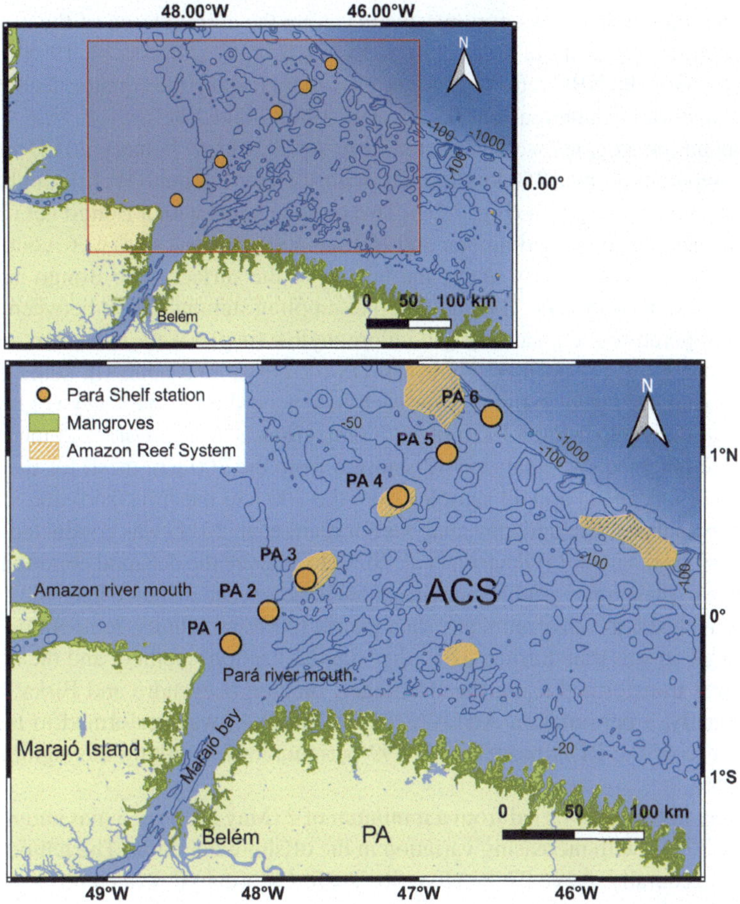

Fig. 7.20 Map of the Amazon Continental Shelf (ACS) showing the sampling stations (PA1–PA6). Additionally, depicted are existing reefs and mangrove forests occurring in the region

atmospheric humidity due to its proximity to the Intertropical Convergence Zone (ITCZ) (Bittencourt 2004). The ITCZ regulates the rainfall regime in the equatorial region and defines the dry and rainy periods in the Amazon. For instance, from January to June, heavy rains are recorded due to the southward displacement of the ITCZ. From September to November, the rains are less frequent and intense due to the northward displacement of the ITCZ (Silva et al. 2007). This rainfall regime affects other parameters, such as salinity and temperature, in the ACS (Silva et al. 2007).

The dominant oceanic current in the region is the North Brazil Current (NBC), which is generated from the meeting between the North Brazil Undercurrent (NBU), is fiormed by the northern bifurcation of the South Equatorial Current, and the South Equatorial Central Current (Stramma and Schott 1999). The NBC is a current of strong magnitude that flows northwestwards along the entire region of the northern continental shelf, mainly in areas close to the shelf break (Johns et al. 1998). From July to January, the NBC retroflects to the east after crossing the equator, feeding the North Equatorial Countercurrent (Flagg et al. 1986; Field 2005).

Zooplankton samples were collected from April 2013 to January 2015, totaling seven campaigns along a 250 km transect, with six fixed stations (PA1 through PA6) occupied in each campaign (Fig. 7.20). Two oblique hauls with a Bongo net (2 m in length with 60 cm opening diameter and 200 and 300 μm mesh size) were conducted at each station from slightly above the bottom to the surface. The Bongo net was equipped with flow meters. The depth of the sampling stations varied between 9 and 90 m. Temperature (°C), salinity and chlorophyll-*a* (μg/L) were measured at each location using a CTD profiler (Hydrolab DS 5). Larval and adult (in some cases) densities (ind./m^3) were estimated for each haul according to its water volume. At the end of the campaigns, 84 samples (7 campaigns × 6 locations × 2 mesh size nets) were obtained. We analyzed 42 samples from the 200 μm mesh for the identification of Decapoda and 42 samples from the 300 μm mesh for zoo- and ichthyoplankton analysis. Redundancy analysis (Borcard et al. 2011) was conducted using the 'vegan' package in R (Oksanen et al. 2020) to analyze the decapod community in relation to explanatory variables (environmental factors and distance from the coast). For this analysis, the environmental data matrix (outflow, salinity, temperature, and chlorophyll-*a*) was transformed using the 'standardization' method, and the density matrix was transformed using the 'Hellinger' method (Legendre and Birks 2012). Subsequently, a permutation ANOVA (PERMANOVA) was performed to test the significance of the RDA model, with 999 permutations, adopting the significance level of 5%.

Salinity in the ACS varied from a minimum of 2 (May 2014) to a maximum of 38 (January 2015), with increasing variation in the offshore direction. The temperature remained generally above 27 °C along the ACS. Chlorophyll-*a* ranged from 0.3 to 85.1 μg/L in January 2014 and showed a decreasing trend with increasing distance from the coast. The river flow recorded in the year of collection was higher than the average recorded in the last 30 years (see details in Silva et al. 2021).

7.5.1 Zooplankton

The mesozooplankton community (0.2–20 mm) is commonly represented by most of the existing animal phyla and forms a diverse and abundant group of organisms, mainly in the tropical region. In the ACS, the phylum Arthropoda was dominant and was mainly represented by Copepoda and Decapoda. However, other groups were found frequently but with lower abundances, such as the medusoid stage of hydrozoans, polychaete larvae, holoplanktonic mollusks, chaetognaths and pelagic tunicates. Other groups appeared sporadically in certain regions and periods forming blooms that were probably related to the reproductive phases of some invertebrates belonging to the phyla Bryozoa, Nemertea, Brachyopoda and Echinodermata, and

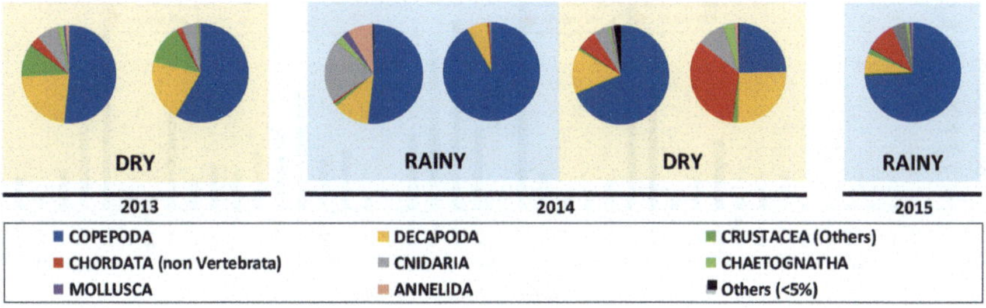

Fig. 7.21 Zooplankton relative abundance for dry and rainy seasons on the Amazon Continental Shelf from July 2013 to January 2015 on a quarterly basis

some vertebrates were represented mainly by larvae and eggs of fishes. More detailed results on Decapoda composition and ichthyoplankton will be presented later in this section.

The relative abundance of zooplanktonic groups reflected a high variability over the 3-year period analyzed (2013–2015), without observing a specific pattern associated with seasonality (Fig. 7.21). Copepoda represented more than 50% of the organisms in 6 of the 7 months analyzed, characterizing this group as the most abundant in the Amazon Continental Shelf. The relative abundance of Copepoda in May 2014 represented 92% of the total community. The only campaign in which Copepoda was not dominant was in October 2014, where planktonic tunicates were the most abundant group, with 33.37%, followed by Decapoda (25.7%) and copepods (24.6%) (Fig. 7.21).

Appendicularians were responsible for the highest density of planktonic tunicates at a sampling station close to the shelf break (PA5). High appendicularian densities have been recorded in other studies in the Amazon basin (Neumann-Leitão et al. 2018). *Oikopleura dioica* is an indicator species of the Amazon River plume (Neumann-Leitão et al. 2018). Decapoda larvae were also a frequent group (98%) and abundant in the ACS, with relative abundance values ranging from 6.5 to 25.7%.

The total density of mesozooplankton during the analyzed period presented values between 4.34 and 6399 ind./m^3 (Fig. 7.22). The minimum and maximum values of abundances were registered in May 2014 (rainy season). There was a significant difference among campaigns (ANOVA, F = 4.903, $p = 0.00745$). The larval density in July 2013 (beginning of the dry season), 9.22–16.8 ind./m^3, was significantly lower than that in October (dry period) and January (rainy period).

Marked density peaks were always observed during the rainy season at sampling stations PA1 and PA2, which were located closer to the mouth of the Pará River (Fig. 7.21). The most marked peak was recorded in May 2014, corresponding to a bloom of the copepod *Acartia tonsa* at sampling station PA1, reaching a density of 6300 ind./m^3, representing the highest relative abundance and the highest density of the group and consequently of the community. *Acartia tonsa* is an abundant copepod of estuarine regions (Bradford-Grieve et al. 1999). In the Curuçá estuary (Amazonian coast, northeast of Pará State), it was the most abundant copepod, mainly in

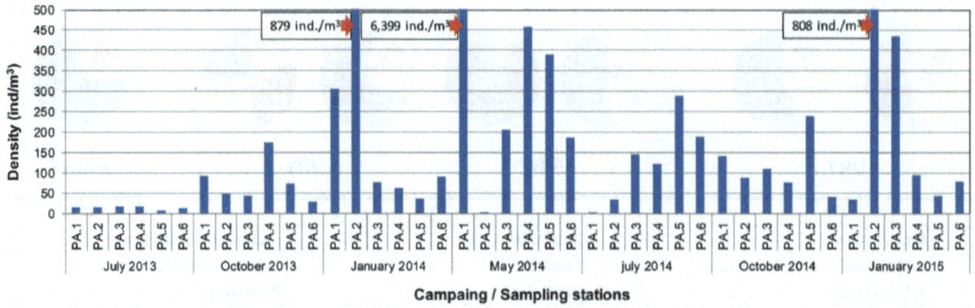

Fig. 7.22 Density (ind./m^3) of the zooplankton community on the Amazon Continental Shelf from July 2013 to January 2015

the rainy season, showing an inverse correlation with salinity and evidencing its preference for estuarine waters (Magalhães et al. 2009). Copepoda show the highest richness, diversity and abundance in almost all coastal and oceanic waters of the world (Bradford-Grieve et al. 1999). In the Amazon region, Copepoda showed itself as a key component of the zooplankton community and of the environmental conditions, being important to establish the patterns of the composition and abundance of species for the region.

7.5.1.1 Decapod Composition

Decapod crustaceans represent an important group for marine ecosystems both in their larval phase, as a component of the plankton and trophic pyramid, and in their benthic or nektonic phase, where they act as omnivores or detritivores that play an important role in nutrient cycling and energy flow, serving as the main energy pathway from primary producers to higher trophic level organisms, such as fish (Lima et al. 2021; Rodrigues-Inoue et al. 2021; Silva et al. 2021). A total of 23 taxa were identified along the ACS: Caridea (not identified families); Luciferidae, Penaeidae, Sergestidae, Sicyoniidae and Solenoceridae (Dendrobranchiata); Axianassidae and Upogebiidae (Gebiidea); Callichiridae and Callianassidae (Axiidea); Diogenidae, Paguridae, Albuneidae, Porcellanidae, Galatheidae (Anomura); Sesarmidae, Pinnotheridae, Calappidae, Grapsidae, Panopeidae, Leucosiidae, Ocypodidae and Portunidae (Brachyura). Except for some taxa, more than one larval stage and sometimes distinct stages were found, with occurrence of the entire larval series (zoea and megalopa) for some groups. There was a remarkable difference between families (Fig. 7.23), and this difference is explained mainly by salinity and chlorophyll-*a* (Fig. 7.24).

Seasonal events such as upwelling and tidal and plume fronts affect the food chain by temporarily changing nutrient inputs (Brandão et al. 2020).

Caridean shrimp present an opposite pattern (closer to the coast) when compared to penaeid shrimps and crabs (away from the coast), and anomurans and talassinoids occur in more specific places and months (Fig. 7.23).

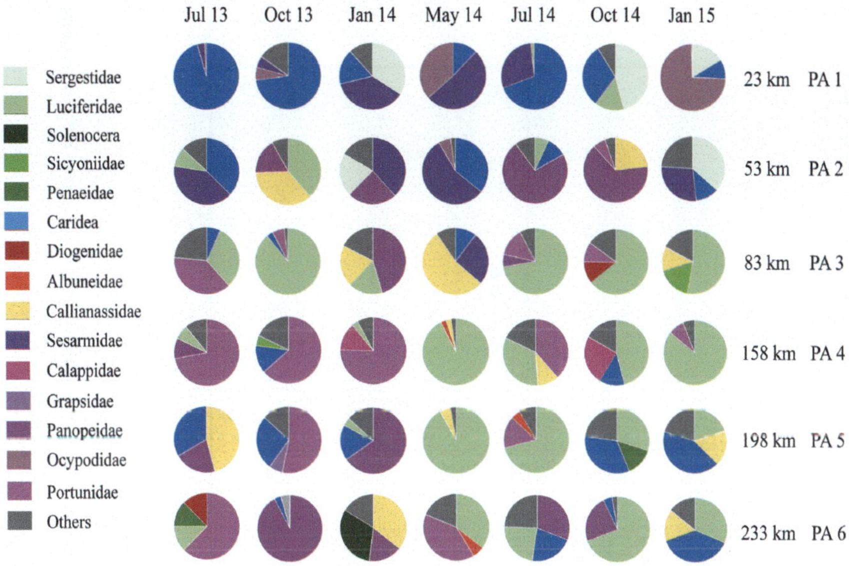

Fig. 7.23 Frequency of each Decapoda family at the six sampling sites from July 2013 to January 2015 surveyed on the Amazon Continental Shelf. See Fig. 7.20 for location

The distribution of Decapoda was significantly correlated with explanatory variables. The RDA model explained $R^2 = 19\%$ ($R^2_{\text{nonadjusted}} = 29\%$; $F = 2.81$; $p = 0.001$) of the total variation. The first two RDA axes together explained 79.3% of the variability; the first axis explained 55.4% of the variability and was positively related to salinity (0.94) and coastal distance (0.69) and negatively related to chlorophyll-*a* (−0.67). The second axis explained 23.9% of the variability and was positively correlated with temperature (0.70).

7.5.2 Ichthyoplankton Community

In general, there was a low density of fish eggs in the study area. The highest densities were observed during the rainy period, with 2342.67 eggs/100 m^3 (January 2014) and 1128.78 eggs/100 m^3 (January 2015), mainly, at most sample sites (Fig. 7.25).

This high egg density in a period of higher rainfall is probably associated with the species that seek the coast to spawn due to the low salinities caused by the influence of the Amazon River discharge in the region (Barletta-Bergan et al. 2002a, b; Zacardi et al. 2014). Fish larval densities were also found to be low, except at the coastal stations, with 1,035.56 larvae/100 m^3 during the rainy season (January 2014) and 1,043.38 larvae/100 m^3 during the dry season (July 2014). These peaks of larval density are associated with the important presence of three fish families, Gobiidae and Engraulidae in the period of higher rainfall and Clupeidae in the month of low

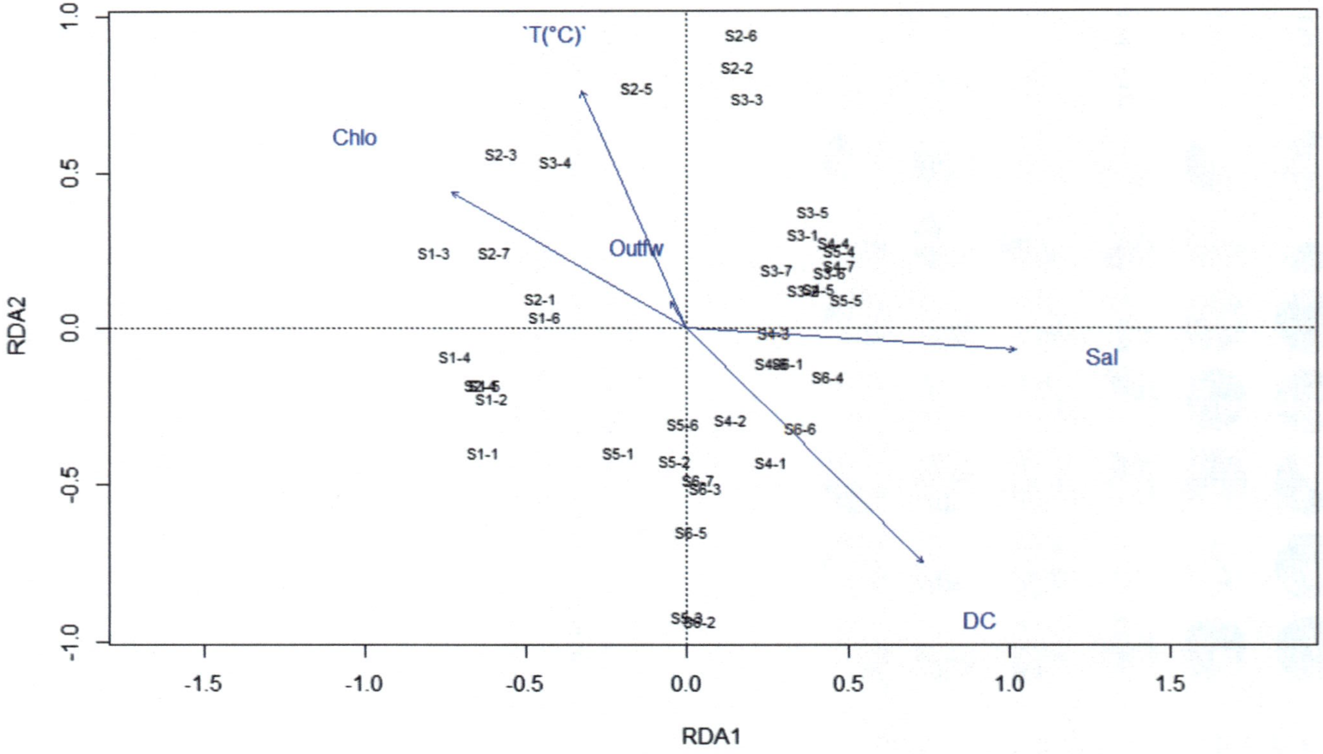

Fig. 7.24 Redundancy analysis for sampling sites and environmental parameters: T (°C) = temperature; Chlo = chlorophyll-*a*; Sal = salinity; Outfw = outflow and DC = distance from the coast (km)

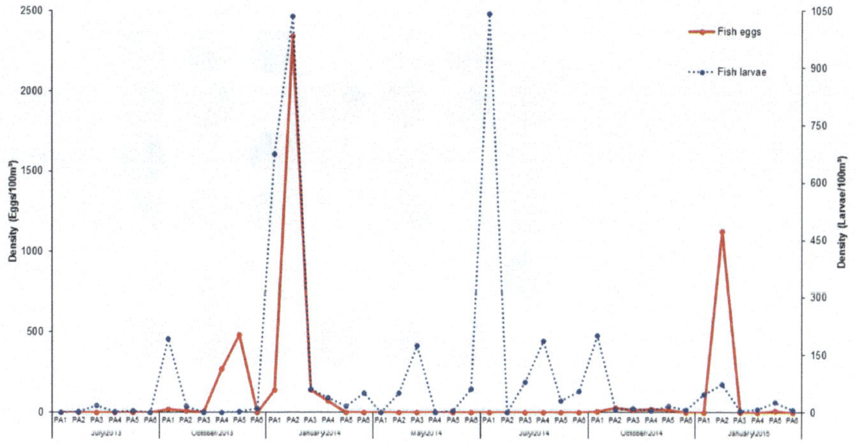

Fig. 7.25 Density (ind./100 m^3) of the ichthyoplankton community from the Amazon Continental Shelf, Brazil, from July 2013 to January 2015

rainfall in the region, which is probably related to the reproductive period of these groups on the northern coast of Brazil.

Larvae of 42 families of fishes were registered: Elopidae, Nettastomatidae, Congridae, Ophichthidae, Clupeidae, Engraulidae, Phosichthyidae, Synodontidae, Bregmacerotidae, Ophididae, Hemiramphidae, Exocoetidae, Belonidae, Holocentridae, Syngnathidae, Trichiuridae, Serranidae, Pomatomidae, Carangidae, Bramidae, Lutjanidae, Haemulidae, Gerreidae, Pomacanthidae, Pomacentridae, Mugilidae, Scaridae, Callyonimidae, Gobiidae, Ptereleotridae, Eleotridae, Blenniidae, Sparidae, Acanthuridae, Sciaenidae, Scombridae, Istiophoridae, Achiridae, Bothidae, Paralichthyidae, Monacanthidae and Tetraodontidae. Clupeidae, Engraulidae, Gobiidae, Sciaenidae, Carangidae, Haemulidae and Pomatomidae were the most abundant families in the study area, differing between the sampling periods (Fig. 7.26).

A clear pattern of spatial distributions of the abundance of families was observed (Fig. 7.26). The families Engraulidae and Sciaenidae were more abundant at the more nearshore stations, while Gobiidae, Carangidae and Haemulidae stood out at the stations farther from the coast. This is probably related to the fact that Sciaenidae, Clupeidae and Engraulidae are considered estuarine-dependent families, present with high frequency and in some cases at high densities in tropical and subtropical estuarine ecosystems (Blaber 2000; Barletta-Bergan et al. 2002a, b), using mangroves extensively as nursery sites. Although the family Gobiidae is regarded as a resident in most estuaries in northern and northeastern Brazil (Andrade et al. 2020), high densities of this family (individuals in later larval stages) were observed at stations farther from the coast, showing that gobiids have different ecological habits, most of them being estuarine dependent (Gomes et al. 2014).

Carangid fishes inhabit marine and estuarine waters, and their larvae are commonly found in studies of ichthyoplankton in coastal environments along the

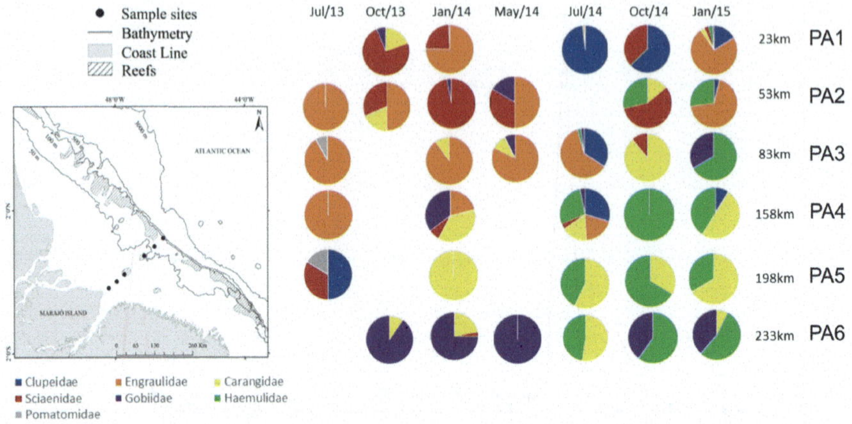

Fig. 7.26 Relative abundance (%) of the main fish families based on the distribution of their larvae on the Amazon continental shelf between July 2013 and January 2015. GARS shapefile provided by Moura et al. (2016)

Brazilian coast. This is because they use different spawning strategies to reduce the co-occurrence of the larvae, which differ between species (Souza and Mafalda Jr 2008). Mangroves and seagrass beds also serve as nursery grounds and provide refuge from predators and an abundance of food for many larvae and juvenile fishes of the Haemulidae family, which are typically found on coral reefs as adults (Burkepile and Hay 2008).

The results presented above support the hypothesis that in the Western Tropical Atlantic, temporal and spatial differences in both composition and abundance appear to be associated with differences in geological and environmental conditions. However, the relationship between zooplankton groups and the several factors that influence the ACS still needs to be clarified. The key groups in the region are the holoplanktonic Copepoda and Tunicata and Decapoda larvae and fish eggs and larvae. These last two groups are of great economic importance, making the ACS an important fishing ground in the country. Despite the low density of fish eggs and larvae of ichthyoplankton, a clear spawning pattern is observed during periods of higher rainfall in the region, strongly caused by the influence of the Amazon River discharge, in addition to a significant dominance of coastal/estuarine fish families. We suggest long-term studies of the ACS aimed to survey larval ecology and taxonomy to contribute to the planning of conservation measures on the Amazon shelf.

7.6 Mesoscale Distribution of Zooplankton off the Oceanic Islands in the Tropical Western Atlantic

In this section, we describe and compare the zooplankton and zooneuston communities from three pelagic environments off oceanic islands in the Western Tropical Atlantic comprising the marine protected areas off FNA, RA and SPSPA (Fig. 7.27). SPSPA exhibits a dry season from June to September and a rainy season from January to May (Souza et al. 2013), whereas FNA and RA have a dry season from August to January and a rainy season from March to July (Lira et al. 2014, 2017; Assunção et al. 2016; Santana et al. 2018). The waters off FNA, RA and SPSPA are oligotrophic, flowing mostly westward under the influence of the north branch of the South Equatorial Current (nSEC) around SPSPA and the central branch of the South Equatorial Current (cSEC) around FNA and RA (Travassos et al. 1999; Araujo and Cintra 2009; Tchamabi et al. 2017). The interaction between the topography of the oceanic islands and these currents results in thermohaline disturbances, vortices, wakes and possibly local mechanisms of upwelling (Araujo and Cintra 2009; Tchamabi et al. 2017).

Samples were collected during three oceanographic campaigns (July and August 2010, September and October 2012 and July and September 2014) aboard the oceanographic vessel "Cruzeiro do Sul/H38" as part of the "Camadas Finas" project along two transects (upstream-north transect and downstream-south transect of the areas) (Fig. 7.27) during day and night. The exact locations of the sampling stations (up- and downstream) were defined on board based on ADCP surface current vector data. Oblique hauls were performed from 200 m depth up to the surface, maintaining

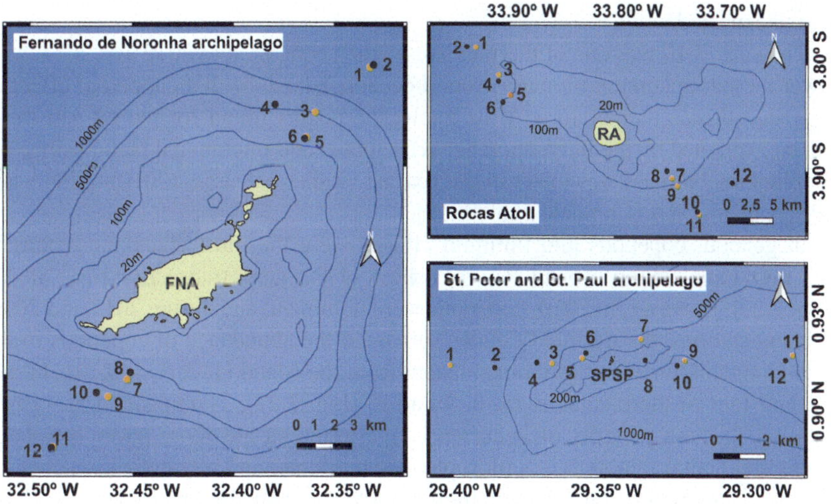

Fig. 7.27 Sampling stations in the Western Tropical Atlantic Ocean at three tropical island environments Saint Peter and Saint Paul Archipelago (SPSP); Rocas Atoll (RA) and Fernando de Noronha Archipelago (FNA)

a wire angle close to 45°, using a conical bongo plankton net. The plankton nets had 64 μm (microzooplankton) and 500 μm (macrozooplankton) mesh sizes, with a 0.6 m mouth opening diameter and 2.5 m length.

Zooneuston was collected with a David-Hempel aluminum catamaran (Hydro-Bios, Kiel, Germany) equipped with two superposed nets, each with a rectangular mouth (29.3 cm wide × 15.2 cm deep) and 500 μm mesh size. The haul duration was 20 min with a speed of 2–3 knots. The upper net was structured to be exactly centered at the air-water interface, thus sampling the epineuston (surface) layer from the surface to a depth of 7.6 cm, while the lower net sampled the hyponeuston (subsurface) layer from a depth of 7.6–22.8 cm. For all Bongo and Catamaran plankton nets, a flowmeter (Hydro-Bios, Kiel, Germany) was used to estimate the filtered volume. All samples were preserved with 4% formaldehyde and buffered with sodium tetraborate (0.5 g L^{-1}) (Lira et al. 2014).

7.6.1 Total Microzooplankton Abundance and Biomass at Rocas Atoll

In general, mainly in 2010, higher total zooplankton abundance and biomass were found at stations closer to the islands during nighttime (Fig. 7.28). These higher values near islands can probably be explained by the island mass effect (Doty and Ogury 1956). During 2012, a peak of total abundance and biomass was found in samplings more distant and upstream from the islands, which was surprising and deviates from the expected island mass effect. The highest values during 2010 were also unexpected.

The year 2010 was one of thermal stress when compared to 2012 and 2014, with lower mesozooplankton biomass around oceanic islands with an apparent effect of high sea surface temperature. Most likely, the results found for the 64-μm mesh can be explained by the interaction between spatial and interannual factors since the zooplankton biomass was lower in the year under thermal stress (2010) in FNA and SPSPA than in RA (Campelo et al. 2019).

In general, copepods and tintinnid ciliates were the most abundant groups in the microzooplankton around RA regardless of the year, transect or time of day (Fig. 7.29). However, in 2010, copepods were dominant, and during 2012 and 2014, at many stations, the tintinnid ciliates surpassed the abundance of copepods in the microzooplankton at some stations distant from the island (1, 2, 3, 4, 9, 10, 11 and 12) and four onshore stations (5, 6, 7 and 8) (Fig. 7.27). Thompson et al. (1999) mention that the patterns of occurrence of Tintinnina are more strongly related to environmental factors (i.e., chlorophyll-a, nutrients, and phytoplankton), which might be an explanation for their high abundance at the nearshore stations. This taxon forms aggregations and assumes strategic behavior when needed, increasing the reproductive rate and persisting in favorable areas (Nogueira et al. 2008).

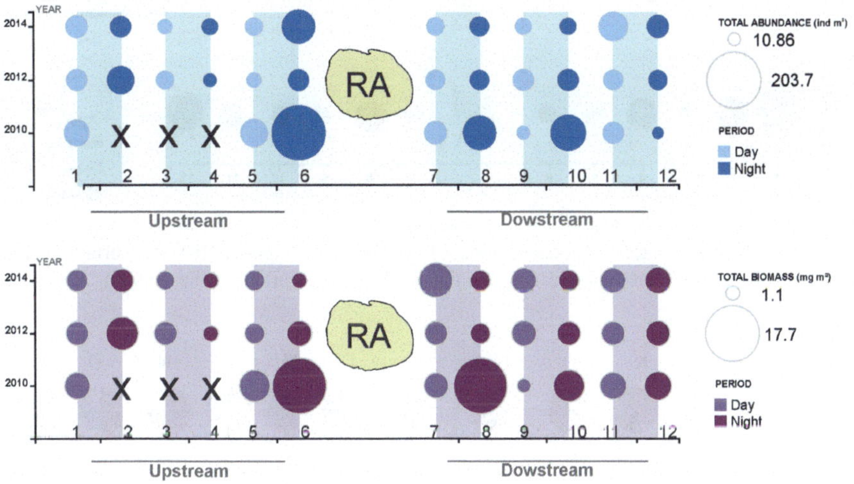

Fig. 7.28 Zooplankton total abundance and biomass for the years 2010, 2012 and 2014 around Rocas Atoll (RA). The surface area of each circle is proportional to abundance (or biomass). X: no samples collected. See Fig. 7.27 for the locations of the sampling stations

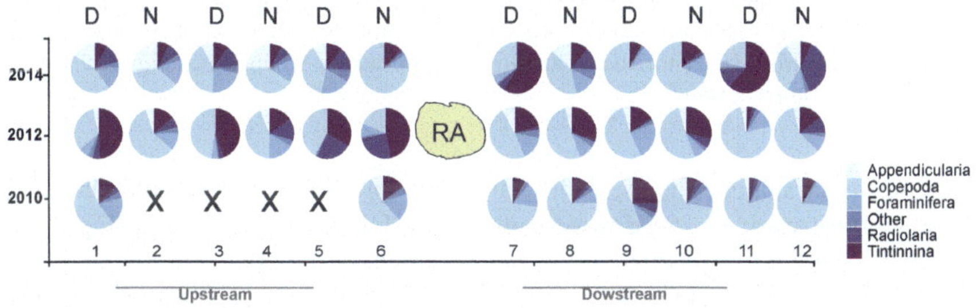

Fig. 7.29 Relative abundance (%) of total microzooplankton collected with a bongo net (64 μm mesh size) between 0–200 m off RA (July and August 2010, September and October 2012 and July and September 2014). See Fig. 7.27 for the locations of the sampling stations

7.6.2 *Macrozooplankton Relative Abundance at the Fernando de Noronha Archipelago*

The total zooplankton abundance around the FNA did not show significant differences in abundance between the up- and downstream areas or between the daytime and nighttime stations (Fig. 7.30). However, some groups, such as fish eggs, were significantly more abundant downstream, indicating the existence of a biomass island effect (Doty and Ogury 1956; Meerhoff et al. 2018).

The dominant zooplankton group, in terms of abundance, was calanoid copepods (Fig. 7.31, Calan, pink color) (upstream: 66.1%, downstream: 52.1%), mainly during the night period (N). Furthermore, gelatinous organisms (mainly chaetognaths,

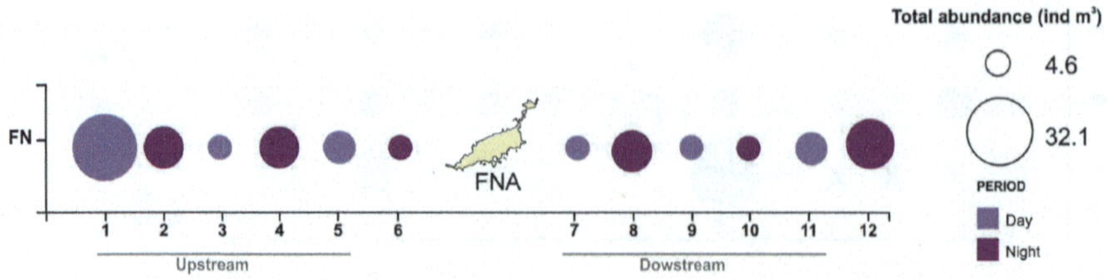

Fig. 7.30 Zooplankton total abundance around FNA (July and August 2010). The surface area of each circle is proportional to abundance. See Fig. 7.27 for the locations of the sampling stations

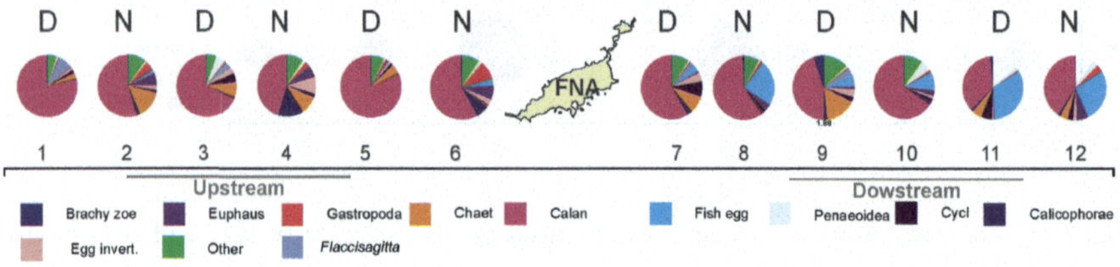

Fig. 7.31 Relative abundance (%) of total macrozooplankton collected with a bongo net (500 μm mesh size) between 0 and 200 m off the Fernando de Noronha Archipelago (July and August 2010). D: Day; N: Night. Zooplankton: Brachy zoeae (Brachyuran zoeae); Calan: (copepods Calanaoida); Calicophorae; Chaet. (Chaetognatha) Egg invert. (eggs of invertebrates); Euphaus (Euphauseacea); *Flaccisagitta*; Gastropoda; Penaeiodea; Other (other zooplankton groups). See Fig. 7.27 for the locations of the sampling stations

hydromedusae, siphonophores, thaliaceans, and appendicularians) were also very abundant (upstream: 15.2%, downstream: 8.6%), followed by fish eggs (Fig. 7.31, blue color), which were particularly abundant downstream (upstream: 0.5%, downstream: 20.9%). García et al. (2005) argue that for fish spawning areas (downstream), aggregation may be due to the interaction of strong currents with the island topography, generating strong lateral frictional contact that can induce mesoscale eddies and convergent fronts. In FNA, we observed a similar process and wakes in downstream areas (Tchamabi et al. 2017). Studies using oceanic current models to understand this larval island effect are crucial off FNA to support management of the stocks of local populations and because FNA seems to be an important source of larvae for the tropical Atlantic.

7.6.3 *Zooneuston Abundance*

The total zooneuston abundance around the islands did not show significant differences in abundance between up- and downstream areas (Fig. 7.32). However, there

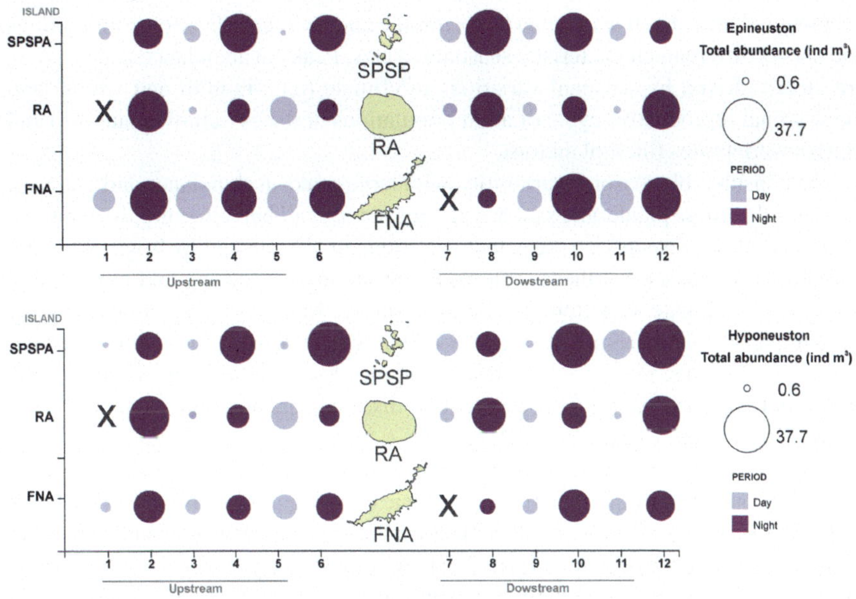

Fig. 7.32 Zooneuston total abundance for Saint Peter and Saint Paul archipelago (SPSP); Rocas Atoll (RA) and Fernando de Noronha archipelago (FNA). The surface area of each circle is proportional to abundance. See Fig. 7.27 for the locations of the sampling stations

was significantly higher abundance during nighttime for both layers of neuston, indicating Diel vertical migration (DVM) for both layers (Meerhoff et al. 2018). Similar results were recorded by Lira et al. (2014) for the zooneuston community off FNA.

7.7 The Future of Pelagic Ecosystems off Northern and Northeastern Brazil in the Context of Climate Change—Scenarios, Challenges, Research Gaps, and Perspectives

The transect results have brought new insights into the seasonal and spatial variability of zoo- and ichthyoplankton in unique and very different Brazilian tropical marine ecosystems. These study areas include two extremely productive ecosystems: the wide and highly productive Pará and Maranhão shelves under the influence of the North Brazil current, with macrotidal mangroves and huge continental runoff, with estuarine plumes on scales of dozens to hundreds of kilometers. Conversely, all other ecosystems analyzed here are characterized by extreme oligotrophy, such as the Brazil Current system off Salvador (Bahia), the oligotrophic system at the narrow continental shelf off Tamandaré (Pernambuco) and the oceanic islands in the Western

Tropical Atlantic. Each study area bears unique and different plankton communities and shows different, characteristic seasonal cycles. Peaks in abundance and biomass are clearly driven by seasonal variations in climate (e.g., rainfall and winds) and the seasonal reproductive cycle of adult populations of fishes (ichthyoplankton) and macroinvertebrates (meroplankton).

The considerable seasonal variability, with large peaks in abundance and biomass, indicates that these planktonic systems are not stable at all but rather highly dynamic, with very strong responses to seasonal variations in climate and hydrography. This observation indicates that these systems are prone to show strong and exacerbated responses to climate variations in the near future, functioning as "amplifiers" of climate signals. Such responses may be drastic changes in the productivity of the ecosystem, but they may also manifest as unprecedented shifts in the timing of the peaks and blooms, leading to a deleterious disarrangement in the food webs, the "trophic mismatch" (Thackeray 2012).

The expected ecosystem responses are highly complex and totally different depending on the study area. In tropical oceanic ecosystems (e.g., off oceanic islands and in the Brazil current waters off Salvador), the expected warming and deepening of the upper mixed layer (Roch et al. 2021) will most likely lead to increased stratification in the layers above the permanent thermocline and thus to a reduction in primary (Gittings et al. 2018) and secondary productivity in the next decades, with deleterious consequences for carbon sequestration, oceanic fish stocks (e.g., tuna and mackerels), seabirds, and other upper trophic levels.

For tropical pelagic ecosystems on the continental shelf, the situation is completely different, since they usually do not have a permanent thermocline. Here, the tides and wind-driven turbulence usually break up any strong thermal stratification, except for areas with estuarine plumes. For the nearshore shelf off Tamandaré (Pernambuco—Northeastern Brazil), coastal and estuarine processes are very important, especially continental runoff. Extreme events, such as very strong rainfall, have drastic consequences for these nearshore ecosystems. Such extreme events off northeastern Brazil (and off southeastern and southern Brazil) have considerably increased in intensity and frequency in recent decades, probably due to increased heat content in the Western Tropical Atlantic Warm Pool (Fonseca et al. 2017; Espinoza et al. 2021). Thus, it is very likely that such events will continue to increase in frequency and intensity under global warming. Such events with extremely strong rainfall have catastrophic consequences for vulnerable communities in coastal cities (e.g., due to landslides and flooding) and for nearshore marine ecosystems (e.g., by killing off organisms due to peaks in turbidity, sediment loads and low salinity in coastal coral reef ecosystems).

Understanding the complex pelagic ecosystems at mid-shelf and on the offshore shelves and continental slopes off northeastern Brazil is still a major challenge, and the prediction of the responses of these systems to future climate change is still beyond the scope of current understanding. These key offshore ecosystems, where important fisheries occur, have been investigated only in sporadic cruises and clearly deserve regular and intensive investigation.

For the Pará and Maranhão shelves, the dynamics of extensive mangroves, the physical and biogeochemical dynamics of the muddy shelf ecosystems and future rainfall variability in the whole Amazon and Tocantins-Araguaia basins are paramount to define the extent and dynamics of the complex basin-scale estuarine plumes. Additionally, the intensive fisheries in these areas, which are among the most important fishing grounds in Brazil, may play an important role in regulating these North Brazilian shelf ecosystems.

Continuous plankton monitoring time series (in the context of current and future research projects), including in diverse shelf areas and oceanic islands, are paramount for better predictions of the responses of tropical pelagic ecosystems to future climate change.

Acknowledgements The authors are greatly indebted to all fellow scientists, students and crew of small and medium-sized vessels for invaluable help during fieldwork at the transects across the continental shelves of Pará, Maranhão, Pernambuco and Bahia. The authors also thank the Brazilian Instituto Chico Mendes de Conservação da Biodiversidade (ICMBio) for SISBIO licenses no. 34067, 37482 and 44803. Many thanks to Brazilian CNPq, Brazilian CAPES, FACEPE (Pernambuco State) and FAPESB (Bahia State) for the fellowships granted to several authors. Many thanks to the Brazilian Navy for support on board NHO Cruzeiro do Sul (H-38) during the "Camadas Finas" campaigns off oceanic islands. This work was supported by the INCT AmbTropic—National Institute on Science and Technology for Tropical Marine Environments, CNPq/FABESB (565054/2010-4, 8936/2011, 465634/2014-1 and inc004/2019).

References

Advincula ACC (1999) Distribuição espacial do macrozooplâncton nas áreas do Arquipélago de São Pedro e São Paulo, Arquipelago de Fernando de Noronha e Cadeia Norte Brasileira. Master Thesis (Universidade Federal de Pernambuco, CCB. Biologia Animal), 61p

Albieri RJ (2009) Biologia reprodutiva da tainha *Mugil liza* Valenciennes e do Parati *Mugil curema* Valenciennes (Actinopterygii, Mugilidae) na Baía de Sepetiba, RJ, Brasil. M.Sc. thesis. Universidade Federal do Rio de Janeiro, Brazil

Almeida ZS (2008) Os Recursos pesqueiros marinhos e estuarinos do Maranhão: Biologia, Tecnologia Socioeconômico, estado de arte e manejo. Ph.D. thesis. Universidade Federal do Pará, Brazil

Almeida ZS, Santos NB, Sousa HL et al (2016) Biologia reprodutiva da pescada amarela (*Cynoscion acoupa*) capturada na Baía de São Marcos, Maranhão, Brasil. Biota Amazônia, Macapá 6(1):46–54. https://doi.org/10.18561/2179-5746/biotaamazonia.v6n1p46-54

Andrade A (2000) Ictionêuston do arquipélago de São Pedro e São Paulo (Programa Arquipélago). Monography. Universidade Federal Rural de Pernambuco, Brazil

Andrade ALRH de, Campos SS, Teixeira SF (2020) Population structure of Gobiidae in a tropical urban estuary. J Env Anal Progr 5:381–390. https://doi.org/10.24221/jeap.5.4.2020.3165.381-390

Angel MV (1985) Vertical migrations in the oceanic realm: possible causes and probable effects. In: Rankin MA (ed) Migration: mechanisms and adaptive significance. Port Aransas: Department of Zoology, University of Texas, Suppl 27, pp 45–70

Angel MV (1993) Biodiversity of the pelagic ocean. Conserv Biol 7:760–772

Araújo EM (2006) Siphonophora (Cnidaria) do Arquipélago São Pedro e São Paulo e da Cadeia Norte/Rocas/Noronha (Brasil). M.Sc. thesis. Universidade Federal de Pernambuco, Brazil

Araujo M, Cintra M (2009) Modelagem matemática da circulação oceânica na região equatorial do Arquipélago de São Pedro e São Paulo. In: Viana DL, Hazin FHV, Souza, MAC (eds) Arquipélago de São Pedro e São Paulo: 10 anos de Estação Científica. SECIRM, Brasília, pp 107–114

Araújo MCB, Costa MF (2003) Análise quali-quantitativa do lixo deixado na baía de Tamandaré—PE—Brasil, por excursionistas. RGCI 3:58–61

Araújo EP, Teles MGL, Lago WJS (2009) Delimitação das bacias hidrográficas da Ilha do Maranhão a partir de dados SRTM. In: Proceedings of XIV Simpósio Brasileiro de Sensoriamento Remoto. Instituto Nacional de Pesquisas Espaciais, Brazil, pp 25–30

Araújo-Lima CARM, Silva JVV, Petry P et al (2001) Diel variation of larval fish abundance in the Amazon and Rio Negro, Brazil. Braz J Biol 61:357–362. https://doi.org/10.1590/S1519-69842001000300003

Assunção RV, Silva AC, Martins J, Flores Montes M (2016) Spatial-temporal variability of the thermohaline properties in the coastal region of Fernando de Noronha archipelago, Brazil. J Coast Res 75:512–516. https://doi.org/10.2112/SI75-103.1

Badú MLAS (2019) Dinâmica ictioplanctônica em ambientes hipersalinos do semiárido brasileiro. M.Sc. thesis. Universidade Estadual da Paraíba, Brazil

Barletta M, Barletta-Bergan A (2009) Endogenous activity rhythms of larval fish assemblages in a mangrove-fringed estuary in North Brazil. Open Fish Sci Res 2:15–24. https://doi.org/10.2174/1874401X00902010015

Barletta M, Barletta-Bergan A, Saint-Paul U et al (2005) The role of salinity in structuring the fish assemblages in a tropical estuary. J Fish Biol 66:45–72. https://doi.org/10.1111/j.0022-1112.2005.00582.x

Barletta M, Blaber SJM, Craig JF (2016) Fish and aquatic habitat conservation in South America. J Fish Biol 89:1–3. https://doi.org/10.1111/jfb.13032

Barletta-Bergan A, Barletta M, Saint-Paul U (2002a) Structure and seasonal dynamics of larval fish in the Caeté River estuary in North Brazil. Estuar Coast Shelf Sci 54:193–206. https://doi.org/10.1006/ecss.2001.0842

Barletta-Bergan A, Barletta M, Saint-Paul U (2002b) Community structure and temporal variability of ichthyoplankton in North Brazilian mangrove creeks. J Fish Biol 61:33–51. https://doi.org/10.1111/j.1095-8649.2002.tb01759.x

Becker EC (2014) Comunidade planctônica, especialmente copépodos, da Plataforma Continental de Santa Catarina (26–29° S): associação com processos oceanográficos e estrutura de tamanho de fitoplâncton. M.Sc. thesis. Universidade Federal de Santa Catarina, Brazil

Bernal AR (1990) Variaciones nictemerales del zooplâncton y su relación con factores ambientales en la Bahia de Santa Marta, Caribe Colombiano. Monografia. Universidad Nacional de Colombia, Colombia

Beys-Da-Silva WO, Santi L, Dias JAG (2014) Mangroves: a threatened ecosystem under-utilized as a resource for scientific research. J Sustain Dev 7:40–51. https://doi.org/10.5539/jsd.v7n5p40

Bittencourt SCS (2004) Composição, distribuição e abundância do ictioplâncton da ZEE Norte coletado na região da costa do Amapá e plataforma do Amazonas. MSc thesis. Universidade Federal do Pará, Brasil

Bittencourt SCS, Zacardi DM, Souza SP et al (2007) Composição e distribuição de larvas de peixes na Zona Neustônica da Costa do Amapá (REVIZEE/SCORE Norte-Op. Norte II e III). Bol Téc Cient CEPNOR 7:41–54. https://doi.org/10.32519/tjfas.v7i1.1222

Björnberg TKS (1963) On the free-living Copepods of Brazil. Bol Inst Oceanogr 13(1):1–142

Björnberg TKS, Forneris L (1956) On the uneven distribution of the Copelata of the Alcatrazes area. Boletim do Instituto Oceanográfico 7:113–118

Blaber SJM (2000) Tropical estuarine fishes: ecology, exploitation and conservation. Blackwell Science, London

Bloch ME, Schneider JG (1801) M.E. Blochii, Systema Ichthyologiae iconibus cx illustratum. Post obitum auctoris opus inchoatum absolvit, correxit, interpolavit Jo. Gottlob Schneider, Saxo.

Berolini. Sumtibus Auctoris Impressum et Bibliopolio Sanderiano Commissum, pp i-lx + 1-584, Pls. 1-110. Available online at page(s): 32. https://www.biodiversitylibrary.org/bibliography/5750#/summary

Boltovskoy D (ed) (1981) Atlas del zooplancton del Atlantico Sudoocidental y métodos de trabajos con el zooplancton marino. INIDEP, Mar del Plata

Boltovskoy D (ed) (1999) South Atlantic Zooplankton, vol 2. Backhuys Publishers, Leiden

Boltovskoy D, Correa N, Boltovskoy A (2002) Marine zooplanktonic diversity: a view from the South Atlantic. Oceanologica Acta 25(5):271–278

Bonecker ACT, Castro MS (2018) Larval fish assemblages in the Foz do Amazonas Basin. Pan-Am J Aquat Sci 13(2):114–120

Bonecker ACT, Castro MS, Namiki AP et al (2007) Larval fish composition of a tropical estuary in northern Brazil (2° 18′–2° 47′ S/044° 20′–044° 25′ W) during the dry season. Pan-Am J Aquat Sci 2(3):235–241

Borcard B, Gillet F, Legendre P (2011) Numerical ecology with R. In: Gentleman R, Hornik K, Parmigiani GG (eds) Use R! series. Springer Science+Business Media, New York, pp 153–225

Böttger-Schnack R (1996) Vertical structure of small metazoan plankton, especially noncalanoid copepods. I. Deep Arabian sea. J Plankton Res 18:1073–1101. https://doi.org/10.1093/plankt/18.7.1073

Bradford-Grieve JM, Markhaseva EL, Rocha CEF et al (1999) Copepoda. In: Boltovskoy D (ed) South Atlantic Zooplankton, vol 2. Backhuys Publishers, Leiden, pp 869–1098

Brandão MC, Garcia CAE, Freire AS (2020) Meroplankton community structure across oceanographic fronts along the South Brazil Shelf. J Mar Syst. https://doi.org/10.1016/j.jmarsys.2020.103361

Brito-Lolaia M, Santos GS, Neumann-Leitão S et al (2020) Micro- and mesozooplankton at the edges of coastal tropical reefs (Tamandaré, Brazil). Helgol Mar Res. https://doi.org/10.1186/s10152-020-00539-4

Burkepile DE, Hay ME (2008) Coral Reefs. In: Jørgensen SE, Fath BD (eds) Encyclopedia of ecology. Elsevier, Amsterdam, pp 784–796

Callède J, Cochonneau G, Alves F, Guyo JL, Guimarães V, Oliveira E (2010) Les apports eneau de l'Amazone à l'Océan Atlantique. Rev Sci Eau 23:247–273. https://doi.org/10.7202/044688ar

Campelo RPDS, Diaz XFG, Santos G, Melo PAMDC, Melo Junior M, Figueiredo LGP, Neumann-Leitão S et al (2018) Small-scale distribution of the mesozooplankton in a tropical insular system. Braz J Oceanogr 66:15–29

Campelo RPS, Bonou FK, Melo Júnior M et al (2019) Zooplankton biomass around marine protected islands in the tropical Atlantic Ocean. J Sea Res 154:101810. https://doi.org/10.1016/j.seares.2019.101810

Campos CC, Garcia TM, Neumann-Leitão S et al (2017) Ecological indicators and functional groups of copepod assemblage. Ecol Indic 83:416–426. https://doi.org/10.1016/j.ecolind.2017.08.018

Cardoso RL (2014) Variabilidade espaço-temporal da comunidade ictioplanctônica em um perfil da plataforma continental maranhense. Monography. Universidade Federal do Maranhão, Brazil

Cardoso RC (2016) Mapeamento do Ictioplâncton na Costa Maranhense, Brasil. Monography. Universidade Federal do Maranhão, Brazil

Cardoso RL, Silveira PCA, Costa DSN (2021) Comunidade ictioplanctônica da zona de arrebentação das praias do Araçagy e Panaquatira, ilha do Maranhão, Maranhão, Brasil. Lat Am J Devt 3:1783–1799. https://doi.org/10.46814/lajdv3n4-008

Castro MF, Medeiros T, França EJ, Severi W (2008) Occurrence of early life stages of *Hirundichthys affinis* (Günther, 1866) and *Cheilopogon* sp. (Beloniformes, Exocoetidae) in a tropical estuary, Northeastern Brazil. Rev Bras Zoociênc 10:139–143

Castro Filho BM, Miranda LB (1998) Physical oceanography of the western Atlantic continental shelf located between 4° N and 34° S. In: Robinson R, Brink KH (eds) The sea, vol 11. Wiley, New York, pp 209–251

Cavalcanti EAH (2002) Macrozooplâncton da Zona Econômica Exclusiva do Nordeste do Brasil (REVIZEE NEII e NEIII) com ênfase em Copepoda. M.Sc. thesis. Universidade Federal de Pernambuco, Brazil
Cavalcanti EAH, Larrazábal MEL (2004) Macrozooplâncton da Zona Econômica Exclusiva do Nordeste do Brasil (segunda expedição oceanográfica—REVIZEE/NE II) com ênfase em Copepoda (Crustacea). Rev Bras Zoo 21:467–475. https://doi.org/10.1590/S0101-81752004000300008
Chuecas L (1998) Programa de monitoreo versus Programa de biomonitoreo del ambiente acuático. In: Arcos D (ed) Mineria del cobre, ecología y ambiente costero. Editorial Anibal Pinto, Chile, pp 407–450
Cialdi M, Secchi PA (1865) Sur la transparence de la mer. C R Hebd Séances Acad Sci 61:100–104
Cirano M, Mata MM, Campos EJ et al (2006) A circulação oceânica de larga-escala na região oeste do Atlântico Sul com base no modelo de circulação global OCCAM. Rev Bras Geofís 24(2):209–230. https://doi.org/10.1590/S0102-261X2006000200005
Codina JCU (2010) O Zooplâncton associado aos máximos subsuperficiais de clorofila na Plataforma Continental sudeste do Brasil. Ph.D. thesis. Universidade Federal do Paraná, Brazil
Conceição JL (2016) Composição e distribuição espacial da comunidade ictioplanctônica na baía do Arraial, Maranhão. Monography, Universidade Federal do Maranhão, Brazil
Conceição LR, Souza CS, Mafalda PO Jr et al (2021) Copepods community structure and function under oceanographic influences and anthropic impacts from the narrowest continental shelf of Southwestern Atlantic. Reg Stud Mar Sci 47:101931. https://doi.org/10.1016/j.rsma.2021.101931
Correia EP (2014) Migração vertical do microzooplâncton do Arquipélago de São Pedro e São Paulo. Master's thesis, Universidade Federal de Pernambuco
Costa DSN (2013) Influência da Temperatura, Salinidade e Oxigênio dissolvido sobre a Abundância e Distribuição das Larvas de Peixes na Área Portuária do Itaqui-Ma, Brasil. Monography, Universidade Federal do Maranhão, Brazil
Costa DSN (2017) Diversidade e zonação do Ictioplâncton em um perfil da Plataforma Maranhense. M.Sc. thesis. Universidade Federal do Maranhão, Brazil
Costa KG, Pereira LCC, Costa RM (2008) Short and long-term temporal variation of the zooplankton in a tropical estuary (Amazon region, Brazil). Bol Mus Para Emílio Goeldi Ciênc Nat 3:127–141
Costa AJG, Costa KG, Pereira LCC et al (2011) Dynamics of hydrological variables and the fish larva community in an Amazonian estuary of northern Brazil. J Coast Res SI 64:1–5
Costa AESF, Santana JR, Neumann-Leitão S (2018) Changes in microplanktonic protists assemblages promoted by the thermocline induced stratification around an oceanic archipelago. An Acad Bras Ciênc 90:2249–2266. https://doi.org/10.1590/0001-3765201820170607
Costa AESF, Santana JR, Melo PAMC et al (2019) Polycystine radiolarians within oligotrophic waters: higher abundance closer to tropical oceanic islands. Trop Ecol 60:261–270. https://doi.org/10.1007/s42965-019-00030-3
Cummings JA (1984) Habitat dimensions of calanoid copepods in the western Gulf of Mexico. J Mar Res 42:163–188. https://doi.org/10.1357/002224084788506121
Dantas PSB (2019) Microplâncton no entorno da ilha oceânica do Atol das Rocas, Atlântico Sudoeste-Brasil. Doctoral thesis. https://repositorio.ufpe.br/handle/123456789/38028
Dias JF (1989) Estimativa da Fecundidade Instantânea de *Sardinella brasiliensis* (Steindachner, 1879). In: Abstracts of I Simpósio sobre Oceanografia, Instituto Oceanográfico da Universidade de São Paulo, São Paulo
Dias CO, Bonecker SLC (2009) The Copepod Assemblage (Copepoda: Crustacea) on the inner Continental Shelf adjacent to Camamu Bay, Northeast Brazil. Zoologia 26:629–640. https://doi.org/10.1590/S1984-46702009000400007
Dominguez JM (2009) The coastal zone of Brazil. In Geology and geomorphology of holocene coastal barriers of Brazil. Springer, Berlin, Heidelberg, pp 17–51
Domingos-Nunes R, Resgalla C Jr (2012) The zooplankton of Santa Catarina continental shelf in southern Brazil with emphasis on Copepoda and Cladocera and their relationship with physical

coastal processes. Lat Am J Aquat Res 40:893–913. https://doi.org/10.3856/vol40-issue4-fulltext-7
Domingues EC, Schettini CAF, Truccolo EC et al (2017) Hydrography and currents on the Pernambuco Continental Shelf. RBRH. https://doi.org/10.1590/2318-0331.0217170027
Doty MS, Oguri M (1956) The island mass effect. ICES J Mar Sci 22:33–37. https://doi.org/10.1093/icesjms/22.1.33
Ekau W, Knoppers B (1996) Sedimentation process and productivity in the continental shelf waters off East and Norteast Brazil. JOPS-II–Cruise Report and First Results. Center for Tropical Marine Ecology, Bremen
Ekau W, Knoppers B (1999) An introduction to the pelagic system of the Northeast and East Brazilian shelf. Arch Fish Mar Res 47(2–3):5–24
Espinoza NS, dos Santos CAC, Silva MT et al (2021) Landslides triggered by the May 2017 extreme rainfall event in the East Coast Northeast of Brazil. Atmosphere 2021:12. https://doi.org/10.3390/atmos12101261
Ferreira BP, Maida M (2006) Monitoramento dos recifes de coral do Brasil: situação atual e perspectivas. Série biodiversidade, vol 18. MMA, Brasília, Brazil, p 250
Ferreira BP, Maida M, Cava F et al (2003) Interações entre a pesca artesanal e o turismo em Tamandaré, APA Costa dos Corais. In: Abstracts of IX Congresso da Associação Brasileira de Estudos do Quaternário/II Congresso sobre Planejamento e Gestão das Zonas Costeiras dos Países de Expressão Portuguesa/II Congresso do Quaternário dos Países de Língua Ibéricas. Associação Portuguesa dos Recursos Hídricos, Recife, Brazil, 12–19 Oct 2003
Ferreira SCG, Silva AC, Mangas AP et al (2014) Distribuição espacial de ictiopâncton em águas neríticas e oceânicas da Costa Norte do Brasil. Bol Téc Cient CEPNOR 14:39–45. https://doi.org/10.32519/tjfas.v14i1.1869
Field A (2005) North Brazil current rings viewed by TRMM Microwave Imager SST and the influence of the Amazon Plume. Deep-Sea Res I Oceanogr Res Pap 52:137 160. https://doi.org/10.1016/j.dsr.2004.05.013
Figueiredo GGAA, Schwamborn R, Bertrand A et al (2020) Body size and stable isotope composition of zooplankton in the western tropical Atlantic. J Mar Syst. https://doi.org/10.1016/j.jmarsys.2020.103449
Flagg CN, Gordon R, Mcdowell S (1986) Hydrographic and current observations on the continental slope and shelf of the western equatorial Atlantic. J Phys Oceanogr 16(8):1412–1429. https://doi.org/10.1175/1520-0485(1986)016%3c1412:HACOOT%3e2.0.CO;2
Franco-Gordo C, Godínez-Domínguez E, Suárez-Moralez E (2002) Larval fish assemblages in waters off central Pacific coast of Mexico. J Plankton Res 24:775–784. https://doi.org/10.1093/plankt/24.8.775
Fonseca P, Veiga JA, Correia FW et al (2017) (2004) An analysis of precipitation extremes in the Northern South America and their behaviors for future climate based on A1B scenario. Rev Bras Climatol 20:117–134
Gallienne CP, Robins DB, Woodd-Walker RS (2001) Abundance, distribution and size structure of zooplankton along a 20 west meridional transect of the northeast Atlantic Ocean in July. Deep Sea Res Part II: Topical Stud Oceanogr 48(4–5):925–949
Galvão IB (2000) O nanofitoplâncton e o microplâncton da Zona Econômica Exclusiva do Nordeste do Brasil (REVIZEE – NEII). Master Thesis (Universidade Federal de Pernambuco), Recife, 160 p
García A, Alemany F, Velez-Belchí P et al (2005) Characterization of the Blue fin Tuna Spawning Habitat off the Balearic Archipelago in relation to key hydrographic. Col Vol Sci Pap ICCAT 58:535–549
Garcia-Díaz XFG (2007) Zooplâncton do Arquipélago de São Pedro e São Paulo. M.Sc. thesis. Universidade Federal de Pernambuco, Brazil
García-Díaz XFG, Gusmão LMO, Neumann-Leitão S (2009) Biodiversidade e dinâmica espaço-temporal do zooplâncton. In: Hazin FHV (ed) O arquipélago de São Pedro e São Paulo: 10 anos de estação científica. SECIRM, Brasília, Brazil, pp 138–147

Gaughan DJ, Neira FJ, Beckley LE et al (1990) Composition, seasonality and distribution of the Ichthyoplankton in the lower Swan estuary, south-western Australia. Mar Freshw Res 41:529–543. https://doi.org/10.1071/MF9900529
Gibbs RJ (1970) Mechanisms controlling world water chemistry. Science 170:1088–1090. https://doi.org/10.1126/science.170.3962.1088
Gittings JA, Raitsos DE, Krokos G et al (2018) Impacts of warming on phytoplankton abundance and phenology in a typical tropical marine ecosystem. Sci Rep 8. https://doi.org/10.1038/s41598-018-20560-5
Gkanasos A, Somarakis S, Tsiaras K et al (2019) Development, application and evaluation of a 1-D full life cycle anchovy and sardine model for the North Aegean Sea (Eastern Mediterranean). PLoS ONE 14(8):e0219671. https://doi.org/10.1371/journal.pone.0219671
Gomes EAP, Campos PN, Bonecker ACT (2014) Occurrence of Gobiidae larvae in a tropical Brazilian estuary, with particular emphasis on the use of size classes to categorize species guilds. J Fish Biol 84:996–1013. https://doi.org/10.1111/jfb.12340
Gorsky G, Ohman MD, Picheral M et al (2010) Digital zooplankton image analysis using the ZooScan integrated system. J Plankton Res 32:285–303. https://doi.org/10.1093/plankt/fbp124
Gove J, McManus M, Neuheimer A, Polovina JJ, Drazen JC, Smith CR, Merrifield MA, Fiedlander AM, Ehses JS, Young CW, Dillon AK, Williams GJ (2016) Near-island biological hotspots in barren ocean basins. Nat Commun 7:10581. https://doi.org/10.1038/ncomms10581
Grego CKS, Feitosa FAN, Silva MH et al (2009) Fitoplâncton do ecossistema estuarino do rio Arinquindá (Tamandaré, Pernambuco, Brasil): variáveis ambientais, biomassa e produtividade primária. Atlântica 31:183–198. https://doi.org/10.5088/atl.2009.31.2.183
Gross MG, Gross E (1996) Oceanography, a view of earth. Pretence Hall, New Jersey, 472p
Gusmão LMO (1986) Chaetognatha planctônicos de províncias nerítica e oceânica do nordeste do Brasil. Doctoral dissertation, Dissertação de Mestrado, Universidade Federal de Pernambuco, 192 p
Gusmão LMO (2000) Comunidade zooplanctônica nas províncias nerítica e oceânica do Estado de Pernambuco—Brasil (Latitude 7° 32,98′ a 8° 41,51′ S—Longitude 34° 04,47′ a 35° 1,51′ W). Ph.D. thesis. Universidade Federal de Pernambuco, Brazil
Gusmão LMO, Neumann-Leitão S, Nascimento-Vieira DA, Silva TA, Silva AP, Porto Neto FF, Moura MCO (1997) Zooplâncton oceânico entre os Estados do Ceará e Pernambuco, Brasil. Trabalhos Oceanográficos da Universidade Federal de Pernambuco 25:17–30
Haney JF (1988) Diel patterns of Zooplankton behavior. Bull Mar Sci 43:583–603
Harzsch S, Müller CHG, Perez Y (2015) Chaetognatha. In: Wanninger A (ed) Evolutionary developmental biology of invertebrates. Springer, Vienna, pp 215–240
Haury LR, Yamazaki H, Itsweire EC (1990) Effects of turbulent shear flow on zooplankton distribution. Deep-Sea Res I Oceanogr Res Pap 37:447–461. https://doi.org/10.1016/0198-0149(90)90019-R
Hernández-León S (1991) Accumulation of mesozooplankton in a wake area as a causative mechanism of the "island-mass effect". Mar Biol 109(1):141–147
Heywood KJ, Barton ED, Simpson JH (1990) The effects of flow disturbance by an oceanic island. J Mar Res 48:55–73. https://doi.org/10.1357/002224090784984623
Hosoe K (1956) Chaetognaths from the Isles of Fernando de Noronha. Universidade de Sao Paulo
Ibáñez AL, Gutiérrez Benítez O (2004) Climate variables and spawning migrations of the striped mullet and white mullet in the north-western area of the Gulf of Mexico. J Fish Biol 65(3):822–831
Isaac-Nahum VJ, Vazzoller AEA, Zaneti-Prado EM (1983) Estudos sobre a estrutura, o ciclo de vida e o comportamento da *Sardinella brasiliensis* (Steindachner, 1879) na área entre 22° S e 28° S, Brasil. 3—Morfologia e histologia de ovários e escala de maturidade. Bol Inst Oceanogr 32:1–16. https://doi.org/10.1590/S0373-55241983000100001
Johns WE, Lee TN, Beardsley R et al (1998) Annual cycle and variability of the North Brazil Current. J Phys Oceanogr 28:103–128. https://doi.org/10.1175/1520-0485(1998)028%3c0103:ACAVOT%3e2.0.CO;2

Kanaeva IP (1960) The distribution of plankton along the 30° W meridian in the Atlantic (April–May 1959). Soviet Fisheries Investigation in North European Seas (Moscow), pp 173–183

Kineke GC, Sternberg RW, Cacchione DA et al (1991) Distribution and characteristics of suspended sediment on the Amazon shelf. Oceanogr 4:21–26. https://doi.org/10.5670/oceanog.1991.17

Kirby RR, Beaugrand G, Kleparski L et al (2021) Citizens and scientists collect comparable oceanographic data: measurements of ocean transparency from the Secchi disk study and science programmes. Sci Rep. https://doi.org/10.1038/s41598-021-95029-z

Klein VLM, Moreira ICP (1977) Algumas Observações sobre o plâncton da região costeira de Paracuru (Ceará-Brasil)

Krumme U, Giarrizzo T, Pereira R et al (2015) Airborne synthetic-aperture radar (SAR) imaging to help assess impacts of stationary fishing gear on the north Brazilian mangrove coast. ICES J Mar Sci 72:939–951. https://doi.org/10.1093/icesjms/fsu188

Lacepède BGE (1801) Histoire naturelle des poissons. 3: i-lxvi + 1-558, Pls. 1-34, page(s): 540, 546

Lalli C, Parsons TR (eds) (1997) Biological oceanography: an introduction, 2nd edn. Elsevier Butterworth-Heinemann

Larrazábal MED, Oliveira VSD (2003) Thecosomata e Gymnosomata (Mollusca, Gastropoda) da cadeia Fernando de Noronha, Brasil. Revista Brasileira de Zoologia 20:351–360

Lavagnino AC, Bastos AC, Amado Filho CM et al (2020) Geomorphometric seabed classification and potential megahabitat distribution in the Amazon continental margin. Front Mar Sci. https://doi.org/10.3389/fmars.2020.00190

Legendre P, Birks HJB (2012) From classical to canonical ordination. In: Birks HJB, Lotter AF, Juggins S et al (eds) Tracking environmental change using lake sediments. Data handling and numerical techniques, vol 5. Springer, Dordrecht, pp 201–248 (Chapter 8)

Legendre P, Gallagher ED (2001) Ecologically meaningful transformations for ordination of species data. Oecologia 129:271–280. https://doi.org/10.1007/s004420100716

Legendre P, Legendre L (1998) Numerical ecology, 2nd edn. Elsevier, Amsterdam

Leite AAC, Queiroz AS, Morais FS et al (2020) Zooplankton at the Northern Brazilian Coast: evaluation and gaps. J Coast Res 95:12–17. https://doi.org/10.2112/SI95-003.1

Leite GR, Silva VV, Freitas CE (2006) Abundância e distribuição das larvas de peixes no Lago Catalão e no encontro dos rios Solimões e Negro, Amazonas, Brasil. Acta Amazon 36:557–562. https://doi.org/10.1590/S0044-59672006000400018

Leite RG, Cañas C, Forsberg B et al (2007) Larvas dos Grandes Bagres Migradores. Walter H. Wust Ediciones, Lima, p 127

Lessa RPT, Mafalda P Jr, Advíncula R et al (1999) Distribution and abundance of ichthyoneuston at seamounts and islands off North-Eastern Brazil. Arch Fish Mar Res 47(2/3):239–265

Lessa, GC, Cirano M, Tanajura CAS et al (2009) Oceanografia Física. In: Hatje V, Andrade JB (org) Baía de Todos os Santos: Aspectos Oceanográficos. Ed. Salvador. EDUFBA, Brazil

Levinton JS (1995) Marine biology: function, biodiversity, ecology. Oxford University Press, New York

Lima IF (2019) Variação nictemeral da comunidade de larvas de peixes sob influência de parâmetros físicos nas marés enchente e vazante no sistema estuarino da Raposa-MA-Brasil. Monography, Universidade Federal do Maranhão, Brazil

Lima AC, Araújo-Lima CARM (2004) The distribution of larval and juveniles fishes in Amazonian rivers of the different nutrients status. Freshw Biol 49:787–800. https://doi.org/10.1111/j.1365-2427.2004.01228.x

Lima ARA, Barletta M, Costa MF (2016) Seasonal-dial shifts of ichthyoplankton assemblages and plastic debris around an Equatorial Atlantic Archipelago. Front Environ Sci. https://doi.org/10.3389/fenvs.2016.00056

Lima FA, Martinelli-Lemos JM (2019) Checklist of the Brachyura of the brazilian Amazon coastal zone and knowledge status of their larval development. Zootaxa 4646:301–321. https://doi.org/10.11646/zootaxa.4646.2.6

Lima FA, Butturi-Gomes D, Martinelli-Lemos JM (2021) Megalopa bloom of *Panopeus lacustris* (Decapoda: Panopeidae) on the Amazon Continental Shelf. Reg Stud Mar Sci. https://doi.org/10.1016/j.rsma.2021.101960

Linnaeus C (1758) Systema Naturae per regna tria naturae, secundum classes, ordines, genera, species, cum characteribus, differentiis, synonymis, locis. Editio decima, reformata, 10th revised edn, vol 1, 824 p. Laurentius Salvius, Holmiae. Available online at page(s): 316. https://biodiversitylibrary.org/page/726886

Lins-Silva N, Marcolin C, Kessler F et al (2021) A fresh look at microplastics and other particles in the tropical coastal ecosystems of Tamandaré, Brazil. Mar Environ Res. https://doi.org/10.1016/j.marenvres.2021.105327

Lira SMA, Teixeira IA, Lima CDM et al (2014) Spatial and nycthemeral distribution of the zooneuston off Fernando de Noronha, Brazil. Braz J Oceanogr 62:35–45. https://doi.org/10.1590/s1679-87592014058206201

Lira SMA, Santana CS, Lima CDM et al (2017) New records of the larval forms *Cerataspis monstrosa* and *Amphionides reynaudii* (Crustacea: Decapoda) from the western tropical Atlantic. Zootaxa 4237:335–346. https://doi.org/10.11646/zootaxa.4237.2.7

Longhurst AR (1985) Relationship between diversity and the vertical structure of the upper ocean. Deep-Sea Res I Oceanogr Res Pap 32:1535–1570

Longhurst A, Pauly D (1987) Ecology of tropical oceans. Academic Press, Inc., New York, 407 p

Longhurst A, Williams R (1979) Materials for plankton modelling: vertical distribution of Atlantic zooplankton in summer. J Plankton Res 1:1–28. https://doi.org/10.1093/plankt/1.1.1

Lopes RM, Brandini FP, Gaeta SA (1999) Distribution patterns of epipelagic copepods off Rio de Janeiro (SE Brazil) in summer 1991/1992 and winter 1992. Hydrobiologia 411:161–174. https://doi.org/10.1023/A:1003859107041

Lucas OAP (2008) Variação sazonal do zooplâncton no Estuário dos Rios Botafogo e Siriji, Litoral Norte de Pernambuco, Brasil. Trop Oceanogr 36:28–39. https://doi.org/10.5914/tropocean.v36i1-2.5086

Macedo-Soares LCP, Freire AS, Muelbert JH (2012) Small-scale spatial and temporal variability of larval fishes assemblages at an isolate oceanic island. Mar Ecol Progr Ser 444:207–222

Mafalda PO Jr, Rubín JP (2006) Interannual variation of larval fish assemblages in the Gulf of Cádiz (SW Iberian Peninsula) in relation to summer oceanographic conditions. Braz Arch Biol Technol 49:287–296. https://doi.org/10.1590/S1516-89132006000300014

Mafalda PO Jr, Sinque C, Muelbert H (2006) Associações de Larvas de Peixes na Costa Norte da Bahia. Atlântica 28(19):5–11

Mafalda PO Jr, Sinque C, Muelbert JH et al (2004) Distribuição e abundância do ictioplâncton na costa norte da Bahia, Brasil. Trop Oceanogr 32:69–88. https://doi.org/10.5914/tropocean.v32i1.5036

Mafalda PO Jr, Souza CS, Velame MPB (2008) Fish larvae assemblage of a coastal area under influence of petrochemical activities, in Todos os Santos Bay, Bahia, Brazil. Aquat Ecosyst Health Manag 11:457–464. https://doi.org/10.1080/14634980802533933

Magalhães A, Leite NR, Silva JGS, Pereira LCC, Costa RM (2009) Seasonal variation in the copepod community structure from a tropical Amazon estuary, Northern Brazil. An Acad Bras Ciênc 81(2):187–197

Mangas AP, Silva AC, Ferreira SCG et al (2013) Ictioplâncton da baía do Guajará e do estuário do Rio Pará, ilha do Marajó, Pará, Brasil. Bol Tec Cient CEPNOR 13:43–54. https://doi.org/10.32519/tjfas.v13i1.1619

Manso VAV, Corrêa ICS, Guerra NC (2003) Morfologia e Sedimentologia da Plataforma Continental Interna entre as Praias Porto de Galinhas e Campos—Litoral Sul de Pernambuco, Brasil. Pesq Geociências 30:17–25. https://doi.org/10.22456/1807-9806.19587

Marcolin CR, Schultes S, Jackson GA et al (2013) Plankton and seston size spectra estimated by the LOPC and ZooScan in the Abrolhos Bank ecosystem (SE Atlantic). Cont Shelf Res 70:74–87. https://doi.org/10.1016/j.csr.2013.09.022

Marques V, Silva-Falcão E, Severi W (2015) Estrutura da assembleia ictioplanctônica em dois estuários tropicais de Pernambuco (Brasil), sujeitos a diferentes condições hidrológicas. Rev Bras Ciênc Agrár 10:304–314. https://doi.org/10.5039/agraria.v10i2a5012

Matos IP, Lucena F (2006) Descrição da pesca da pescada-amarela, Cynoscion acoupa, da costa do Pará. [Description of the yellow weakfish fishery, Cynoscion acoupa, of the coast of Pará]. Arq Ciên Mar 39:66–73

Medeiros GF, Björnberg TKS (1978) Levantamento preliminar do zooplâncton das águas das ilhas da região de Natal e das águas de Fernando de Noronha. Ciência e Cultura 30(3):348–349

Meerhoff E, Ramos M, Yannicelli B, Bravo L, Zambra D, Varela C (2018) Meroplankton distribution in South Pacific islands, implications for larval connectivity. J Plankton Res 40(1):77–90

Melo M Jr (2005) Padrões dinâmicos de transporte e migração do zooplâncton, com ênfase nos Decapoda planctônicos, da Barra de Catuama, Pernambuco—Brasil. M.Sc. thesis. Universidade Federal de Pernambuco, Brazil

Melo PAMC (2009) Dinâmica em microescala do mesozooplâncton do Arquipélago de São Pedro e São Paulo, NE, Brasil. Master Thesis (Universidade Federal de Pernambuco), 50p

Melo M Jr, Almeida VLS, Neumann-Leitao S et al (2007) O estado da arte da biodiversidade de rotíferos planctônicos de ecossistemas límnicos de Pernambuco. Biota Neotrop 7:1–9. https://doi.org/10.1590/S1676-06032007000300013

Melo PAMC, Silva TA, Neumann-Leitão S et al (2010) Demersal zooplankton communities from tropical habitats in the southwestern Atlantic. Mar Biol Res 6:530–541. https://doi.org/10.1080/17451000903426557

Melo PAMC, Diaz XFG, Macedo SJD, Neumann-Leitão S (2012) Diurnal and spatial variation of the mesozooplankton community in the Saint Peter and Saint Paul Archipelago, equatorial Atlantic. Mar Biodivers Rec 5:1–14. https://doi.org/10.1017/S1755267212001054

Melo NFAC, Neumann-Leitão S, Gusmão LMO et al (2014a) Distribution of the Planktonic Shrimp *Lucifer* (Thompson, 1829) (Decapoda, Sergestoidea) of the Amazon. Braz J Biol 74:S45-51. https://doi.org/10.1590/1519-6984.20612

Melo PAMC, Melo Júnior M, Macedo SJ et al (2014b) Copepod distribution and production in a mid-Atlantic ridge archipelago. An Acad Bras Ciênc 86:1719–1733. https://doi.org/10.1590/0001-3765201420130395

Melo DCM, Lira SMA, Moreira APB et al (2020) Genetic diversity and connectivity of Flaccisagitta enflata (Chaetognatha: Sagittidae) in the tropical Atlantic Ocean (northeastern Brazil). PLoS ONE. https://doi.org/10.1371/journal.pone.0231574

Moreno D (2017) Composição e Variação Temporal e Nictemeral da Fauna Zooplanctônica na Zona de Arrebentação em Praia Grande, SP. M.Sc. thesis. Universidade Estadual Paulista, Brazil

Moser HG, Smith PE (1993) Larval fish assemblages of the California current Region and their horizontal and vertical distributions across a front. Bull Mar Sci 53:645–691

Moura RT, Passavante JZO (1993) Taxa de assimilação do fitoplâncton da baía de Tamandaré—Rio Formoso—PE—Brasil. Bol Tec Cient CEPENE 1:17–23

Moura RL, Amado-Filho GM, Moraes FC et al (2016) An extensive reef system at the Amazon River mouth. Sci Adv. https://doi.org/10.1126/sciadv.1501252

Nascimento-Vieira DAN, Sant'anna E, Luz BRA, Neumann-Leitão S (1990) Zooplâncton nerítico e oceânico dos estados de Alagoas e Pernambuco (Brasil). Trabalhos Oceanográficos da Universidade Federal de Pernambuco 21:81–101

Nascimento-Vieira DA, Neumann-Leitão S, Porto Neto FF et al (2010) Mesozooplâncton de área recifal do Atlântico sudoeste Tropical. Trop Oceanogr 38(1):45–59. https://doi.org/10.5914/1679_3013.2014.0099

Neumann-Leitão S, Matsumura-Tundisi T (1998) Dynamics of a perturbed estuarine zooplanktonic community: Port of Suape, PE, Brazil. Verh Int Ver Theor Angew Limnol 26(4):1981–1988. https://doi.org/10.1080/03680770.1995.11901089

Neumann-Leitão S, Gusmão LMO, Nogueira-Paranhos JD et al (1991/1993) Zooplâncton da Plataforma Continental Norte de Pernambuco (Brasil). Trab Oceanogr 22:97–116. https://doi.org/10.5914/tropocean.v22i1.2659

Neumann-Leitão S, Gusmão LMO, Silva TA et al (1999) Mesozooplankton biomass and diversity in coastal and oceanic waters off north-eastern Brazil. Arch Fish Mar Res 47:153–165
Neumann-Leitão SN, Eskinazi-Santanna EM, Gusmão LMO et al (2008a) Diversity and distribution of the Mesozooplankton in the tropical Southwestern Atlantic. J Plankton Res 30:795–805. https://doi.org/10.1093/plankt/fbn040
Neumann-Leitão S, Sant′Anna EME, Gusmão LMO et al (2008b) Diversity and distribution of the mesozooplankton in the tropical Southwestern Atlantic. J Plankton Res 30:795–805. https://doi.org/10.1093/plankt/fbn040
Neumann-Leitão S, Feitosa FAN, Mayal E, Schwamborn R, Silva-Cunha MGG, Silva TA, Melo NF, Porto Neto FF (2009) The plankton from Maracajaú reef ecosystem (Brazil)—offshore coral reefs under multiple human stressors. WIT Trans Ecol Environ 122:173–182
Neumann-Leitão S, Melo PAMC, Schwamborn R et al (2018) Zooplankton from a Reef system under the influence of the Amazon River Plume. Front Microbiol. https://doi.org/10.3389/fmicb.2018.00355
Neumann Leitão S, Melo M Jr, Porto Neto FF et al (2019a) Connectivity between Coastal and Oceanic Zooplankton from Rio Grande do Norte in the Tropical Western Atlantic. Front Mar Sci. https://doi.org/10.3389/fmars.2019.00287
Neumann-Leitão S, Melo M Jr, Porto-Neto FF (2019b) Connectivity between Coastal and Oceanic Zooplankton from Rio Grande do Norte in the Tropical Western Atlantic. Front Mar Sci. https://doi.org/10.3389/fmars.2019.00287
Nixon SW, Oviatt CA, Frithsen J, Sullivan B (1986) Nutrients and the productivity of estuarine and coastal marine ecosystems. J Limnol Soc South Africa 12:43–71. https://doi.org/10.1080/03779688.1986.9639398
Nogueira EMDS, Sassi R (2011) Nychthemeral variations of Tintinnina (Ciliata: Oligotrichida) near the Rocas atoll (South Atlantic) and relationships with other microzooplanktonic components
Nogueira EMDS, Sassi R, Cordeiro TA (2005) A família Rhabdonellidade (Tintinnina: Oligotrichida) do Atol das Rocas e Arquipélago de Fernando de Noronha, Atlântico Sul Ocidental, Brasil
Nogueira EMS, Sassi R, Cordeiro T (2008) Estrutura da comunidade dos tintinnina na região do atol das rocas (Rio Grande do Norte) e 454 arquipélago de Fernando de Noronha (Pernambuco). Biota Neotrop 8:135–140. https://doi.org/10.1590/s1676-06032008000300134
Nonaka RH, Matsuura Y, Suzuki K (2000) Seasonal variation in larval fish assemblages in relation to oceanographic conditions in the Abrolhos Bank region off eastern Brazil. Fish Bull 98:767–784
Nunes YBS, Diniz TS, Figueiredo MB (2019) Análise socioeconômica e caracterização dos sistemas pesqueiros. Bol Lab Hidrobiol 29:19–25. https://doi.org/10.18764/1981-6421e2019.3
Oksanen J, Blanchet FG, Friendly M et al (2020) Vegan: community ecology package. R package version 2.5-7. Available at https://CRAN.R-project.org/package=vegan
Olavo G, Costa PAS, Martins AS (2005) Prospecção de grandes peixes pelágicos na região central da ZEE brasileira entre o Rio Real-BA e o Cabo de São Tomé-RJ, In: Costa PAS, Martins AS, Olavo G (eds) Pesca e potenciais de exploração de recursos vivos na região central da Zona Econômica Exclusiva brasileira (Série Livros No. 13), Museu Nacional, Rio de Janeiro, Brazil, pp 167–202
Oliveira LP (2009) Análise comparativa da distribuição das famílias Salpidae e Dioliolidae em relação ao zooplâncton total na Plataforma Continental Sudeste do Brasil por meio de técnicas semi-automaticas de identificação e contagem. M.Sc. thesis. Universidade de São Paulo, Brazil
Omori M, Ikeda T (1984) Methods in marine zooplankton ecology. Wiley, New York
Paes E, Dos R, Souza, SNG et al (2020) Abundância e densidade das larvas de peixes e sua relação com nível das águas na confluência dos rios Solimões e Negro e no lago Catalão, Amazonas-Brasil. Biota Amazônia 10:20–24. https://doi.org/10.18561/2179-5746/biotaamazonia.v10n2p20-24
Palacios DM (2002) Factors influencing the island-mass effect of the Galápagos Archipelago. Geophys Res Lett 29(23):49–51

Paranaguá MN (1966) Sobre o plâncton da região compreendida entre 3º Lat. S e 13º Lat. S, ao largo do Brasil. Trabalhos do Instituto de Oceanografia da Universidade Federal de Pernambuco 5(6):125–139

Paranaguá MN (1967/69) Estudo da Plataforma Continental na área do Recife (Brasil) IIIb. Composição e variação do zooplâncton. Trabalhos Oceanográficos da Universidade Federal de Pernambuco 18:193–206

Paranaguá MN, Nascimento-Vieira DA, Gusmão LMO et al (2004) Estrutura da comunidade zooplanctônica. In: Eskinazi-Leça E, Neumann-Leitão S, Costa MF (orgs) Oceanografia: um cenário tropical. Bagaço/UFPE, Recife, pp 441–459

Pereira MAG, Lessa GC (2009) Varying patterns of water circulation in canal de Cotegipe, Baía de Todos os Santos. Rev Bras Geofis 27:103–119. https://doi.org/10.1590/S0102-261X2009000100009

Pereira RC, Soares-Gomes A (2002) Biologia Marinha. Interciência, Rio de Janeiro

Pinheiro HT, Macena BCL, Francini-Filho RB (2020) Fish biodiversity of Saint Peter and Saint Paul's Archipelago, Mid-Atlantic Ridge, Brazil: new records and a species database. J Fish Biol 97:1143–1153. https://doi.org/10.1111/jfb.14484

Pinto NCT, Mafalda P, Santos AT (1997) Caracterização do zooplâncton da Reserva Biológica do Atol das Rocas, na Campanha de Março-1991 (verão). Trabalhos Oceanográficos da Universidade Federal de Pernambuco 25:31–46

Piorski N M, Serpa S, Nunes JLS (2009) Análise comparativa da pesca de curral na Ilha de São Luís, Estado do Maranhão, Brasil. Arq Ciênc 42:65–71. https://doi.org/10.32360/acmar.v42i1.6040

Platt T (1985) Structure of marine ecosystems: its allometric basis. In: Ulanowicz RE, Platt T (eds) Ecosystem theory for biological oceanography, vol 213. Department of Fisheries and Oceans, Canada, pp 55–64

PORTOBRAS (1988) Relatório de apresentação das medições meteorológicas observadas em Pontal do Sul, Paranaguá-PR, período set. 1982 a dez. 1986. INPH, Rio de Janeiro, Brazil

Pugh PR, Boxshall GA (1984) The small-scale distribution of plankton at a shelf station off the Northwest African coast. Cont Shelf Res 3(4):399–423

Quevedo M, Viesca L, Anadón R, Fernández E (2003) The protistan microzooplankton community in the oligotrophic North-Eastern Atlantic: large-and mesoscale patterns. J Plankton Res 25(5):551–563

Rakocinski CF, Baltz DM, Fleeger JW (1995) Correspondence between environmental gradients and the community structure of marsh-edge fishes in a Louisiana estuary. Mar Ecol Prog Ser 80:135–148. https://doi.org/10.3354/meps080135

Roch M, Brandt P, Schmidtko S et al (2021) Southeastern tropical Atlantic changing from subtropical to tropical conditions. Front Mar Sci 26. https://doi.org/10.3389/fmars.2021.748383

Rodrigues BCC (2021) Influência da defasagem temporal de parâmetros ambientais sobre as larvas de peixes do sistema estuarino da Raposa, Maranhão-Brasil. Monography, Universidade Federal do Maranhão, Brazil

Rodrigues-Inoue ACM, Santos A, Martinelli Lemos JM (2021) Distribution patterns of Anomura, Axiidea and Gebiidea (Crustacea, Decapoda) larvae at the Amazon shelf. Reg Stud Mar Sci. https://doi.org/10.1016/j.rsma.2021.101946

Röpke A (1993) Do larvae of mesopelagic fishes in the Arabian Sea adjust their vertical distribution to physical and biological gradients? Mar Ecol Prog Ser 101:223–235

Sankarankutty C, Medeiros GF, Santos NQ, Silva IJ, Rego FA (1990) On zooplankton of the coastal waters of Rio Grande Do Norte sampled from oil producing platform. J Mar Biol Assoc India 32(1 & 2):254–259

Sankarankutty C, Oliveira JE, Cunha KMF, Silva ACC, Barroca EVG (1995) Further observations on zooplankton of the Potengi Estuary (Natal, Rio Grande do Norte, Brazil) with special reference to the larvae of Brachyura (Crustacea, Decapoda). Revista Brasileira de Zoologia 12:273–301

Santana CS, Schwamborn R, Neumann-Leitão S et al (2018) Spatio-temporal variation of planktonic decapods along the leeward coast of the Fernando de Noronha archipelago, Brazil. Braz J Oceanogr 66:1–14. https://doi.org/10.1590/s1679-87592018147206601

Santana CS, Lira SM, Varona HL et al (2020a) Amazon river plume influence on planktonic decapods in the tropical Atlantic. J Mar Syst 212:103428. https://doi.org/10.1016/j.jmarsys.2020.103428m

Santana JR, Costa AEF, Neumann Leitão S et al (2020b) Spatial variability of the ichthyoneuston around oceanic islands at the tropical Atlantic. J Sea Res. https://doi.org/10.1016/j.seares.2020.101928

Santos RVS, Severi W (2019) Dynamics of early life-history stages of fish along an estuarine gradient. Fish Oceanogr 28:402–418. https://doi.org/10.1111/fog.12420

Santos GS, Stemmann L, Lombard F et al (2019) Are tropical coastal reefs sinks or sources of mesozooplankton? A case study in a Brazilian marine protected area. Coral Reefs 38:1107–1120. https://doi.org/10.1007/s00338-019-01860-2

Sarpedonti V, Anunciação SEM, Bordalo AO (2008) Ichthyoplankton variations in two mangrove creeks of the Curuçá estuary, Pará, Brasil. Ecotropicos 21(1):1–12

Sarpedonti V, Anunciação SEM, Bordalo AO (2013) Spatio-temporal distribution of fish larvae in relation to ontogeny and water quality in the oligohaline zone of a North Brazilian estuary. Biota Neotrop 13:55–63. https://doi.org/10.1590/S1676-06032013000300007

Sato K, Matsuno K, Arima D et al (2015) Spatial and temporal changes in zooplankton abundance, biovolume, and size spectra in the neighboring waters of Japan: analyses using an optical plankton counter. Zool Stud 54:1–15. https://doi.org/10.1186/s40555-014-0098-z

Schnack-Schiel SB, Mizdalski E, Cornils A (2010) Copepod abundance and species composition in the Eastern subtropical/tropical Atlantic. Deep Sea Res Part II: Top Stud Oceanogr 57(24–26):2064–2075

Schott FA, Fischer J, Stramma L (1998) Transports and pathways of the upper-layer circulation in the western tropical Atlantic. J Phys Oceanogr 28:1904–1928. https://doi.org/10.1175/1520-0485(1998)028%3c1904:TAPOTU%3e2.0.CO;2

Schwamborn R (1997) The influence of mangroves on community structure and nutrition of macrozooplankton in northeast Brazil. ZMT Contrib 4:1–77. https://epic.awi.de/id/eprint/13526/

Schwamborn R, Bonecker ACT (1996) Seasonal changes in the transport and distribution of meroplankton into a Brazilian estuary with emphasis on the importance of floating mangrove leaves. Braz Arch Biol Technol 39:451–462

Schwamborn R, Neumann-Leitão S, Almeida T et al (2001) Distribution and dispersal of decapod crustacean larvae and other zooplankton in the Itamaracá estuarine system, Brazil. Trop Oceanogr 29:1679–3013. https://doi.org/10.5914/tropocean.v29i1.2834

Schwamborn R, Melo Júnior M, Neumann-Leitão S et al (2008) Dynamic patterns of zooplankton transport and migration in Catuama Inlet (Pernambuco, Brazil), with emphasis on the decapod crustacean larvae. Lat Am J Aquat Res 36:109–113

Schwamborn R, Mildenberger TK, Taylor MH (2019) Assessing sources of uncertainty in length-based estimates of body growth in populations of fishes and macroinvertebrates with bootstrapped ELEFAN. Ecol Modell 393:37–51. https://doi.org/10.1016/j.ecolmodel.2018.12.001

Severi W, Urach B, Castro M (2008) Occurrence of *Microdesmus bahianus* and *M. longipinnis* (Teleostei: Microdesmidae) larvae and juveniles in estuaries of the State of Pernambuco, Brazil. Rev Bras Cienc Agrárias 3:360–364

Sigman DM, Hain MP (2012) The biological productivity of the ocean. Nat Educ Knowl 3(10):21

Signorini SR, Mcclain CR, Dandonneau Y (1999) Mixing and phytoplankton bloom in the wake of the Marquesas Islands. Geophys Res Lett 26:3121–3124. https://doi.org/10.1016/10.1029/1999GL010470

Silva AC (2006) An analysis of water properties in the western tropical Atlantic using observed data and numerical model results. Ph.D. thesis. Universidade Federal de Pernambuco, Brazil

Silva TA, Neumann-Leitão S, Schwamborn R et al (2003) Diel and seasonal changes in the macrozooplankton community of a tropical estuary in Northeastern Brazil. Rev Bras Zool 20:439–446. https://doi.org/10.1590/S0101-81752003000300012

Silva ACD, Araujo M, Pinheiro LDS (2007) Caracterização hidrográfica da plataforma continental do Maranhão a partir de dados oceanográficos medidos, remotos e modelados. Rev Bras Geofis 25:281–293. https://doi.org/10.1590/S0102-261X2007000300005

Silva AC, Araújo M, Bourlès B (2010) Seasonal variability of the Amazon River plume during REVIZEE program. Trop Oceanogr 38:76–87. https://doi.org/10.5914/tropocean.v38i1.5162

Silva AC, Mangas AP, Palheta GDA et al (2011a) Caracterização do ictioplâncton de quatros estuários da Microregião do Salgado do estado do Pará-Brasil. Bol Tec Cient CEPNOR 11:33–44. https://doi.org/10.32519/tjfas.v11i1.1201

Silva AC, Mangas AP, Palheta GDA et al (2011b) Variação intermareal na composição do ictioplâncton no estuário do Rio Guajará-Mirim (Vigia de Nazaré-Pará) em período de alta pluviosidade. Bol Tec Cient CEPNOR 1:21–32. https://doi.org/10.17080/1676-5664/btcc.v11n1p21-32

Silva AC, Mangas AP, Silva SC et al (2017) Variação de ictioplâncton em relação ao ciclo de maré e à distribuição espaço sazonal no estuário Guajará-Mirim, Pará, Brasil. Bol Cient CEPNOR 17(1):47–57. https://doi.org/10.32519/tjfas.v17i1.2140

Silva LS, Cavalcante-Braga DV, Lourenço CB et al (2021) Factors affecting the seasonal variability of planktonic shrimps (Dendrobranchiata) along an estuary-ocean gradient on the Amazon continental shelf. J Mar Biol Assoc 101:331–342. https://doi.org/10.1017/S0025315421000308

Silva-Falcão EC, Severi WE, Rocha AAF (2007) Dinâmica espacial e temporal de zoeas de Brachyura (Crustacea, Decapoda) no estuário do Rio Jaguaribe, Itamaracá, Pernambuco, Brasil. Iheringia Ser Zool 97(4):434–440. https://doi.org/10.1590/S0073-47212007000400012

Silva-Falcão E, Severi W, Araújo M (2013) Spatial–temporal variation of Achirus larvae (Actinopterygii: Achiridae) in mangrove, beach and reef habitats in north-eastern Brazil. J Mar Biol Assoc 93:381–388. https://doi.org/10.1017/S0025315411001706

Silveira PCA (2003) Ictionêuston da Zona Econômica exclusiva referente ao Estado do Maranhão (Região Norte do Brasil). M.Sc. thesis. Universidade Federal de Pernambuco, Brazil

Silveira PCA (2008) Impact of oceanographic conditions on distribution and abundance of larval fish in northern Brazil. Ph.D. thesis. Bremen University

Silveira PCA (2011) Distribuição e abundância das larvas de peixes na Zona Econômica Exclusiva do Estado do Maranhão, Brasil. In: Nunes JLS, Piorski NM (org) Peixes Marinhos e Estuarinos do Maranhão. Café & Lápis/FAPEMA, São Luís, Brazil, pp 5–25

Smith SL, Vallina SM, Merico A (2016) Phytoplankton size-diversity mediates an emergent trade-off in ecosystem functioning for rare versus frequent disturbances. Sci Rep 6:34170. https://doi.org/10.1038/srep34170

Soares FMM (2019) Variação espaço-sazonal das larvas de peixes no estuário do Rio Perizes, Maranhão, Brasil. Monography, Universidade Federal do Maranhão, Brazil

Soares RMT (2021) Caracterização da comunidade de larvas de peixes, sob variáveis físicas e químicas, durante um ciclo nictemeral, nos períodos de chuva e estiagem no Complexo Estuarino da Raposa-MA, Brasil. Monography, Universidade Federal do Maranhão, Brazil

Soares RDB, Cutrim MVJ, Silveira PCA (2014a) Comunidade ictioplanctônica da bacia hidrográfica do rio Bacanga na cidade de São Luís, Brasil. Rev Ciênc Ambient 8:37–48. https://doi.org/10.18316/1435

Soares RDB, Cutrim MVJ, Silveira PCA (2014b) Diversidade de Larvas de Peixes da Bacia Hidrográfica do Rio Bacanga (Maranhão Brazil). Rev Ciênc Ambient 8:5–19. https://doi.org/10.18316/1436

Soares RA, Ribeiro Júnior JCM, Silveira PCA et al (2020) Fish larval distribution in a macro-tidal regime: an in situ study in São Marcos Bay (Amazon Coast, Brazil). Res Soc Dev. https://doi.org/10.33448/rsd-v9i10.9238

Souza CS, Mafalda PO Jr (2008) Distribution and Abundance of Carangidae (Teleostei, Perciformes) Associated with oceanographic factors along the Northeast Brazilian exclusive economic zone. Braz Arch Biol Technol 51:1267–1278. https://doi.org/10.1590/S1516-89132008000600023

Souza CS, Mafalda PO Jr (2019) Large-scale spatial and temporal variability of larval fish assemblages in the tropical Atlantic ocean. An Acad Bras Ciênc. https://doi.org/10.1590/0001-3765201820170567

Souza CS, Luz JA, Macedo S et al (2013) Chlorophyll a and nutrient distribution around seamounts and islands of the tropical south-western Atlantic. Mar Freshwater Res 64:168–184. https://doi.org/10.1071/MF12075

Souza CS, Luz JAG, Mafalda PO Jr (2014) Relationship between spatial distribution of chaetognaths and hydrographic conditions around seamounts and islands of the tropical southwestern Atlantic. An Acad Bras Ciênc 86(3):1151–1165. https://doi.org/10.1590/0001-3765201420130101

Souza CS, Conceicao LR, Freitas TSS et al (2020) Size spectra modeling of Mesozooplankton over a tropical continental shelf. J Coast Res 36:795–804. https://doi.org/10.2112/JCOASTRES-D-19-00102.1

Stramma L, Schott F (1999) The mean flow field of the tropical Atlantic ocean. Deep-Sea Res II Top Stud Oceanogr 46:279–303. https://doi.org/10.1016/S0967-0645(98)00109-X

Tchamabi CC, Araujo M, Silva M et al (2017) A study of the Brazilian Fernando de Noronha island and Rocas atoll wakes in the tropical Atlantic. Ocean Model 111:9–18. https://doi.org/10.1016/j.ocemod.2016.12.009

Ter Braak, CJF, Smilauer P (2002) CANOCO Reference manual and CanoDraw for windows user's guide: software for canonical community ordination (version 4.5) (microcomputer power). Available at www.canoco.com. https://www.wur.nl/en/Research-Results/Research-Institutes/show/Canoco-for-visualization-of-multivariate-data.htm

Thackeray SJ (2012) Mismatch revisited: what is trophic mismatching from the perspective of the plankton? J Plankton Res 34:1001–1010. https://doi.org/10.1093/plankt/fbs066

Thompson GA, Alder VA, Boltovskoy D et al (1999) Abundance and biogeography of tintinnids (Ciliophora) and associated microzoo plankton in the southwestern Atlantic Ocean. J Plankton Res 21:1265–1298. https://doi.org/10.1093/plankt/21.7.1265

Tinoco IM (1965) Foraminiferos do Atol das Rocas. Trabalhos do Instituto Oceanográfico da Universidade Federal de Pernambuco 7(1):91–114. https://doi.org/10.5914/tropocean.v7i1.2503

Tosetto EG, Bertrand APA, Neumann-Leitao S et al (2021) Spatial patterns in planktonic cnidarian distribution in the western boundary current system of the tropical South Atlantic Ocean. J Plankton Res 43:270–287. https://doi.org/10.1093/plankt/fbaa066

Travassos P, Hazin FHV, Zagaglia JR et al (1999) Thermohaline structure around seamonts and islands off north-eastern Brazil. Arch Fish Mar Res 47:211–222

Vannucci M (1958) Considerações em torno das Hydromedusae da região de Fernando de Noronha. Boletim do Instituto Oceanográfico 9:03–12

Vannucci M, Hosoe K (1956) Pterosagitta besnardi Van. & Hosoe 1952 synonym of P. draco (Krohn 1853). Boletim do Instituto Oceanográfico 7:195–199

Vannucci M, Queiroz D (1963) Plankton collected during cruise No. 7 of the N/E ALMIRANTE SALDANHA. Contribuicoes Avulsas do Instituto Oceanografico 4:1–9

Vazzoler AEAM, Rossi-Wongtschowski CLDB (1976) *Sardinella brasiliensis*: spawning, fecundity and relative spawning power. I. Area between 23° 40′ S and 24° 20′ S. Brazil Bol Inst Oceanogr 25:131–155. https://doi.org/10.1590/S0373-55241976000200001

Venekey V, Fonsêca-Genevois VG, Santos PJP (2011) Influence of the tidal and rainfall cycles on the population structure and density of *Mesacanthion hirsutum* Gerlach (nematoda, thoracostomopsidae) on a tropical sandy beach (Tamandaré Bay, Pernambuco, Brazil). Braz J Oceanogr 59:253–258. https://doi.org/10.1590/S1679-87592011000300005

Wickstead J (1979) Zooplancton marino. Editorial Omega, Barcelona

Yoklavich MM, Stevenson M, Cailliet GM (1992) Seasonal and spatial patterns of ichthyoplankton abundance in Elkhorn Slough, California. Estuar Coast Shelf Sci 34:109–126. https://doi.org/10.1016/S0272-7714(05)80099-X

Zacardi DM (2015) Variação e abundância do ictioplâncton em canais de maré no Extremo Norte do Brasil. Biota Amazônia 5(1):43–52. https://doi.org/10.18561/2179-5746/biotaamazonia.v5n1p43-52

Zacardi DM, Bittencourt SCD (2017) Caracterização morfológica de larvas de peixes capturadas no complexo estuarino dos rios Pará e Paracauarí (estado do Pará—Brasil). ACTAFISH 5:78–102. https://doi.org/10.2312/ActaFish.2017.5.2.78-102

Zacardi DM, Bittencourt SCS, Rawietsch AK et al (2008) Ictioplâncton marinho da Plataforma Continental e águas adjacentes à Foz do Rio Amazonas (Operação Norte-REVIZEE SCORE NORTE). Bol Tec Cient CEPNOR 8:9–20. https://doi.org/10.32519/tjfas.v8i1.1208

Zacardi DM, Sobrinho AF, Silva LMA (2014) Composition and distribution of larval fishes of an urban tributary the mouth of Amazon River, Brazil. Acta Fish Aquat Res 2(2):1–16. https://doi.org/10.2312/ActaFish.2014.2.2.1-16

Zacardi DM, Bittencourt SCS, Nakayama N (2016) O ictioplâncton e sua relação com a variação diária e os ciclos de marés no estuário amazônico. Biota Amazônia 6:32–40. https://doi.org/10.18561/2179-5746/biotaamazonia.v6n2p32-40

Zacardi DM, Ponte SC, Ferreira LC et al (2017) Diversity and spatio-temporal distribution of the ichthyoplankton in the lower Amazon River, Brazil. Biota Amazônia 7:12–20. https://doi.org/10.18561/2179-5746/biotaamazonia.v7n2p12-20

Zacardi DM, Bittencourt SCD, De Queiroz HL (2020) Spatial and seasonal variability of ichthyoplankton in the middle Solimões and lower Japurá rivers, Central Amazon, Brazil. Pan-Am J Aquat Sci 15(4):258–271

Chapter 8
Circulation, Biogeochemical Cycles and CO_2 Flux Variability in the Tropical Atlantic Ocean

Moacyr Cunha de Araújo Filho, Syumara Queiroz, Carlos Esteban Delgado Noriega, Gbekpo Aubains Hounsou-Gbo, Julia Martins de Araújo, Alex Costa da Silva, Leticia Cotrim da Cunha, and Helen Michelle de Jesus Affe

Abstract The tropical Atlantic (TA) is a very complex region where ocean–atmosphere interactions, oceanic currents, and phenomena such as wave propagations and mesoscale activities occur. The region is considered oligotrophic due to relatively strong static stability, with a well-marked thermocline. Seasonality is mostly governed by the Intertropical Convergence Zone (ITCZ). The TA is the second-largest oceanic source of CO_2 to the atmosphere subject to equatorial upwelling, seasonal variations, and large river discharges, which drive the exchange of CO_2 between the sea and air. Despite the high net CO_2 emissions, zones of net atmospheric CO_2 uptake exist in the TA. This is mainly linked to the seasonal SST cycle and its associated thermodynamic effects on the partial pressure of CO_2 in the North Equatorial Current, as well as freshwater sources, such as the Amazon River flow. The Amazon River continuum and the North Brazilian chain of islands and sea mounts compose the equatorial broadband. This region has biogeochemical characteristics that distinguish it from eastern Northeast Brazil and sustain most productivity and fisheries of the northern and northeastern regions. This chapter summarizes the main characteristics of the circulation, biogeochemical cycles and CO_2 flux variability in the tropical Atlantic Ocean.

M. C. de Araújo Filho (✉) · S. Queiroz · C. E. D. Noriega · A. C. da Silva
Federal University of Pernambuco (UFPE), Recife, Pernambuco, Brazil
e-mail: moa@ufpe.br

M. C. de Araújo Filho · C. E. D. Noriega · L. C. da Cunha · H. M. de J. Affe
Brazilian Research Network on Global Climate Change (Rede CLIMA), São José dos Campos, São Paulo, Brazil

L. C. da Cunha · H. M. de J. Affe
State University of Rio de Janeiro (UERJ), Rio de Janeiro, Rio de Janeiro, Brazil

J. M. de Araújo
University of São Paulo (USP), São Paulo, São Paulo, Brazil

G. A. Hounsou-Gbo
Federal University of Ceará (UFC), Fortaleza, Ceará, Brazil

J. M. Landim Dominguez et al. (eds.), *Tropical Marine Environments of Brazil*,
The Latin American Studies Book Series, https://doi.org/10.1007/978-3-031-21329-8_8

Keywords Tropical Atlantic · Western boundary circulation · Biogeochemical cycles · Climate variability

8.1 The Western Boundary Systems of Brazil: General Overview

The tropical Atlantic (TA) is a very complex region where ocean–atmosphere interactions, oceanic currents, and phenomena such as propagation of gravitational waves and mesoscale phenomena occur. Climate seasonality in this region is governed by the Intertropical Convergence Zone (ITCZ), which reaches its southernmost position between February and March (austral summer) (Lumpkin and Garzoli 2005; Hounsou-Gbo et al. 2015, 2016). The warmest season occurs between March and June when SST reaches values above 28 °C. The coldest season occurs between August and November, with SST values of approximately 26.5 °C (Hounsou-Gbo et al. 2015; Domingues et al. 2017).

The tropical Atlantic is characterized by permanent oligotrophic conditions outside the areas affected by river discharge (Da Cunha and Buitenhuis 2013) due to the relatively strong static stability with a well-marked thermocline, which is seasonally modulated by the meridional displacement of the ITCZ, controlling the regime of precipitation and trade winds (Araujo et al. 2011; Assunção et al. 2020). Southeast trade winds blow from June to November, while northeast trade winds are dominant from December to May (Geyer et al. 1996; Silva et al. 2005, 2010). The permanent thermocline restricts vertical mixing and nutrients up to photic layers and constraints biological productivity (Araujo et al. 2019).

8.1.1 Ocean Circulation

The upper circulation in the TA comprises the equatorial system (ES) and the Western Boundary System (WBS). The first is represented by the zonal flows that connect the eastern to the western boundary, while the latter is represented by predominantly meridional flows.

The ES starts with the South Equatorial Current (SEC), which ranges from 20° S to 4° N. This current is divided into the North (nSEC), Central (cSEC) and South (sSEC) branches (Fig. 8.1) (Lumpkin and Garzoli 2005). The nSEC flows in the Northern Hemisphere (Lumpkin and Garzoli 2005), and the cSEC flows at approximately 4° S and presents seasonal variability with a decrease in intensity at the beginning of April and November and intensification in the middle of July (Lumpkin and Garzoli 2005).

The sSEC extends from 8° S to 20° S and reaches the Brazilian continental margin between 12° and 20° S (Veleda et al. 2011; Dossa et al. 2021). The bifurcation of this branch on the Brazilian coast is controlled by wind stress variability and initiates

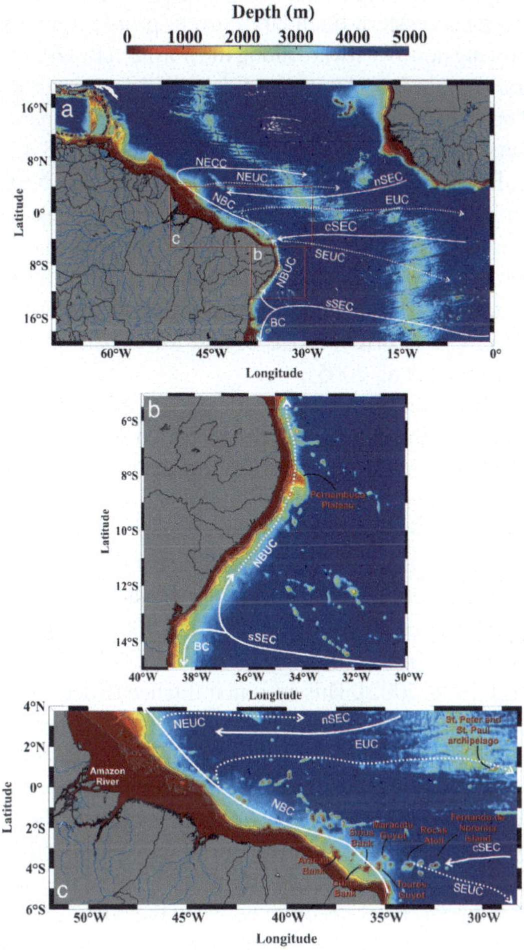

Fig. 8.1 Tropical Atlantic (TA) main currents (**a**), eastern Northeast Brazil (**b**) and equatorial broadband (**c**). The white continuous lines denote surface currents, and the dotted lines denote undercurrents. BC: Brazil Current; nSEC/cSEC/sSEC: north/central/south South Equatorial Current; NBC: North Brazil Current; NECC: North Equatorial Counter Current; NBUC: North Brazil Undercurrent; NEUC: North Equatorial Undercurrent; and SEUC: South Equatorial Undercurrent

the WBS currents: the North Brazil Undercurrent (NBUC) and Brazil Current (BC) (Silva et al. 2009a) (Fig. 8.1). The BC runs southward, and the NBUC runs northward along the northeastern Brazilian coast (Schott et al. 2005). The period of greatest (lowest) NBUC intensity occurs when the position of the sSEC bifurcation is further south (north) between May and July (October and December) (Silva et al. 2009a, b; Veleda et al. 2011). Additionally, recent studies have pointed out that the position of the bifurcation controls current-driven upwelling in Salvador Canyon (13° S) (Aguiar et al. 2018).

At approximately 4° S, the NBUC joins the cSEC initiating the North Brazil Current (NBC), which is another WBS current. The shelf off Calcanhar Cape (5° S) has been considered the boundary between the Northeast and North Brazilian regions. North of Calcanhar Cape, the NBUC-cSEC-NBC current system presents a two-core profile (in surface and subpycnocline levels), which evolves then into a typical western boundary current profile with a single surface core—the NBC—flowing northwestward along the northern Brazilian coast. The NBC has its maximum transport in July/August and its minimum transport in April/May (Johns et al. 1998). The dynamics on the northern Brazilian continental margin are also influenced by the seasonality of the Amazon River (see Sect. 8.4.2). At approximately 2°–3° N, the nSEC merges with the North Brazil Current (NBC) (Lumpkin and Garzoli 2005), which flows northward to the Caribbean Sea (Schott et al. 2002; Cintra et al. 2015) (Fig. 8.1a). The NBC also has strong seasonality, and the strength of the NBC retroflection peaks in boreal summer and fall (Richardson and Reverdin 1987). The water from the retroflection was observed to be distributed between the North Equatorial Undercurrent (NEUC) and lower part of the Equatorial Undercurrent (EUC) (Stramma et al. 2005).

The equatorward flow of the NBUC-NBC contributes to the upper branch of global thermohaline circulation, the so-called Atlantic Meridional Overturning Circulation (AMOC), and wind-driven subtropical cells (Schott et al. 2004, 2005; Brandt et al. 2006), which connect the southern warm surface waters to the Northern Hemisphere, crossing the TA (Gordon 1986).

The South Equatorial Undercurrent (SEUC)—a permanent countercurrent in the subthermocline layer and at approximately 4.5° S (Cochrane et al. 1979)—is found at depths that vary between 200 and 500 m with west–east flow and presents greater intensities in the western portion of the Atlantic between July and September (Schott et al. 1998, 2002). This current influences Rocas Atoll (RA) and Fernando de Noronha Island (FN) (Fig. 8.1c) subsurface dynamics and, similar to the NBUC-NBC, is part of the superior branch of the AMOC (Bourlès et al. 1999; Zhang et al. 2019).

Many authors have confirmed the existence of equatorial current recirculation feeding the westward flow and have observed the presence of oxygen-rich waters—typically Southern Hemisphere waters—in the SEUC (Arhan et al. 1998). However, there is still no consensus on the origins of the SEUC. Some authors have claimed that there is no connection between the NBUC-NBC and SEUC (Brandt et al. 2006), and others have affirmed that there is a connection between the western boundary currents and SEUC (Silveira et al. 1994; Schott et al. 1998, 2002). Schott et al. (2002) found that high oxygen values in the SEUC suggest a western boundary origin, while Schott et al. (1998) did not observe any contribution of the NBUC to the SEUC and rather suggested that the SEUC originates from SEC recirculation. Dossa et al. (2021) found that north of 3.8° S, at 35° W, the core of the SEUC extends from 120 to 330 m depth with velocities up to 0.2 m/s and transports 2.6 ± 1.2 Sv. Furthermore, the water mass at 3.8° S, 35° W, corresponding to the SEUC, is oxygen-depleted in the subsurface in the fall (April–May), which is the opposite pattern of the NBUC (Dossa et al. 2021). However, in spring (September–October), the NBUC can retroflect to feed the SEUC (Goes et al. 2005; Costa da Silva et al. 2021). In summary, little is known about the space–time variability of the SEUC, and the origin of this current remains unclear.

8.1.2 Ocean–Atmosphere Teleconnections and Climate Variability

The climate of South America is primarily driven by the variability of the tropical Atlantic and Pacific oceans. The El Niño Southern Oscillation (ENSO), which peaks in boreal winter in the equatorial Pacific, is known to highly influence the climatic variability in South America. During El Niño (La Niña), northern South America experiences dry (wet) events, while southeastern South America experiences wet (dry) events (Rao and Hada 1990; Grimm et al. 2000).

In the Atlantic, the interhemispheric mode is the principal mode of variability that influences the Northeast Brazilian climate. A positive (negative) north–south gradient is associated with a dry (wet) rainy season in Northeast Brazil. The equatorial mode or the Atlantic Niño mode, which peaks in boreal summer, is the second mode of variability of the TA. Previous studies have shown that this equatorial mode interacts with ENSO through atmospheric bridges (Rodriguez-Fonseca et al. 2009; Keenlyside et al. 2013). The Atlantic Niño (ATL3 index), calculated from SST anomalies in the 3° N–3° S, 20° W–0° region, is negatively correlated with the winter Pacific ENSO (Niño3 index), calculated from SST anomalies in the 5° N–5° S, 150° W–90° W region, with the Atlantic leading the Pacific by one season. Positive SST anomalies in the equatorial Atlantic in summer are associated with negative SST anomalies in the eastern equatorial Pacific several months later (Losada and Rodríguez-Fonseca 2016; Cai et al. 2019). This connection was strong during the first and last decades of the twentieth century and weak between these periods (Martín-Rey et al. 2015; Lübbecke et al. 2018).

In addition to the summer Atlantic Niño, the equatorial Atlantic is also characterized by a second Atlantic Niño mode in boreal winter (hereafter called winter Atlantic Niño), which is much weaker than that of the summer. Hounsou-Gbo et al. (2020) investigated the influence of both summer and winter Atlantic Niño on ENSO and South American climate variability for the 1905–2014 period. The results indicated that the winter Atlantic Niño is also negatively correlated with ENSO, with the Atlantic leading the Pacific by 2–3 seasons (Hounsou-Gbo et al. 2020). This winter Atlantic Niño leads to an early development of ENSO from boreal summer onward, with a noticeable multidecadal modulation of the lead time. A nearly 1-year leading connection between winter Atlantic Niño and the following ENSO was generally observed in the mid-twentieth century, mainly when the summer Atlantic Niño teleconnection with the subsequent winter ENSO was weak. Furthermore, Hounsou-Gbo et al. (2020) suggested that the same mechanism of the Atlantic–Pacific Niño teleconnection operates during both summer ATL3 and winter ATL3 events. Positive/negative SST anomalies in the equatorial Atlantic affect the Walker circulation with anomalous ascending/descending branches in the Atlantic and descending/ascending branches in the central equatorial Pacific. These perturbations in the zonal atmospheric circulation drive anomalous easterly/westerly surface winds favorable for ENSO development (Keenlyside et al. 2013; Losada and

Rodríguez-Fonseca 2016). However, the lead time of the strongest negative correlation between the winter Atlantic Niño and the following winter Pacific ENSO is clearly higher than that previously identified (i.e., summer ATL3 through subsequent winter ENSO) (Fig. 8.2).

Both the summer and winter Atlantic Niño events can be relevant for Pacific ENSO predictability (Hounsou-Gbo et al. 2020). The advantage of the winter Atlantic Niño is that it could bridge the spring barrier to ENSO prediction. Nevertheless, the mechanisms of this delayed oceanic response in the equatorial Pacific are still not well known. The Atlantic–Pacific teleconnection is associated with variability in each basin at a multidecadal timescale that can be modulated by the Atlantic Multidecadal Oscillation (AMO). Recent studies have indicated that the AMO can modulate the multidecadal variability of the Atlantic equatorial mode (Martín-Rey et al. 2018), although there are still several uncertainties regarding the mechanisms that control the variability of the equatorial Atlantic from seasonal to decadal timescales. Some authors have indicated that the variance in the Atlantic Niño depends primarily on thermodynamic feedback rather than on the dynamic interaction between the ocean and atmosphere (Nnamchi et al. 2015). Other studies have supported the important contribution of dynamic processes to the variability of the equatorial Atlantic (Jouanno et al. 2017). Other mechanisms, such as meridional advection of SST anomalies from the north to the equator and reflection of Rossby waves in the western boundary, have also been suggested to generate the Atlantic Niño (Lübbecke and McPhaden 2012; Lübbecke et al. 2018).

Summer Atlantic events indirectly impact South American rainfall variability during boreal winter and spring through their teleconnection with Pacific ENSO (Hounsou-Gbo et al. 2019, 2020). The winter Atlantic events also influence the seasonal precipitation of South America in two ways. First, these events affect the spring northern Northeast Brazil (NEB) rainfall through evolution into the meridional mode of the tropical Atlantic (Okumura and Xie 2006; Hounsou-Gbo et al. 2020). Winter Atlantic events can also influence South American precipitation through their connection to the following year's ENSO (Hounsou-Gbo et al. 2020).

The influence of the positive/negative phase of the Atlantic interhemispheric mode and El Niño/La Niña are mainly strong over the northern part of NEB in late boreal winter and early spring. In eastern NEB (5°–11° S; 34.5°–37° W), the rainy season, which peaks in austral winter from May to July (MJJ), is mainly linked to events occurring in the tropical South Atlantic, such as easterly disturbance activities (Kouadio et al. 2012; Silva et al. 2018). The seasonal establishment of the southern ITCZ (Grodsky and Carton 2003) also coincides with the rainy season in eastern NEB. This southern ITCZ is associated with atmospheric convection and rainfall over warm waters in the southwestern Atlantic warm pool, around June–July.

Hounsou-Gbo et al. (2015, 2019) investigated how earlier (nearly 6 months of lead time) oceanic–atmospheric conditions in the tropical South Atlantic influence the climate of eastern NEB. They identified a significant positive relationship between the eastern NEB rainfall anomalies and the SST anomalies in the southeastern tropical Atlantic (SETA), when the SETA leads rainfall by 4–6 months (Fig. 8.3a). This positive relationship passes through the oceanic portion near the eastern coast of

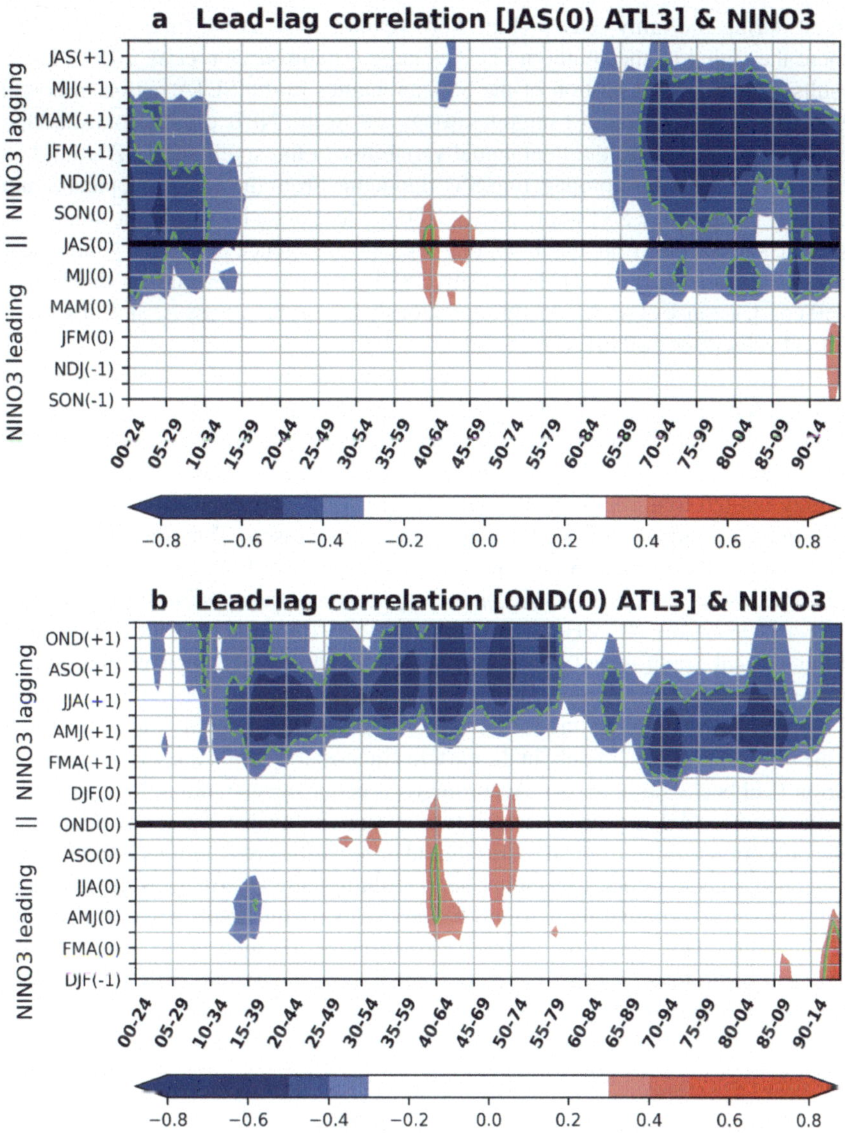

Fig. 8.2 Lead–lag correlation, with a 25-year running window, between the Niño3 (5° N–5° S, 150°–90° W) SST index and ATL3 (3° N–3° S, 20° W–0°) SST index in **a** July–September [JAS(0)] and **b** October–December [OND(0)]. The horizontal black lines at JAS(0) in **a** and OND(0) in **b** indicate the zero-lag correlation between ATL3 and Niño3. The values on the *y*-axis are the 3-month mean of Niño3. Values below horizontal black lines in **a** and **b** indicate Niño3 leading the ATL3. Values above the horizontal black lines in **a** and **b** indicate Niño3 lagging ATL3. The green contours show correlations significant at the 95% confidence level using a t test. SST data are from HadISST (Hadley Center Global Sea Ice and Sea Surface Temperature) for 1900–2018

Brazil during the rainy season (Fig. 8.3b) (Cintra et al. 2015; Silva et al. 2018), where the signal of the southern ITCZ is large (Hounsou-Gbo et al. 2019). These results indicate that a portion of the SST anomalies in the SETA region in boreal winter (November–January) migrates westward to the NEB coast (Hounsou-Gbo et al. 2015). Therefore, the interannual variability in the southern ITCZ should be related to the westward propagating SST anomalies from the eastern tropical South Atlantic. The positive SST anomalies in NDJ are also associated with an anomalous cyclonic surface atmospheric circulation (i.e., a convergence of surface winds) near the coast in May–July (MJJ). The positive SST anomalies, located near the eastern NEB coast, are also associated with anomalously low surface pressure and an upward vertical velocity at 500 hPa (Fig. 8.3c).

The predictability of rainfall over several months in NEB, a largely semiarid region, is fundamental for water resource decision-making, on which several millions of people depend. Currently, several climate forecast centers use the Pacific Niño indices and the Atlantic "dipole" index to forecast the seasonal rainfall in NEB (mainly limited to northern NEB), but this forecast is provided for relatively short lead times, i.e., just before and/or during the rainy season. Therefore, the surface conditions in the tropical South Atlantic, and especially SST anomalies near the African coast, may be relevant for predicting seasonal precipitation several months in advance in eastern NEB. Hounsou-Gbo et al. (2019) indicated that the combination of the tropical South Atlantic index with ENSO could clearly improve the forecast ability of eastern NEB rainfall over several months.

Finally, current climate model simulations present large warm systematic errors in the tropical Atlantic, mainly in the eastern equatorial Atlantic and in the Benguela region (Lübbecke et al. 2018). Accordingly, a better understanding of the mechanisms that drive Atlantic Niño variability from seasonal to decadal timescales and its teleconnection to other tropical basins should improve the predictability of ENSO and associated climate impacts for future years. Additionally, in situ measurements and long-term monitoring systems would be very relevant for South American climate predictability. This could be achieved, for instance, through an extension of the "Prediction and Research Moored Array in the Tropical Atlantic" PIRATA network (Bourlès et al. 2019).

8.2 Biogeochemical Cycles and CO_2 Flux

The tropical Atlantic is the second-largest oceanic source of CO_2 to the atmosphere after the tropical Pacific (Takahashi et al. 2009; Araujo et al. 2019). This oceanic region is subject to equatorial upwelling (Andrié et al. 1986), seasonal variations (warming/cooling, seasonal migration of the Intertropical Convergence Zone), interannual climatic variability (Lefèvre et al. 2013), and large river discharges (Jacobson et al. 2007), which drive sea–air CO_2 exchanges. Takahashi et al. (2009) estimated the total sea–air CO_2 flux in this region to be 0.10 Pg C yr^{-1} (2000 reference year).

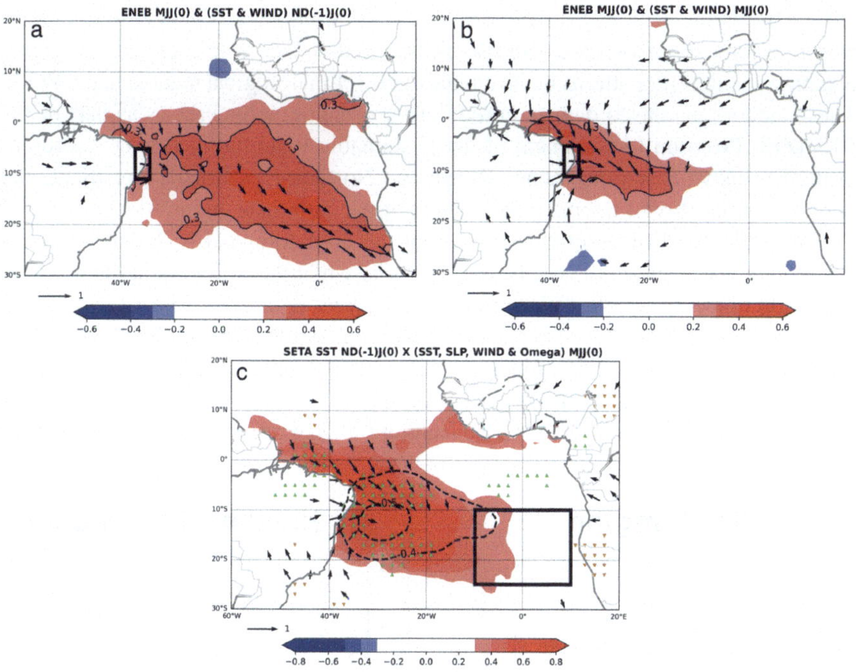

Fig. 8.3 Maps of the lagged linear correlation between the May–July(0) [MJJ(0)] rainfall anomalies in eastern NEB (black box in **a** and **b**, from Global Precipitation Climatology Center—GPCC) and the gridded SST anomalies (color, from the Hadley Center Global Sea Ice and Sea Surface Temperature—HadISST) in **a** November–December(−1)–January(0) [ND(−1)J(0)] and **b** May–July(0) [MJJ(0)]. The vectors represent the linear correlation (significant at the 95% confidence level according to the t test) between the rainfall anomalies in eastern NEB and the surface wind vectors (u and v; from the NCEP anomalies). The contours in **a** and **b** indicate regions of significant correlation at the 95% confidence level for SST anomalies. **c** Map of the lagged linear correlation between the SSTA inside the southeastern tropical Atlantic (SETA; 10° S–25° S, 10° W–10° E; oceanic black box) in ND(−1)J(0) and the gridded SST (shaded), surface wind (vectors), 500 hPa vertical velocity (triangles; green triangle-up indicates upward motion and brown triangle-down indicates downward motion) and sea level pressure (contours) anomalies within the entire tropical Atlantic in MJJ(0) for the 1960–2018 period. Only the correlations that are significant at a 95% confidence level, according to the t test, are plotted for all variables in (**c**)

The western tropical Atlantic Ocean (WTA) is an area whose participation in global inorganic carbon cycling is not completely understood. Despite the net CO_2 outgassing recorded in the tropical Atlantic, zones of net atmospheric CO_2 uptake exist, which are mainly linked to the seasonal cycle of sea surface temperature (SST) and its associated thermodynamic effects on the partial pressure of CO_2 (pCO_2) in the North Equatorial Current (Ibánhez et al. 2015), as well as freshwater sources to the basin.

Recent research efforts in this area have constrained the relationships between carbon parameters and temperature (SST), salinity (SSS) and chlorophyll-*a* (Bonou et al. 2016; Bruto et al. 2017; Araujo et al. 2019). Araujo et al. (2019) used a combination of long series of SST, SSS and chlorophyll-*a* data from regional and global databases to identify patterns in the annual cycle (climatological means) in the WTA. These authors also used data acquired during oceanographic cruises as part of the REVIZEE Program (Assessment of the Sustainable Potential of Living Resources in the Exclusive Economic Zone—North and Northeastern Brazil) to estimate the temporal variability of the main parameters of the carbonate and nutrient system.

The 1955–2012 time series showed that the climatological SST data exhibit a strong seasonal cycle in the northeastern region, while the SSS data in the northern region do not vary significantly throughout the year (Fig. 8.4). The largest annual SSS variations were observed in the northern region, while in the northeastern region, the range of variation in SSS values was lower than 2 units (Fig. 8.4).

Satellite records and in situ observations showed that the northern region is characterized by average chlorophyll-*a* concentrations that are one order of magnitude

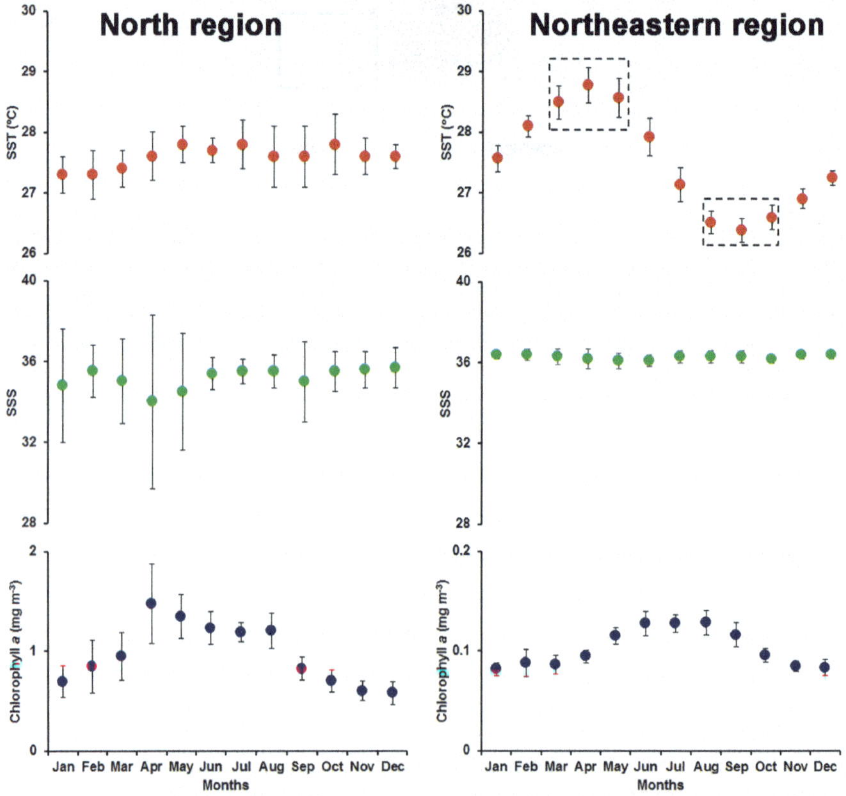

Fig. 8.4 SST and SSS (monthly mean ± SD; $n = 57$) climatology in the tropical Atlantic (1955–2012). Chlorophyll-*a* (monthly mean ± SD; $n = 11$) (2°–12° S; 30°–40° W, 4 km resolution). The dashed-line boxes in the figure indicate the periods with the highest and lowest observed SST values. Araujo et al. (2019)

higher than those in the northeastern region (with average values of 1.0 ± 0.3 and 0.1 ± 0.02 mg m^{-3}, respectively) (Fig. 8.4).

According to Araujo et al. (2019), the strong fluctuations observed in SSS were a result of high discharge in the northern region. The amplitude of SST was influenced by mixtures of diverse water masses, such as the South Equatorial Current (SEC) and the southern branch of the South Equatorial Current (sSEC) (Fig. 8.5). Additionally, these authors considered differences in time periods between the different cruises, with the seasonal SSS distribution in the northern region related to the periods of maximum (April–June) and minimum (October–December) discharge of the Amazon River. The lowest SSS values (< 15 units) were found from April to June (N3), and the highest SSS values (> 35 units) were found between October and December (N2) (Fig. 8.5f, e respectively). In the NE region, the recorded SSS showed a smaller amplitude than that in the N region (35.3–37.4), which was attributed to the lack of large freshwater inputs (Fig. 8.5e, g).

The concentrations of dissolved inorganic nitrogen (DIN), PO_4^- and SiO_2^- were higher in the N region than in the NE region, mainly due to fluvial input (Araujo et al. 2019). Nitrogen: phosphorus (N:P) ratios of < 16 and oxygen supersaturation were observed within the Amazon plume (Araujo et al. 2019). The concentrations of riverine nutrients in the N region support primary production occurring in the offshore plume area (Fig. 8.6a, e, i).

Estimates of the main parameters of the carbonate system (pCO_2, alkalinity and CO_2 fluxes) at the western edge of the tropical Atlantic Ocean showed significant statistical variations between the region near the Amazon River (north) and the oceanic region of northeastern Brazil (Araujo et al. 2019) (Fig. 8.7e, f).

The pCO_2 values observed in the NE region (average: 379 ± 16 µatm) were higher than those observed in the N region (average: 355 ± 52 µatm). The calculated pCO_2 values reached 423 µatm in the oceanic region offshore of the NE region during the warmer period (Fig. 8.7b, NE3 campaign) (Araujo et al. 2019). The pCO_2 values presented an average amplitude of ~ 23 µatm between the two opposite periods [high SST (NE2 and NE3)–low SST (NE1 and NE4)], which is concomitant with a difference in SST of 1.4 °C observed between both periods.

The average sea–air CO_2 fluxes estimated ranged from -1.2 ± 2.6 to $+2.0 \pm 1.0$ mmoles m^{-2} d^{-1}. The N campaigns showed negative average fluxes (range of -1.2 ± 2.6 to $+0.5 \pm 2.1$ mmoles m^{-2} d^{-1}), while in the NE region, the estimated average CO_2 flux ranged from $+0.3 \pm 0.3$ to $+2.0 \pm 1.0$ mmoles m^{-2} d^{-1} (Araujo et al. 2019) (Fig. 8.7e–h).

The largest negative sea–air CO_2 fluxes in the N region occurred during the period of high discharges of the Amazon River (Fig. 8.7f). The NE region showed positive CO_2 fluxes (acting as a CO_2 source to the atmosphere) during all periods (Fig. 8.7e–h). The highest CO_2 fluxes were associated with periods of high SST in the study area (Araujo et al. 2019) (Fig. 8.7e–h).

The variability in the sea–air CO_2 fluxes in the N and NE regions was explained by variations in biological activity and the thermodynamic effect of temperature, respectively. The surface water pCO_2 values showed a positive temporal trend ($+1.10 \pm 0.2$ µatm yr^{-1}) in the NE region during the 1987–2010 period (Araujo et al. 2019).

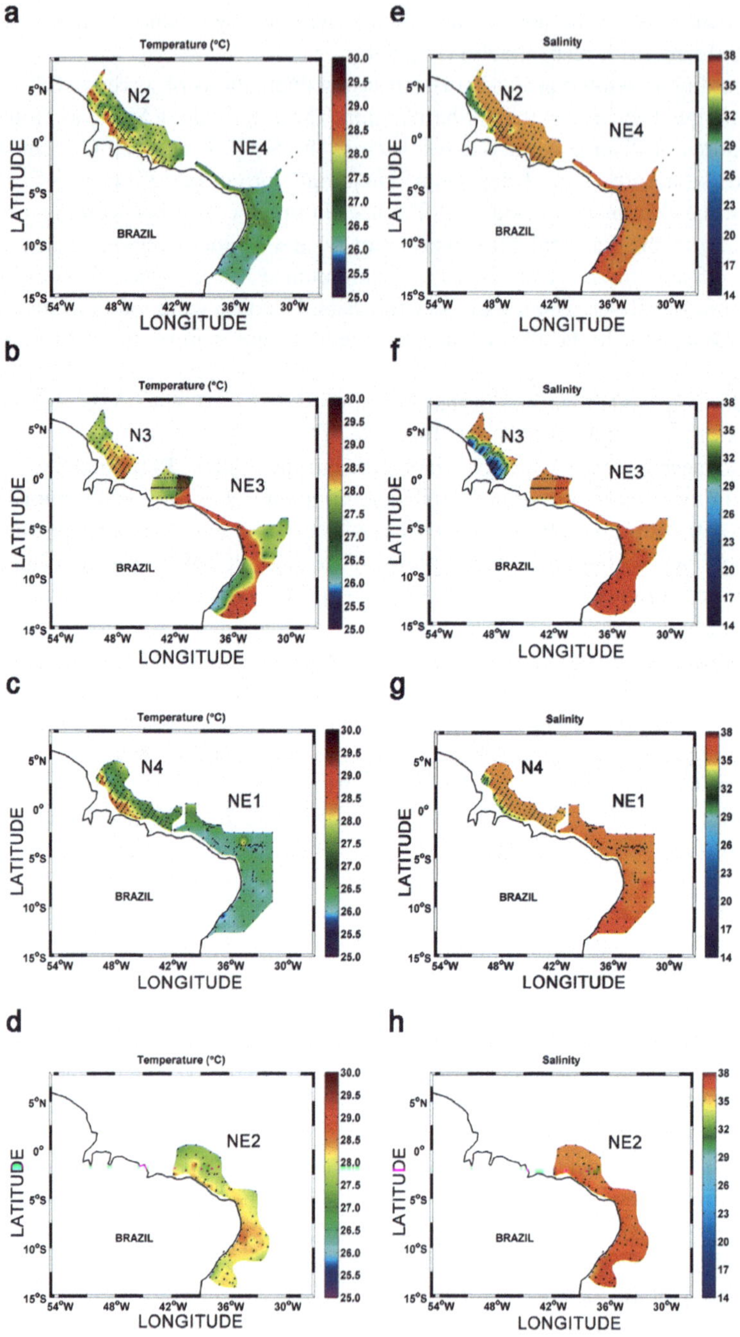

Fig. 8.5 **a**–**d** SST and **e**–**h** SSS (in the tropical Atlantic during REVIZEE cruises). The campaigns are grouped by climatic periods (N2 and NE4—October to December; N3 and NE3—April to June; N4 and NE1—July to September; and NE2—January to March). Araujo et al. (2019)

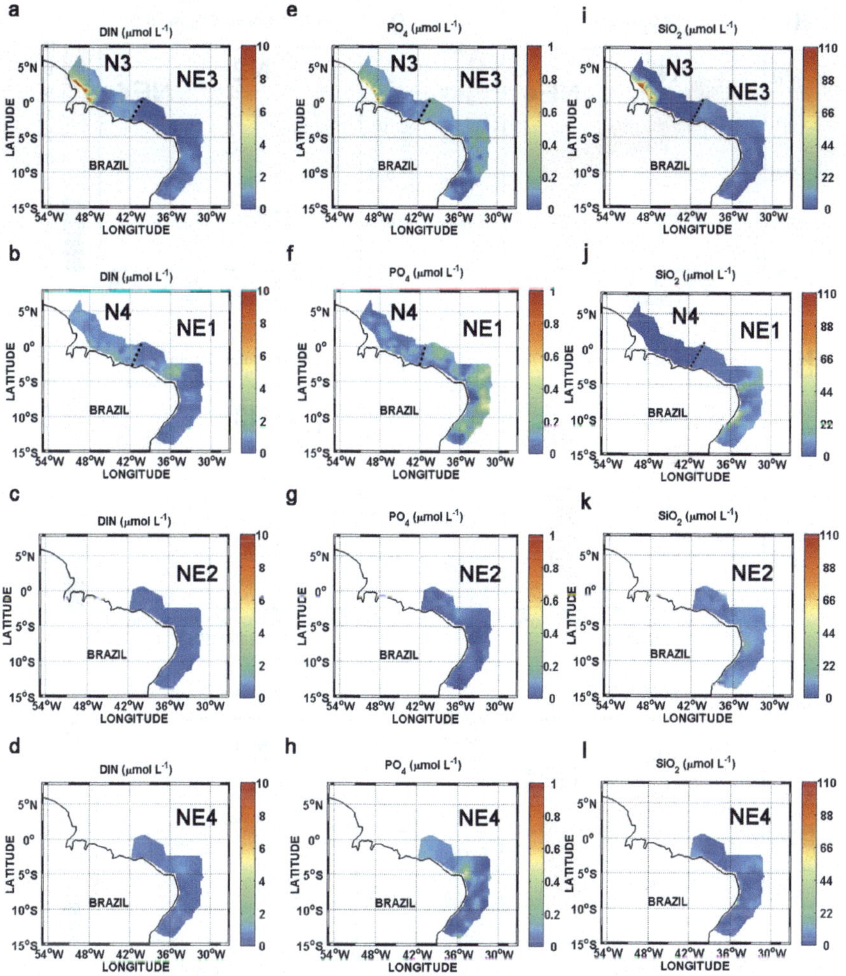

Fig. 8.6 **a–d** DIN, **e–h** PO4 and **i–l** SiO_2 concentrations in the western tropical Atlantic during REVIZEE cruises. The campaigns were grouped by climatic periods (N2 and NE4—October to December; N3 and NE3—April to June; N4 and NE1—July to September; and NE2—January to March). Araujo et al. (2019)

This rate of increase was lower than that verified to have occurred in the atmosphere (+1.72 ± 0.01 µatm yr^{-1}) during the same period.

The spread of the Amazon River waters in the western tropical Atlantic is known to have developed an extensive area of enhanced primary production and associated CO_2 undersaturation; it thus acts as an atmospheric CO_2 sink of global relevance (Ibánhez et al. 2015, 2016; Araujo et al. 2019). The systematic overlooking of this significant atmospheric CO_2 sink has resulted in the current sea–air CO_2 balances

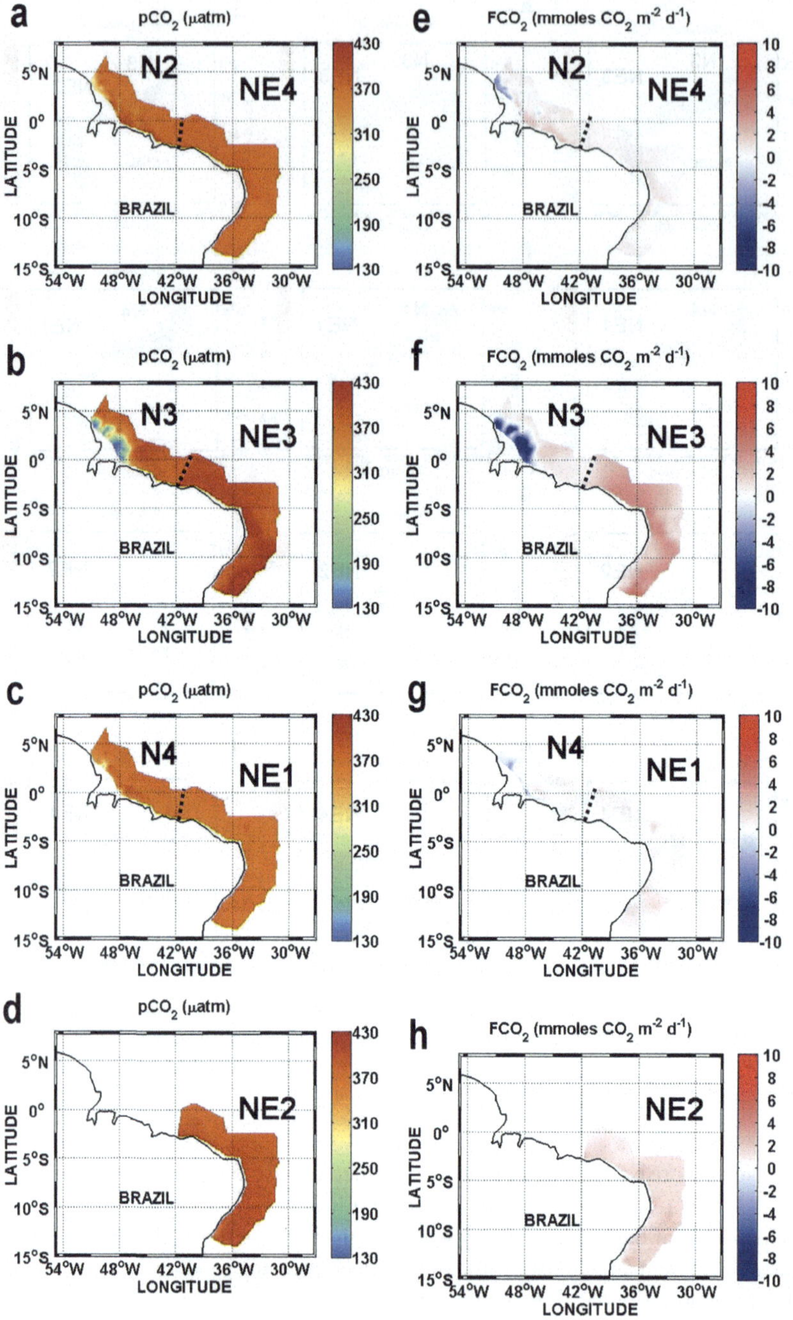

Fig. 8.7 **a–d** pCO_2 values in the western tropical Atlantic during REVIZEE cruises. **e–h** CO_2 fluxes in the western tropical Atlantic during REVIZEE cruises. The campaigns were grouped by seasonality (N2 and NE4—October to December; N3 and NE3—April to June; N4 and NE1—July to September; and NE2—January to March). Araujo et al. (2019)

for the equatorial Atlantic being overestimated by approximately 10% (Ibánhez et al. 2016).

Rivers and estuaries play an important role in the transport and transformation of terrestrial carbon to the adjacent coastal zone and typically act as sources of CO_2 to the atmosphere (Noriega and Araujo 2014). In addition, anthropogenic input can change the biogeochemical composition of rivers and estuaries, changing the quality of coastal waters and leading to eutrophication processes. Very little is known about the seasonal and interannual variability of the CO_2 flux at the air–water interface in rivers and estuaries in the western tropical Atlantic (WTA), particularly in the northeastern region of Brazil.

The global exchange of CO_2 between inner estuaries and the atmosphere was recently evaluated using data from 106 systems (Chen et al. 2012). The estimated global emissions of 0.26 Pg C yr^{-1} from these systems are lower than previous estimates, which have been in a range of 0.27–0.60 Pg C yr^{-1} (Borges 2005; Laruelle et al. 2010). However, all recent studies have reported higher values than those first reported (0.1 Pg C yr^{-1}) (Kempe 1984).

Estuaries at low (tropical) latitudes have received less attention, although the total surface area of low latitude estuaries is greater than that of estuaries in middle and high latitude systems (Borges 2005). Chen et al. (2012) estimated that the mean CO_2 flux in estuaries at low latitudes (<23.5°) is 23.9 ± 39.5 mol m^{-2} yr^{-1}. Sarma et al. (2012) reported a mean CO_2 flux of 27 mmol m^{-2} d^{-1} from 27 Indian estuaries. Borges et al. (2005) noted that these current calculations are based on very limited datasets from subtropical and tropical regions. Two studies in tropical estuaries in northeastern Brazil reported values of + 30 and + 35.6 mmoles m^{-2} d^{-1} (Souza et al. 2009; Noriega et al. 2013, respectively).

In the WTA, the primary freshwater contributions originate from the Amazon, Orinoco, São Francisco and Paraíba do Sul rivers (Fig. 8.8) (Araujo et al. 2014). The inflow of the rivers varies seasonally, with a maximum in May and a minimum in November (Geyer et al. 1996; Dai and Trenberth 2002). Another study that included 28 estuaries in northern and northeastern Brazil reported a value of 55 ± 45 mmol m^{-2} d^{-1} (Fig. 8.9) (Noriega and Araujo 2014). Additionally, a negative correlation between dissolved oxygen saturation and pCO_2 was observed, indicating a control by biological processes and especially by organic matter degradation. This led to an increased dissolved CO_2 concentration in estuarine waters, which resulted in pCO_2 values that reached 8638 μatm (Noriega and Araujo 2014), which led these authors to suggest that northern and northeastern Brazilian estuaries act as sources of atmospheric CO_2.

In general, the estuarine systems of the great rivers emptying in the tropical Atlantic Ocean (Fig. 8.8) (Araujo et al. 2014) show lower CO_2 fluxes than those from northeastern Brazil studied by Noriega and Araujo (2014). Noriega and Araujo (2014) detected a strong negative correlation between the saturation of dissolved oxygen (DO%) and pCO_2 ($r2 = 0.92$), suggesting that intense organic matter decomposition in estuaries, either in the water column or the sediment, results in a decrease in pH and enhanced pCO_2 values. Other processes, such as direct groundwater discharge,

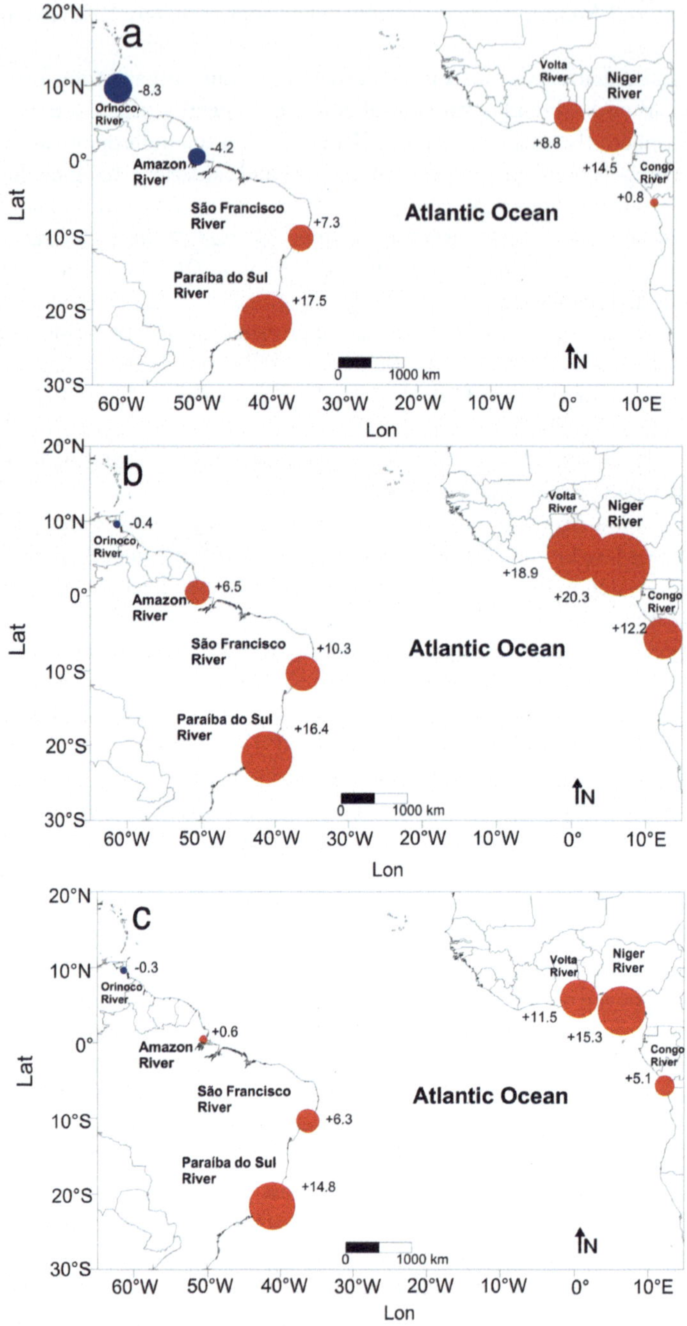

Fig. 8.8 **a** CO_2 flux (mmols m^{-2} d^{-1}) in the estuaries of the main rivers that flow into the tropical Atlantic during the low discharge period of these rivers. **b** CO_2 flux during the period of high flow of these rivers. **c** Annual average values of CO_2 flux. The red circles (+) indicate CO_2 sources to the atmosphere, and the blue circles (−) indicate atmospheric CO_2 sinks. Araujo et al. (2014)

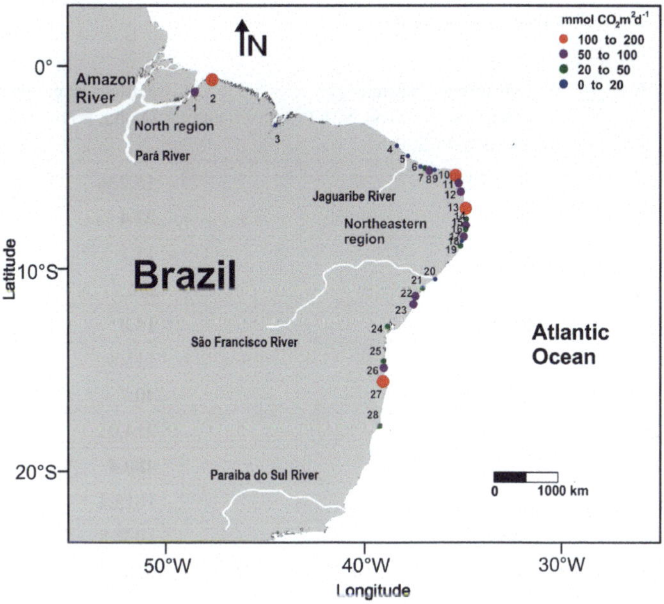

Fig. 8.9 CO_2 fluxes (levels) of 28 northeastern Brazilian estuaries (after Noriega and Araujo 2014). Estuaries: 1. Guajará Bay, 2. Marapanim, 3. São Marcos Bay, 4. Catú, 5. Jaguaribe, 6. Apodí, 7. Conchas, 8. Cavalos, 9. Assu, 10. Ceará-Mirim, 11. Potengi, 12. Guaráiras, 13. Paraíba, 14. Goiana, 15. Timbó, 16. Capibaribe, 17. Barra das Jangadas, 18. Sirinhaem, 19. Una, 20. São Francisco, 21. Poxim, 22. Real, 23. Itapicuru, 24. Paraguaçú, 25. Cachoeira, 26. Cururupe, 27. Acuñpe, and 28. Caravelas

biogeochemical reactivity at seepage faces or benthic organic matter processing, could also influence the observed surface water composition in these estuaries.

Similar to pCO_2, CO_2 fluxes also accompany this association. Noriega et al. (2015) also showed that population density exhibits a strong correlation with positive CO_2 fluxes (source of CO_2 to the atmosphere) when compared to local climate variations such as rainfall. Scattered data suggest that some of these areas are already showing signs of eutrophication.

If we now consider the main characteristics of the large estuarine systems of the tropical Atlantic Ocean (Table 8.1), it can be observed that CO_2 fluxes, dissolved inorganic nitrogen-DIN, dissolved inorganic phosphorus-DIP and dissolved inorganic carbon-DIC are significantly lower in the western estuarine systems than in the eastern estuarine systems of the tropical Atlantic. Among the possible causes, we can mention the following: (i) lithological composition, as African watersheds have a higher percentage of sand and sandstone, while western basins have a higher percentage of shale rock; (ii) the load of organic carbon is higher in the eastern rivers than in the western rivers.

Additionally, according to Amiotte Suchet et al. (2003), variations in hydroclimatic factors (precipitation, runoff and temperature) have a great influence on the

Table 8.1 Statistical analysis of categories

Categories	Spatial	
Parameters	Western river estuaries (Orinoco, Amazon, São Francisco and Paraíba do Sul)	Eastern river estuaries (Volta, Niger and Congo)
Discharge (m^3 s^{-1})	**52,870**	**15,236**
Temperature (°C)	**26.8**	**27.4**
Salinity	**6.9**	**11.1**
DIP (μmol kg^{-1})	**0.6**	**2.2**
DIN (μmol kg^{-1})	**10.4**	**14.8**
Silicate (μmol kg^{-1})	117.6	115.8
Iron (nmol kg^{-1})	**11.3**	**40.7**
DO (μmol kg^{-1})	205.7	214.0
DOC (μmol kg^{-1})	**388.4**	**480.4**
DIC (μmol kg^{-1})	**680.7**	**1013.4**
TA (μmol kg^{-1})	**685.8**	**1037.8**
pCO_2 (μatm)	**501.6**	**675.3**
FCO_2 (mmol m^2 d^{-1})	**+5.3**	**+10.6**

Numbers in bold indicate significant differences ($\alpha = 0.05$). The temporal analysis includes the three wettest months and the three driest months of each river estuary (Araujo et al. 2014)

global CO_2 consumed by rock weathering and on the riverine transport of dissolved elements into the oceans.

8.3 The Eastern Northeast Brazilian Offshore Region: Connecting the Central Atlantic

As presented in the previous section, there is a significant difference between the region under the influence of the Amazon River plume and the eastern Northeast region of Brazil. Eastern Northeast Brazil comprises the continental margin from 4° S to 16° S (Fig. 8.1b).

Eastern Northeast Brazil is under the influence of the western boundary system represented by the BC and the NBUC-NBC (Fig. 8.1a), fed by central Atlantic waters transported and modulated by the sSEC. The region is dominated by South Atlantic water masses: the tropical Atlantic water (TW) is located at the surface (0–100 m) (temperature > 25 °C and isopycnals $\sigma_\theta = 23$–24.5 kg m^{-3}); just below, the subtropical Underwater (SUW) flows between 80 and 150 m (maximum salinity > 36.5; $\sigma_\theta < 25$ kg m^{-3}); then, the South Atlantic Central Water (SACW) flows down to 500 m (temperature between 10 °C and 23 °C; salinity > 35; $24.5 < \sigma_\theta < 27$ kg m^{-3}); and finally, the Antarctic intermediate water (AAIW) flows at greater depths (low salinity and $\sigma_\theta > 27$ kg m^{-3}) (Dossa et al. 2021). The SACW originates in the subtropical South Atlantic and crosses the ocean from the Agulhas leakage; the TW originates from the surface circulation of the tropics and subduction in the tropical–subtropical transition region; and the AAIW originates at the surface of the circumpolar layer (Schott et al. 1998).

Regarding the seasonality of the regional circulation, current measurements acquired during the ABRACOS surveys (Bertrand 2015, 2017) indicate that the NBUC is trapped along the NEB coast between 9° S and 5° S. Its maximum velocity is on average located at approximately 200–300 m depth and could locally reach 1.2 m/s in spring 2015, in agreement with Silveira et al. (1994), among others. The NBUC carries equatorward high-salinity and oxygen-enriched South Atlantic water within its core and exhibits higher transport in spring than in fall (Bourlès et al. 1999). These results provide additional information to previous studies based on both in situ data (Stramma et al. 1995; Schott et al. 2005; Veleda et al. 2012; Hummels et al. 2015; Herrford et al. 2020) and numerical simulations (Rodrigues et al. 2007; Silva et al. 2009a; Veleda et al. 2011; Herrford et al. 2020). These studies reported a seasonal variability in the near-surface sSEC-NBUC-BC system transport associated with changes in the local wind stress curl due to the annual north–south excursion of the ITCZ. Indeed, in austral winter, the sSEC bifurcation latitude reaches its southernmost position, NBUC transport is higher and BC transport is lower (e.g., Rodrigues et al. 2007; Silva et al. 2009a). In contrast, the near-surface sSEC bifurcation latitude is located at its northernmost position during austral summer, when NBUC transport is seasonally lower and BC transport is higher. The wind velocity follows this pattern, with the highest values (> 9 m s^{-1}) between August and December and the lowest velocities (< 8 m s^{-1}) in April–May (Domingues et al. 2017).

The equatorward increase in NBUC strength is usually accompanied by an increase in its core. As previously described (e.g., Schott et al. 1998), this increase is attributed to Ekman drift, which is less intense north of the area (Stramma et al. 1995). Indeed, the westerly driven Ekman drift is south-westward in the region, constraining the equatorward flow to the subsurface from its origin to approximately 5° S, from where it decreases. Another reason for the latitudinal variation in the NBUC along the NEB continental slope is the orographic effect. Dossa et al. (2021) observed, irrespective of the season, a change in the zonal flow, shifting from eastward south of 7.5° S to westward farther north, and the intensification of the equatorward flow between 8° S and 7° S. This consistency suggests that the coastal curvature affects the mean nearshore circulation characteristics. In fact, using a two-layer model, Ou and De Ruijter (1986) pointed out that boundary currents are subjected to a separation where a positive coastal curvature occurs, being dominantly controlled by inertial and beta effects.

An aspect of the local current dynamics that seems to not be directly affected by seasonality is the crucial transition area where the SEUC and the NBC originate and which is located north of 5° S, between 36.5° W and 34.5° W (Schott et al. 1998; Bourlès et al. 1999; Goes et al. 2005). Dossa et al. (2021) provided a comprehensive description of the NBUC/NBC transition, its interaction with the cSEC, and further evidence about the origin of the SEUC. Although the situation described is based only on data from fall 2017, the feature and position of this transition may not significantly vary according to the season. Indeed, surface currents are stronger in this transitional area during fall than in spring, but the position where the geostrophic current increases is similar. In this area, the NBUC and cSEC coalesce to form the NBC as already reported in previous studies (Stramma et al. 1995; Schott et al. 1998). Moving to the west, a clear distinction between the NBC and cSEC is no longer possible in the upper layer.

One overarching question in the literature is the development of the NBUC after crossing 5° S. After the generation of the NBC in the upper layer, a subsurface part of the NBUC continues flowing in the same direction as the NBC. The increase in NBUC transport from 6.1 to 11.5 Sv south of 5° S to 15.8 $\pm$ 4.6 Sv north of this latitude at 35° W implies that the contribution of the cSEC to the western boundary equatorward flow is not restricted to the upper layer. The subsurface cSEC contribution maintains the NBUC flow northwestward after the formation of the NBC in the upper layer (Goes et al. 2005).

Farther south, a study was conducted by Aguiar et al. (2018) to investigate upwelling and uplift events from 2008 to 2012 in the southern branch of the South Equatorial Current bifurcation region in the Salvador Canyon near 13° S. The main results showed a clear seasonality in sea surface temperature, with cold anomalies related to upwelling events occurring preferentially from austral spring to summer (September to March). In turn, uplift events occurred more evenly throughout the year, with a higher frequency during the autumn/winter period. In general, uplift exhibited an approximately two times higher frequency of occurrence and lasted longer than upwelling across the region. Upwelling and uplift were enhanced in Salvador Canyon in comparison with the remaining domain. Both current and wind-driven mechanisms were responsible for the observed upwelling/uplift events (Aguiar et al. 2018). The mechanisms are similar to those of the wind field (E/NE), and the West Boundary Current (WBC), formed when the sSEC is located north of the domain, favors upwelling during the spring/summer period. During autumn and winter, the region is under upwelling-unfavorable conditions. The main difference between upwelling and uplift phenomena in the region is the magnitude of the forcing mechanisms. Wind transport is stronger than WBC transport in the canyon region, and the latter increases southward, overcoming wind transport. Therefore, wind-driven mechanisms predominantly induce upwelling/uplift processes, with the current-driven mechanism providing an additional contribution.

Presently, investigations regarding flow–topographic interactions are still being carried out in eastern Northeast Brazil (Silva et al. 2021a).

8.4 The Equatorial Broadband

The Brazilian equatorial broadband is a region that encompasses the coastal, shelf and oceanic zones adjacent to the northern and northeastern regions of Brazil, extending from 31.30° W to 52° W and including the oceanic islands and seamounts of the North Brazilian Chain (Fig. 8.1c).

8.4.1 The North Brazil Chain

The North Brazil Chain, which includes Rocas Atol (RA), Fernando de Noronha Island (FN), and submarine banks (Fig. 8.1c), notably host high biomass and biodiversity when compared with those of open-water oceanic regions (Morato et al. 2008; Pitcher et al. 2008; Koettker et al. 2010). The presence of these geological features serves not only as shelter and physical substrate for the development of several species but also induces a variety of flow phenomena.

For example, flow topographic interactions (FTIs) can promote turbulence and mixing, localized upwelling, formation of a Taylor column and/or induce an "island mass effect" (IME) (Doty and Oguri 1956). These processes can increase nutrient transfer from deeper to shallower waters (Flagg 1987) and can thus enhance primary production (Cushman-Roisin 1994), reflecting the concentration of chlorophyll-*a* in surface water (Souza et al. 2013). Indeed, some studies have suggested that significant chlorophyll enrichment near ocean islands is linked to the IME (Doty and Oguri 1956).

The IME phenomenon refers to the disturbance induced by an island, which changes the dynamics of ocean circulation around these areas, causing moderately deep water that is rich in nutrients to rise into the photic zone (Palacios 2002; Melo et al. 2012; Lira et al. 2014; Tchamabi et al. 2018). According to Gove et al. (2016) and Tchamabi et al. (2017), the current wakes create eddy vortices downstream of the island, which increases the mixing rate, and transport deep nutrients and CO_2-rich waters to the surface (small-scale upwelling), promoting the increase in productivity. However, these vortices are controlled by several factors, such as the incoming current velocity, island size, surrounding bathymetry and generated internal waves (Gove et al. 2006). FN and RA are both influenced by the westward flow of the cSEC and the deeper eastward SEUC (Fig. 8.1c). Oceanic areas near FN and RA are energetic regions subjected to strong seasonally driven features, such as the complex system of zonal equatorial currents and countercurrents, confluence of water masses, or trade wind systems (Araujo et al. 2011; Tchamabi et al. 2017, 2018; Foltz et al. 2019). Several studies have suggested that the high productivity found in both is a result of locally induced topographic upwelling, which does not reach the sea surface (Ekau and Knoppers 1999; Travassos et al. 1999; Becker 2001). Moreover, it was shown that primary production and ichthyoplankton and neuston abundance appear to increase in areas adjacent to FN and AR (Chaves et al. 2006). Souza et al. (2013) showed that nutrients exhibit an inverse linear relationship with temperature in FN and RA. Cordeiro et al. (2013) showed an increase in the concentration of integrated chlorophyll over a 100 m depth in the wake of FN and RA when compared to the concentration upstream of these islands. Jales et al. (2015) also noticed a significant increase in most environmental variables due to the turbulence downstream of AR.

They found that the concentration of chlorophyll-*a* and nutrients increased when the mixed layer temperature was reduced due to the influence of SACW.

Tchamabi et al. (2017) investigated oceanic current wake-induced disturbances/alterations occurring around FN and AR. Their results supported the IME hypothesis regarding the high productivity of subsurface waters generally observed on the western side of these islands. To account for the topographic effects on ocean dynamics, they performed two different numerical simulations using the Regional Oceanic Modeling System (ROMS). The first included FN and RA (Scenario I), and the second included the artificial removal of the island and atoll (Scenario NI). Scenario I reproduced well the wakes that give rise to the development of eddies downstream of FN and RA. These mesoscale structures had a strong influence on the thermodynamic properties. In the presence of FN and RA, on the western side of both, shoaling of the mixed layer was observed throughout the year. Additionally, the increase in mixing at the base of the mixed layer induced subsurface cooling, which was enhanced in the downstream portion of FN and RA, particularly when the cSEC was stronger (Fig. 8.10a, b).

Tchamabi et al. (2017) also showed that the chlorophyll-*a* concentration was higher downstream of FN and AR (Fig. 8.10c, d) and suggested that this pattern

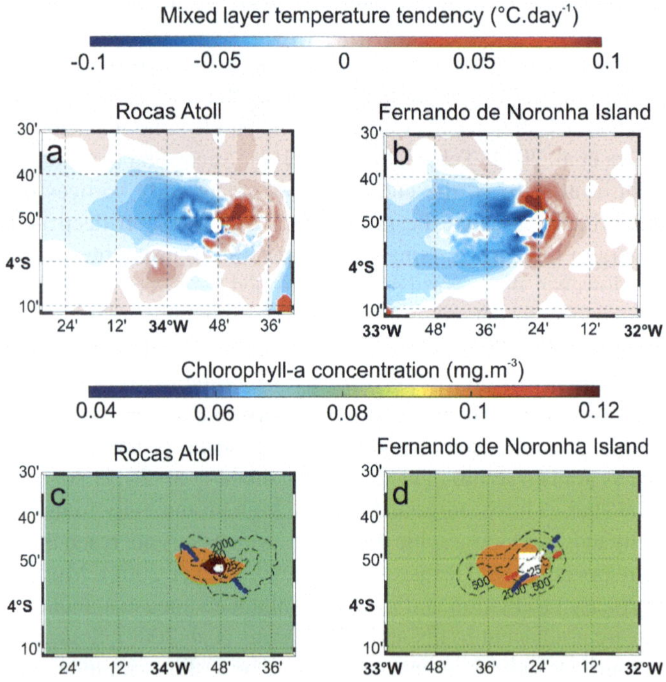

Fig. 8.10 Difference (I-NI) of the mixing term representing mixed layer temperature tendency (°C d^{-1}) for cooling for RA (**a**) and FN (**b**). Chlorophyll concentration (in mg m^{-3}) annual mean (2003–2014) from MODIS surrounding RA (**c**) and FN (**d**). Tchamabi et al. (2017)

was a result of the IME mechanism (Souza et al. 2013; Tchamabi et al. 2017). These authors argued that the impact of the current disruption due to the bathymetry of FN and RA could induce turbulent mixing, facilitating the enrichment of nutrients from the subsurface to the euphotic layer and promoting high productivity. According to Tchamabi et al. (2017), there is no evidence of a cold spot in these surface waters despite the identification of mixing in the water masses. However, the IME can contribute to enhancing biological productivity, and potential fisheries surrounding these islands are considered an "oasis" in oligotrophic environments and may contribute to the global CO_2 budget (Heywood et al. 1996). Biogeochemical modeling studies regarding the wake influence on nutrients and chlorophyll in FN and AR are underway, and further investigation of the dynamics of oceanographic variables still requires additional observational data around these islands and banks from the North Brazilian Chain (Tchamabi et al. 2017; Costa da Silva et al. 2021; Silva et al. 2021b).

Costa da Silva et al. (2021), using current, hydrographic and satellite data, investigated the general oceanic circulation and its modifications induced by FN and RA during spring 2015 and fall 2017. They noticed, in addition to the island wake effects, that the presence of FN strongly perturbs cSEC and SEUC features, with an upstream core splitting and a reorganization of single current core structures downstream of the island (Fig. 8.11). Near the island, the flow disturbances impact the thermohaline structure and biogeochemistry, with a negative anomaly in temperature (−1.3 °C) and salinity (−0.15) between 200 and 400 m depth on the southeastern side of FN, where the fluorescence peak (> 1.0 mg m^{-3}) is shallower than at the other stations around the island.

8.4.2 Amazon River–Ocean Continuum

The tropical Atlantic Ocean adjacent to the northern region of Brazil is a high-energy marine system due to the combined action of high continental water inputs, tidal oscillations, trade winds, and a strong western boundary current (Silva et al. 2005, 2009b, c, 2010; Araujo et al. 2017). The Amazon Continental Shelf (ACS) extends from the mouth of the Parnaíba River to the mouth of the Oiapoque River with a width from 100 km (Parnaíba River mouth) to 330 km (off Marajó Island).

A unique feature of the northern region is the high discharge from the Amazon River, which accounts for approximately 18% of the world's continental waters draining into the ocean (Chen et al. 2012; Araujo et al. 2014). As the Amazon River enters the ACS, low salinity and turbid water masses form a plume that persists both temporally and spatially, covering an extensive area. The freshwater of the Amazon River forms a surface plume area that can exceed 10^6 km^2, thus covering a vast portion of the western tropical Atlantic and reaching longitudes of 30° W (Coles et al. 2013; Ibánhez et al. 2015, 2016). The plume contributes to a great input of organic and inorganic components and sediments into the ocean (Santos et al. 2008; Araujo et al. 2014). The associated nutrient loads are responsible for significant

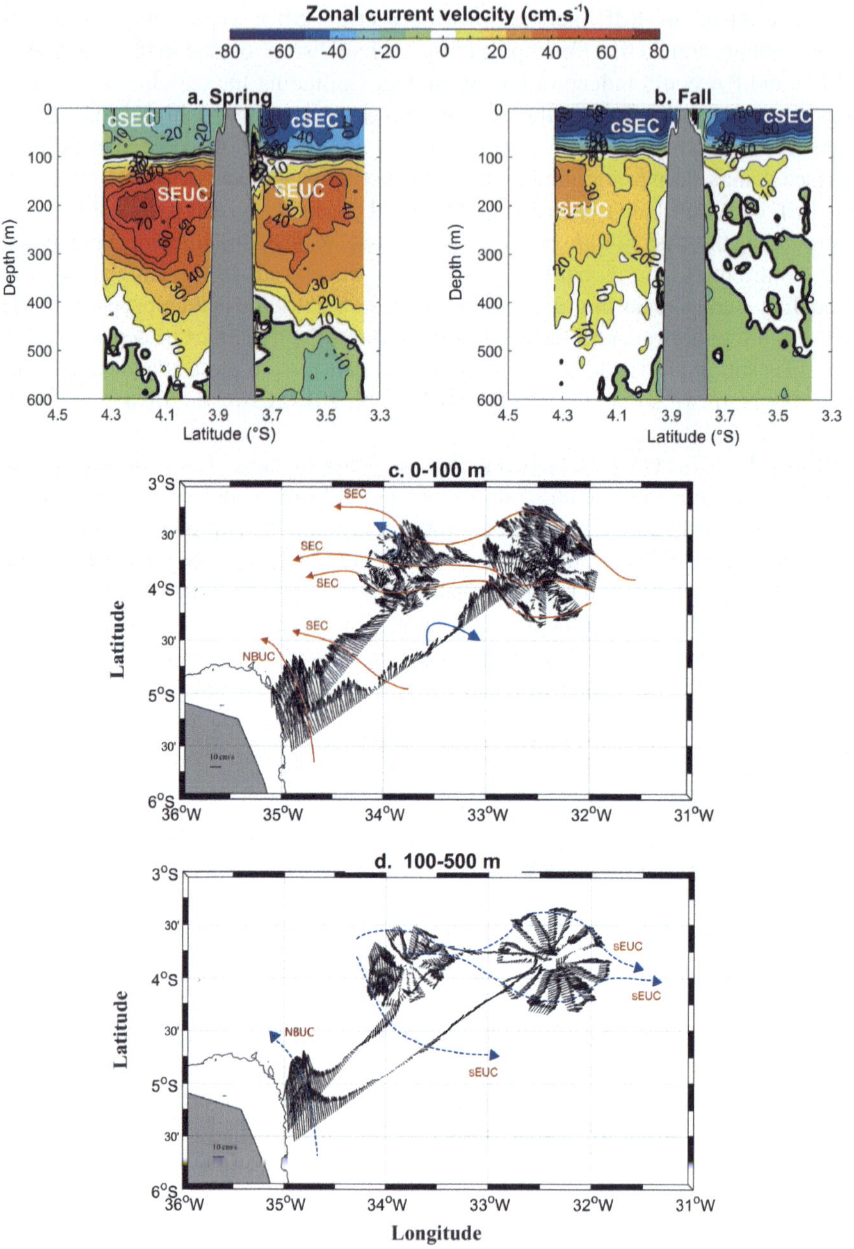

Fig. 8.11 Mean zonal current velocity (in cm s^{-1}) obtained from ABRACOS cruises in spring 2015 (**a**) and fall 2017 (**b**) around FN. Average of the currents observed in the surface layer (0–100 m) (**c**) and subsurface layer (100–500 m) (**d**) during the ABRAÇOS cruises (black arrows) and schematic illustration of the circulation. Red arrows in **c** represent large currents (cSEC and a part of the NBUC), while blue arrows represent mesoscale circulation structures. The blue dashed arrows in **d** represent the subsurface currents (sEUC and NBUC). Costa da Silva et al. (2021)

changes in primary production in the continental shelf and the adjacent oceanic region, even inducing atmospheric nitrogen fixation (N_2; diazotrophy mechanism) and producing an increase in atmospheric CO_2 sequestration (Yeung et al. 2012; Araujo et al. 2014). The high turbidity of this plume is due to the resuspension of sediments from the bottom caused by wind, tides, and convective mixing of the river flows, altering the pelagic physical–chemical parameters (e.g., salinity, light, oxygen, pH, and dissolved nutrients) over the shelf (Silva et al. 2009c, 2010; Francini-Filho et al. 2018). This results in high primary productivity on the coasts of Pará and Amapá (45°–50° W), which counteracts the characteristic oligotrophy of the western edge (REVIZEE 2006).

The ACS is also known for meso- to macrotidal semidiurnal tides (Prestes et al. 2018) and strong induced tidal currents, with M2 as its main semidiurnal component. These tides reach amplitudes greater than 3 m near the mouth of the Amazon River (Beardsley et al. 1995) with sufficient pressure gradients to generate velocities of approximately 1.0 m s^{-1} over the continental shelf, with residual transport perpendicular to the coast (Prestes et al. 2018). In addition to the effect of tides, internal wave interactions with the bottom can also influence the upwelling process of cooler waters (Venayagamoorthy and Fringer 2012; Lamb 2013; Lauton et al. 2021). The generation sites of solitary large-amplitude internal waves commonly occur in coastal areas or in areas with large bathymetric variations, which influence the near-surface current field. Distinct groups of internal waves propagate on the ACS, with most observations being of along-shore waves (Lentini et al. 2016).

The ACS and its adjacent oceanic region are equally under the direct influence of the NBC, one of the strongest western boundary currents in the oceans. The NBC flows near the ACS slope, predominantly in a northwest direction, with speeds ranging from 0.5 to 1.0 m s^{-1} (Schott et al. 1998). The estimated transport values of the NBC adjacent to the Amazonian coast (4° N, 45° W) range between 13 Sv ($1\ Sv = 10^6\ m^3\ s^{-1}$) in April and May and 35 Sv in July and August, with an annual mean of approximately 26 Sv (Johns et al. 1998). Farther north, this strong contour current retroflects seasonally to the east (7° N, 52° W) and feeds the North Equatorial Countercurrent (NECC), thus contributing to the formation of anticyclonic vortices (Fratantoni and Glickson 2002; Jochum and Malanotte-Rizzoli 2003). The ACS and adjacent oceanic area are characterized by the presence of coastal water (CW; $\sigma_\theta \leq 21.6$), associated with the continental contributions of the Amazon River, TW, SACW and AAIW (Silva et al. 2005).

All oceanographic features described above are the result of complex physical–biogeochemical interactions that have been studied, but some mechanisms are not yet well understood. Previous studies have shown, for example, the presence of cold waters near the mouth of the Amazon River (Geyer et al. 1996; Neto and Silva 2014; Araujo and Mezilet 2016). These studies indicate that SST anomalies extended over a large portion of the NBC region, where the vertical temperature distribution showed an elevation of the isotherms to a depth near the surface in the shelf-break area. The upwelling of high-salinity water along the plume axis may contribute to the tongue-shaped structure (Gibbs 1976), and the upwelling may indeed be significant on the ACS (Geyer et al. 1996). Cold water cells (Neto and Silva 2014) and lenses were identified on the ACS in front of the mouth of the Amazon River in a period when the NBC was stronger (Araujo and Mezilet 2016). The presence of a colder surface water cell located on the continental slope (100 m isobath) was found, showing a breakdown of water column stratification and uprise of the isotherms (27 °C and 28 °C) from a depth of 120 m to the surface over the shelf-break. In May 1999 (REVIZEE N3 campaign), however, the surface waters in the region were warmer and influenced by the Amazon River's low salinity waters, and the vertical structure contained waters quite stratified, with isobaths completely parallel and close to each other between 50 and 80 m (Araujo and Mezilet 2016).

Moreover, whether the NBC intensity variations influence the observed upwelling process must still be determined. Seasonally, an intensification of the NBC occurs in the second semester of each year, and a peak is observed in July–August (Johns et al. 1998). This intensification also occurs in August, with mean ACS values of 1.5 m s^{-1} (northwestward) (Araujo and Mezilet 2016). Therefore, the NBC intensification may result in a greater interaction between the circulation and topography of the ACS and its slope, thus increasing the upwelling mechanism.

8.5 Synopsis

The TA is a complex region with two main current systems in play, the Equatorial System (ES) and the Western Boundary System (WBS). The equatorward western boundary North Brazil Current and Undercurrent (NBC-NBUC) contributes to the upper Atlantic Meridional Overturning Circulation (AMOC) and wind-driven subtropical cells. In turn, the NBC-NBUC are fed by the zonal flow of the SEC branches, connecting the western TA with the central Atlantic. The water masses are considered oligotrophic because of the relatively strong static stability with a well-marked thermocline. The currents, winds, precipitation, biogeochemical cycles, and CO_2 flux are seasonally regulated by the Intertropical Convergence Zone (ITCZ). Anomalous rainfall in eastern Northeast Brazil (NEB) is highly correlated with early (nearly 6-month lead time) southeastern tropical Atlantic (SETA) SST anomalies.

On a larger scale, the climate of South America is primarily driven by the variability in the tropical Atlantic and Pacific Oceans. The first variability mode in the Atlantic is the interhemispheric mode, and the second is the equatorial mode or the Atlantic Niño mode. The Pacific El Niño Southern Oscillation (ENSO) is known to influence dry/wet events in South America. Recent studies have shown that the winter Atlantic Niño is also negatively correlated with ENSO, with the Atlantic leading the Pacific by 2–3 seasons (Hounsou-Gbo et al. 2020). SST anomalies in the equatorial Atlantic affect Walker circulation with anomalous ascending/descending branches over the Atlantic and central equatorial Pacific, driving anomalous easterly/westerly surface winds favorable for ENSO development.

The tropical Atlantic is the second-largest oceanic source of CO_2 to the atmosphere after the tropical Pacific (Lefèvre et al. 2010; Ibánhez et al. 2015; Araujo et al. 2019). However, little is known about the seasonal and interannual variability of the CO_2 flux at the air–water interface in rivers and estuaries in the western tropical South Atlantic, particularly in Northeast Brazil. Despite the net CO_2 outgassing recorded in the tropical Atlantic, zones of net atmospheric CO_2 uptake exist, which are mainly linked to the seasonal cycle of SST and its associated thermodynamic effects on the partial pressure of CO_2 (pCO_2) in the North Equatorial Current (Ibánhez et al. 2015). Rivers and estuaries can also play an important role in the transport and transformation of carbon from the continent to the adjacent coastal zone, typically acting as sources of CO_2 to the atmosphere. In the western tropical Atlantic, the primary freshwater contributions originate from the Amazon, Orinoco, São Francisco and Paraíba do Sul rivers (Araujo et al. 2014). The inflow varies seasonally, with a maximum in May and a minimum in November. Recent studies in the northern and northeastern Brazilian major river estuaries have reported a negative correlation between dissolved oxygen saturation and pCO_2, indicating a control by biological processes, especially organic matter degradation, that act as a source of atmospheric CO_2 (Araujo et al. 2019). Population density has shown a strong correlation with positive CO_2 fluxes (source of CO_2 to the atmosphere) when compared to local climate variations, being weakly correlated with rainfall, possibly resulting from constant flushing of organic matter by runoff from urban centers (Araujo et al. 2014). The CO_2 fluxes, average nitrogen concentrations, dissolved inorganic nitrogen-DIN, dissolved inorganic phosphorus-DIP and dissolved inorganic carbon-DIC are significantly lower in the Amazon and Orinoco estuarine systems than in estuarine systems located along the eastern edge of the southern tropical Atlantic (Amiotte Suchet et al. 2003; Araujo et al. 2014). In fact, in low discharge conditions, the spread of the Amazon plume acts as an atmospheric CO_2 sink of global relevance (Lefèvre et al. 2010; Ibánhez et al. 2015). In contrast, the NE region acts primarily as a CO_2 source to the atmosphere with the highest CO_2 fluxes associated with periods of high SST in the area (Araujo et al. 2019). The highest productivity in the equatorial broadband—in comparison with eastern NEB—results from the influence of the Amazon River plume (Araujo et al. 2019) and from the island mass effect and wakes in the North Brazilian chain of islands and sea mounts (Tchamabi et al. 2017; Costa da Silva et al. 2021; Silva et al. 2021b), thus supporting most fisheries in North and Northeast Brazil.

Acknowledgements We express our sincere thanks to the participating scientists, captains, officers and crew of the Brazilian and French ships for carrying out the oceanographic cruises. This work was supported by the INCT AmbTropic—National Institute on Science and Technology for Tropical Marine Environments, CNPq/FABESB (565054/2010-4, 8936/2011, 465634/2014-1 and inc004/2019). MA, CN LC and HA thank the support of the Brazilian Research Network on Global Climate Change—Rede CLIMA (FINEP-CNPq 437167/2016-0). This is a contribution to the IJL-TAPIOCA and to the TRIATLAS project, which has received funding from the European Union's Horizon 2020 research and innovation program under grant agreement No. 817578.

References

Aguiar AL, Cirano M, Marta-Almeida M et al (2018) Upwelling processes along the south equatorial current bifurcation region and the Salvador Canyon (13° S), Brazil. Cont Shelf Res 171:77–96. https://doi.org/10.1016/j.csr.2018.10.001

Amiotte Suchet P, Probst JL, Ludwig W (2003) Worldwide distribution of continental rock lithology: implications for the atmospheric/soil CO_2 uptake by continental weathering and alkalinity river transport to the oceans. Global Biogeochem Cycles 17(2). http://doi.org/10.1029/2002GB001891

Andrié C, Oudot C, Genthon C et al (1986) CO_2 fluxes in the tropical Atlantic during FOCAL cruises. J Geophys Res. https://doi.org/10.1029/jc091ic10p11741

Araujo J, Mezilet Y (2016) Variabilité de la cellule d'upwelling au large de l'embouchure de l'Amazone. Rapport de Stage M1-AOC, UPS-OMP/LEGOS

Araujo M, Limongi C, Servain J et al (2011) Salinity-induced mixed and barrier layers in the southwestern tropical Atlantic ocean off the northeast of Brazil. Ocean Sci 7:63–73. https://doi.org/10.5194/os-7-63-2011

Araujo M, Noriega C, Lefèvre N (2014) Nutrients and carbon fluxes in the estuaries of major rivers flowing into the tropical Atlantic. Front Mar Sci. https://doi.org/10.3389/fmars.2014.00010

Araujo M, Noriega C, Hounsou-Gbo GA et al (2017) A synoptic assessment of the Amazon river-ocean continuum during boreal autumn: from physics to plankton communities and carbon flux. Front Microbiol. https://doi.org/10.3389/fmicb.2017.01358

Araujo M, Noriega C, Medeiros C et al (2019) On the variability in the CO_2 system and water productivity in the western tropical Atlantic off North and Northeast Brazil. J Mar Syst 189:62–77. https://doi.org/10.1016/j.jmarsys.2018.09.008

Arhan M, Mercier H, Bourlès B et al (1998) Hydrographic sections across the Atlantic at 7° 30 N and 4° 30 S. Deep Sea Res I 45:829–872. https://doi.org/10.1016/S0967-0637(98)00001-6

Assunção RV, Silva AC, Roy A et al (2020) 3D characterisation of the thermohaline structure in the southwestern tropical Atlantic derived from functional data analysis of in situ profiles. Prog Oceanogr. https://doi.org/10.1016/j.pocean.2020.102399

Beardsley RC, Candela J, Limeburner R et al (1995) The M2 tide on the Amazon shelf. J Geophys Res 100:2283–2319. https://doi.org/10.1029/94JC01688

Becker M (2001) Hidrologia dos Bancos e Ilhas oceânicas do Nordeste Brasileiro, uma Contribuição ao Programa REVIZEE. Doctoral dissertation Universidade Federal de São Carlos, Brazil

Bertrand A (2015) ABRACOS 1 cruise, Antea R/V. http://doi.org/10.17600/15005600

Bertrand A (2017) ABRACOS 2 cruise, Antea R/V. http://doi.org/10.17600/17004100

Bonou FK, Noriega C, Lefèvre N et al (2016) Distribution of CO_2 parameters in the western tropical Atlantic ocean. Dyn Atmos Ocean 73:47–60. https://doi.org/10.1016/j.dynatmoce.2015.12.001

Borges AV (2005) Do we have enough pieces of the jigsaw to integrate CO_2 fluxes in the coastal ocean? Estuaries 28:3–27. https://doi.org/10.1007/BF02732750

Borges AV, Delille B, Frankignoulle M (2005) Budgeting sinks and sources of CO2 in the coastal ocean: diversity of ecosystems counts. Geophys Res Lett 32:1–6. https://doi.org/10.1029/2005GL023053

Bourlès B, Gouriou Y, Chuchla R (1999) On the circulation in the upper layer of the western equatorial Atlantic. J Geophys Res Oceans 104:21151–21170. https://doi.org/10.1029/1999JC900058

Bourlès B, Araujo M, McPhaden MJ et al (2019) PIRATA: a sustained observing system for tropical Atlantic climate research and forecasting. Earth Space Sci 6:577–616. https://doi.org/10.1029/2018EA000428

Brandt P, Schott FA, Provost C et al (2006) Circulation in the central equatorial Atlantic: mean and intraseasonal to seasonal variability. Geophys Res Lett. https://doi.org/10.1029/2005gl025498

Bruto L, Araujo M, Noriega C et al (2017) Variability of CO_2 fugacity at the western edge of the tropical Atlantic ocean from the 8° N to 38° W PIRATA buoy. Dyn Atmos Ocean 78:1–13. https://doi.org/10.1016/j.dynatmoce.2017.01.003

Cai W, Wu L, Lengaigne M et al (2019) Pantropical climate interactions. Science 363. https://doi.org/10.1126/science.aav4236

Chaves TBC, Mafalda JRP, Santos C et al (2006) Planktonic biomass and hydrography in the exclusive economic zone of Brazilian Northeast. Trop Oceanogr 34:12–30

Chen CTA, Huang TH, Fu YH et al (2012) Strong sources of CO_2 in upper estuaries become sinks of CO_2 in large river plumes. Curr Opin Environ Sustain 4:179–185. https://doi.org/10.1016/j.cosust.2012.02.003

Cintra M, Lentini CAD, Servain J et al (2015) Physical processes that drive the seasonal evolution of the southwestern tropical Atlantic warm pool. Dyn Atmos Ocean 7:1–11. https://doi.org/10.1016/j.dynatmoce.2015.08.001

Cochrane JD, Kelly FJ, Olling CR (1979) Subthermocline countercurrents in the western equatorial Atlantic ocean. J Phys Oceanogr 9:724–738. https://doi.org/10.1175/1520-0485(1979)009%3c0724:scitwe%3e2.0.co;2

Coles VJ, Brooks MT, Hopkins J et al (2013) The pathways and properties of the Amazon river plume in the tropical North Atlantic ocean. J Geophys Res Oceans 118:6894–6913. https://doi.org/10.1002/2013JC008981

Cordeiro TA, Brandini FP, Rosa RS et al (2013) Deep chlorophyll maximum in western equatorial Atlantic—how does it interact with islands slopes and seamounts? Mar Sci 3:30–37. https://doi.org/10.5923/j.ms.20130301.03

Costa da Silva A, Chaigneau A, Dossa AN et al (2021) Surface circulation and vertical structure of upper ocean variability around Fernando de Noronha Archipelago and Rocas Atoll during spring 2015 and fall 2017. Front Mar Sci. https://doi.org/10.3389/fmars.2021.598101

Cushman-Roisin B (1994) Introduction to geophysical fluid dynamics. Prentice Hall International, Hoboken, New Jersey

Da Cunha LC, Buitenhuis ET (2013) Riverine influence on the tropical Atlantic ocean biogeochemistry. Biogeosciences 10:6357–6373. https://doi.org/10.5194/bg-10-6357-2013

Dai A, Trenberth K (2002) Estimates of freshwater discharge from continents: latitudinal and seasonal variations. J Hydrometeorol 3:660–687. https://doi.org/10.1175/1525-7541(2002)003%3C0660:EOFDFC%3E2.0.CO;2

Domingues EC, Schettini CAF, Truccolo EC et al (2017) Hydrography and currents on the Pernambuco continental shelf. Rev Brasileira Recursos Hidricos. https://doi.org/10.1590/2318-0331.0217170027

Dossa AN, Silva AC, Chaigneau A et al (2021) Near-surface western boundary circulation off Northeast Brazil. Prog Oceanogr. https://doi.org/10.1016/j.pocean.2020.102475

Doty MS, Oguri M (1956) The island mass effect. ICES J Mar Sci 22:33–37. https://doi.org/10.1093/icesjms/22.1.33

Ekau W, Knoppers B (1999) An introduction to the pelagic system of the North-East and East Brazilian shelf. Arch Fish Mar Res 47:113–132

Flagg CN (1987) Hydrographic structure and variability. In: Backus H (ed) Georges banks. The MIT Press, Cambridge, MA, pp 108–124

Foltz GR, Brandt P, Richter I et al (2019) The tropical Atlantic observing system. Front Mar Sci. https://doi.org/10.3389/fmars.2019.00206

Francini-Filho RB, Asp N, Siegle E et al (2018) Perspectives on the Great Amazon reef: extension, biodiversity, and threats. Front Mar Sci. https://doi.org/10.3389/fmars.2018.00142

Fratantoni DM, Glickson DA (2002) North Brazil current ring generation and evolution observed with sea WiFS. J Phys Oceanogr 32:1058–1074. https://doi.org/10.1175/1520-0485(2002)032%3c1058:NBCRGA%3e2.0.CO;2

Geyer WR, Beardsley RC, Lentz SJ et al (1996) Physical oceanography of the Amazon shelf. Cont Shelf Res 16:575–616. https://doi.org/10.1016/0278-4343(95)00051-8

Gibbs RJ (1976) Amazon river sediment transport in the Atlantic Ocean. Geology 4:45–48. https://doi.org/10.1130/0091-7613(1976)4%3c45:ARSTIT%3e2.0.CO;2

Goes M, Molinari R, Da Silveira I et al (2005) Retroflections of the North Brazil current during February 2002. Deep-Sea Res Part I 52:647–667. https://doi.org/10.1016/j.dsr.2004.10.010

Gordon AL (1986) Interocean exchange of thermocline water. J Geophys Res Oceans 91:5037–5046. https://doi.org/10.1029/JC091iC04p05037

Gove JM, Merrifield MA, Brainard R (2006) Temporal variability of current-driven upwelling at Jarvis Island. J Geophys Res. https://doi.org/10.1029/2005JC003161

Gove JM, McManus MA, Neuheimer AB et al (2016) Near-island biological hotspots in Barren ocean basins. Nat Commun. https://doi.org/10.1038/ncomms10581

Grimm AM, Barros VR, Doyle ME (2000) Climate variability in Southern South America associated with El Niño and La Niña events. J Clim 13:35–58. https://doi.org/10.1175/1520-0442(2000)013%3c0035:CVISSA%3e2.0.CO;2

Grodsky SA, Carton JA (2003) The intertropical convergence zone in the South Atlantic and the equatorial cold tongue. J Clim 16:723–733. https://doi.org/10.1175/1520-0442(2003)016%3c0723:TICZIT%3e2.0.CO;2

Herrford J, Brandt P, Kanzow T et al (2020) Seasonal variability of the Atlantic meridional overturning circulation at 11° S inferred from bottom pressure measurements. Ocean Sci Discuss 2020:1–37. https://doi.org/10.5194/os-2020-55

Heywood KJ, Stevens DP, Bigg GR (1996) Eddy formation behind the tropical island of Aldabra. Deep-Sea Res Part I Oceanogr Res Pap 43(4):555–578. https://doi.org/10.1016/0967-0637(96)00097-0

Hounsou-Gbo GA, Araujo M, Bourlès B et al (2015) Tropical Atlantic contributions to strong rainfall variability along the Northeast Brazilian coast. Adv Meteorol. https://doi.org/10.1155/2015/902084

Hounsou-Gbo GA, Servain J, Araujo M et al (2016) Oceanic indices for forecasting seasonal rainfall over the northern part of Brazilian Northeast. Am J Clim Change 5:261–274. https://doi.org/10.4236/ajcc.2016.52022

Hounsou-Gbo GA, Servain J, Araujo M et al (2019) SST indexes in the tropical South Atlantic for forecasting rainy seasons in northeast Brazil. Atmosphere. https://doi.org/10.3390/atmos10060335

Hounsou-Gbo A, Servain J, Vasconcelos Junior FC et al (2020) Summer and winter Atlantic Niño: connections with ENSO and implications. Clim Dyn 55:2939–2956. https://doi.org/10.1007/s00382-020-05424-x

Hummels R, Brandt P, Dengler M et al (2015) Interannual to decadal changes in the western boundary circulation in the Atlantic at 11° S. Geophys Res Lett 42:7615–7622. https://doi.org/10.1002/2015GL065254

Ibánhez JSP, Diverrès D, Araujo M et al (2015) Seasonal and interannual variability of sea-air CO_2 fluxes in the tropical Atlantic affected by the Amazon river plume. Global Biogeochem Cycles 29:1640–1655. https://doi.org/10.1002/2015GB005110

Ibánhez JSP, Araujo M, Lefèvre N (2016) The overlooked tropical oceanic CO_2 sink. Geophys Res Lett 43:3804–3812. https://doi.org/10.1002/2016GL068020

Jacobson AR, Mikaloff Fletcher SE, Gruber N et al (2007) A joint atmosphere-ocean inversion for surface fluxes of carbon dioxide: 2. Regional results. Global Biogeochem Cycles. http://doi.org/10.1029/2006GB002703

Jales MC, Feitosa FA, Koening ML et al (2015) Phytoplankton biomass dynamics and environmental variables around the Rocas Atoll Biological Reserve, South Atlantic. Braz J Oceanogr 63(4):443–454. https://doi.org/10.1590/S1679-87592015093906304

Jochum M, Malanotte-Rizzoli P (2003) On the generation of North Brazil current rings. J Mar Res 61:147–173. https://doi.org/10.1357/002224003322005050

Johns WE, Lee TN, Beardsley RC et al (1998) Annual cycle and variability of the North Brazil current. J Phys Oceanogr 28:103–128. https://doi.org/10.1175/1520-0485(1998)028%3c0103:ACAVOT%3e2.0.CO;2

Jouanno J, Hernandez O, Sanchez-Gomez E (2017) Equatorial Atlantic interannual variability and its relation to dynamic and thermodynamic processes. Earth Syst Dynam 8:1061–1069. https://doi.org/10.5194/esd-8-1061-2017

Keenlyside NS, Ding H, Latif M (2013) Potential of equatorial Atlantic variability to enhance El Nino prediction. Geophys Res Lett 40:2278–2283. https://doi.org/10.1002/grl.50362
Kempe S (1984) Sinks of the anthropogenically enhanced carbon cycle in surface fresh waters. Geophys Res Atmos. https://doi.org/10.1029/JD089iD03p04657
Koettker A, Freire A, Sumida P (2010) Temporal, diel and spatial variability of decapod larvae from St Paul's Rocks, an equatorial oceanic island of Brazil. J Mar Biol Assoc UK 90:1227–1239. https://doi.org/10.1017/S0025315409990890
Kouadio YK, Servain J, Machado LAT, Lentini CAD (2012) Heavy rainfall episodes in the eastern northeast Brazil linked to large-scale ocean-atmosphere conditions in the tropical Atlantic. Adv Meteorol. https://doi.org/10.1155/2012/369567
Lamb KG (2013) Internal wave breaking and dissipation mechanisms on the continental slope/shelf. Annu Rev Fluid Mech 46:231–254. https://doi.org/10.1146/annurev-fluid-011212-140701
Laruelle GG, Dürr HH, Slomp CP et al (2010) Evaluation of sinks and sources of CO_2 in the global coastal ocean using a spatially-explicit typology of estuaries and continental shelves. Geophys Res Lett. https://doi.org/10.1029/2010GL043691
Lauton G, Pattiaratchi CB, Lentini CAD (2021) Observations of breaking internal tides on the Australian north west shelf edge. Front Mar Sci. https://doi.org/10.3389/fmars.2021.629372
Lefèvre N, Diverrès D, Gallois F (2010) Origin of CO_2 undersaturation in the western tropical Atlantic. Tellus Ser B Chem Phys Meteorol 62:595–607. https://doi.org/10.1111/j.1600-0889.2010.00475.x
Lefèvre N, Caniaux G, Janicot S et al (2013) Increased CO_2 outgassing in February-May 2010 in the tropical Atlantic following the 2009 Pacific El Niño. J Geophys Res Oceans 118:1645–1657. https://doi.org/10.1002/jgrc.20107
Lentini CAD, Magalhães JM, Da Silva JCB et al (2016) Transcritical flow and generation of internal solitary waves off the Amazon river: synthetic aperture radar observations and interpretation. Oceanography 29:187–195. http://www.jstor.org/stable/24862294
Lira SMA, Teixeira IA, Lima CDM et al (2014) Spatial and nycthemeral distribution of the zooneuston off Fernando de Noronha, Brazil. Braz J Oceanogr 62:35–45. https://doi.org/10.1590/s1679-87592014058206201
Losada T, Rodríguez-Fonseca B (2016) Tropical atmospheric response to decadal changes in the Atlantic equatorial mode. Clim Dyn 47:1211–1224. https://doi.org/10.1007/s00382-015-2897-2
Lübbecke JF, McPhaden MJ (2012) On the inconsistent relationship between Pacific and Atlantic Niños. J Clim 25:4294–4303. https://doi.org/10.1175/JCLI-D-11-00553.1
Lübbecke JF, Rodríguez-Fonseca B, Richter I et al (2018) Equatorial Atlantic variability—modes, mechanisms, and global teleconnections. Wires Clim Change. https://doi.org/10.1002/wcc.527
Lumpkin R, Garzoli SL (2005) Near-surface circulation in the tropical Atlantic ocean. Deep Sea Res I 52:495–518. https://doi.org/10.1016/j.dsr.2004.09.001
Martín-Rey M, Rodríguez-Fonseca B, Polo I (2015) Atlantic opportunities for ENSO prediction. Geophys Res Lett 42:6802–6810. https://doi.org/10.1002/2015GL065062
Martín-Rey M, Polo I, Rodriguez-Fonseca B et al (2018) Is there evidence of changes in tropical Atlantic variability modes under AMO phases in the observational record? J Clim 31:515–536. https://doi.org/10.1175/JCLI-D-16-0459.1
Melo PAMC, Diaz XFG, Macedo SJ et al (2012) Diurnal and spatial variation of the mesozooplankton community in the Saint Peter and Saint Paul Archipelago, equatorial Atlantic. Mar Biodivers Rec. http://doi.org/10.1017/S1755267212001054
Morato T, Varkey D, Damaso C et al (2008) Evidence of a seamount effect on aggregating visitors. Mar Ecol Prog Ser 357:23–32. https://doi.org/10.3354/meps07269
Neto AVN, da Silva AC (2014) Seawater temperature changes associated with the North Brazil current dynamics. Ocean Dyn 64:13–27. https://doi.org/10.1007/s10236-013-0667-4
Nnamchi HC, Li J, Kucharski F et al (2015) Thermodynamic controls of the Atlantic Niño. Nat Commun. https://doi.org/10.1038/ncomms9895
Noriega C, Araujo M (2014) Carbon dioxide emissions from estuaries of northern and northeastern Brazil. Sci Rep. https://doi.org/10.1038/srep06164

Noriega CED, Araujo M, Lefèvre N (2013) Spatial and temporal variability of the CO_2 fluxes in a tropical, highly urbanized estuary. Estuaries Coasts 36:1054–1072. https://doi.org/10.1007/s12237-013-9608-1

Noriega C, Araujo M, Lefèvre N et al (2015) Spatial and temporal variability of CO_2 fluxes in tropical estuarine systems near areas of high population density in Brazil. Reg Environ Change 15:619–630. https://doi.org/10.1007/s10113-014-0671-3

Okumura Y, Xie SP (2006) Some overlooked features of tropical Atlantic climate leading to a new Niño-like phenomenon. J Clim 19:5859–5874. https://doi.org/10.1175/JCLI3928.1

Ou HW, De Ruijter WPM (1986) Separation of an inertial boundary current from a curved coastline. J Phys Ocean 16:280–289. https://doi.org/10.1175/1520-0485(1986)016%3c0280:SOAIBC%3e2.0.CO;2

Palacios DM (2002) Factors influencing the island-mass effect of the Galápagos Archipelago. Geophys Res Lett. https://doi.org/10.1029/2002GL016232

Pitcher TJ, Morato T, Hart PJB et al (2008) Seamounts: ecology, fisheries and conservation. Wiley

Prestes YO, Silva AC, Jeandel C (2018) Amazon water lenses and the influence of the North Brazil current on the continental shelf. Cont Shelf Res 160:36–48. https://doi.org/10.1016/j.csr.2018.04.002

Rao VB, Hada K (1990) Characteristics of rainfall over Brazil: annual variations and connections with the southern oscillation. Theor Appl Climatol 42:81–91. https://doi.org/10.1007/BF00868215

REVIZEE (2006) Programa REVIZEE - Avaliação do Potencial Sustentável dos Recursos Vivos da Zona Econômica Exclusiva. Executive report. Ministry of the Environment, Brazil

Richardson PL, Reverdin G (1987) Seasonal cycle of velocity in the Atlantic north equatorial countercurrent as measured by surface drifters, current meters, and ship drifts. J Geophys Res Oceans 92:3691–3708. https://doi.org/10.1029/JC092iC04p03691

Rodrigues RR, Rothstein LM, Wimbush M (2007) Seasonal variability of the south equatorial current bifurcation in the Atlantic ocean: a numerical study. J Phys Oceanogr 37:16–30. https://doi.org/10.1175/jpo2983.1

Rodriguez-Fonseca B, Polo I, Garcia-Serrano J et al (2009) Are Atlantic Ninos enhancing Pacific ENSO events in recent decades? Geophys Res Lett. https://doi.org/10.1029/2009GL040048

Santos MLS, Muniz KM, Barros-Neto B et al (2008) Nutrient and phytoplankton biomass in the Amazon river shelf waters. An Acad Bras Cienc 80:703–717. https://doi.org/10.1590/S0001-37652008000400011

Sarma VVSS, Viswanadham R, Rao GD et al (2012) Carbon dioxide emissions from Indian monsoonal estuaries. Geophys Res Lett. https://doi.org/10.1029/2011GL050709

Schott FA, Fischer J, Stramma L (1998) Transports and pathways of the upper-layer circulation in the western tropical Atlantic. J Phys Oceanogr 28:1904–1928. https://doi.org/10.1175/1520-0485(1998)028%3c1904:TAPOTU%3e2.0.CO;2

Schott FA, Brandt P, Hamann M et al (2002) On the boundary flow off Brazil at 5–10° S and its connection to the interior tropical Atlantic. Geophys Res Lett. https://doi.org/10.1029/2002GL014786

Schott FA, McCreary JP, Johnson GC (2004) Shallow overturning circulations of the tropical-subtropical oceans. In: Wang C, Xie SP, Carton JA (eds) Earth's climate: the ocean-atmosphere interaction, geophysical monograph series, 147. AGU, pp 261–304. http://doi.org/10.1029/147GM15

Schott FA, Dengler M, Zantopp R et al (2005) The shallow and deep western boundary circulation of the South Atlantic at 5°–11° S. J Phys Oceanogr 35:2031–2053. https://doi.org/10.1175/JPO2813.1

Silva AC, Araujo M, Bourlès B (2005) Variação sazonal da estrutura de massas de água na plataforma continental do Amazonas e área oceânica adjacente. Rev Bras Geofís 23:1–13. https://doi.org/10.1590/S0102-261X2005000200004

Silva AC, Bourlès B, Araujo M (2009a) Circulation of the thermocline salinity maximum waters off the Northern Brazil as inferred from in situ measurements and numerical results. Ann Geophys 27:1861–1873. https://doi.org/10.5194/angeo-27-1861-2009
Silva M, Araujo M, Servain J (2009b) Circulation and heat budget in a regional climatological simulation of the southwestern tropical Atlantic. Trop Oceanogr 37(1–2). http://doi.org/10.5914/tropocean.v37i1-2.5156
Silva M, Araujo M, Servain J et al (2009c) High-resolution regional ocean dynamics simulation in the southwestern tropical Atlantic. Ocean Model 30:256–269. https://doi.org/10.1016/j.ocemod.2009.07.002
Silva AC, Bourlès B, Araujo M (2010) Seasonal variability of the Amazon river plume during REVIZEE program. Trop Oceanogr 38:76–87. https://doi.org/10.5914/tropocean.v38i1.5162
Silva T, Veleda D, Araujo M et al (2018) Ocean-atmosphere feedback during extreme rainfall events in eastern northeast Brazil. J Appl Meteorol Climatol 57:1211–1229. https://doi.org/10.1175/JAMC-D-17-0232.1
Silva AC, Chaigneau A, Dossa NA et al (2021a) Surface circulation and vertical structure of upper ocean variability around Fernando de Noronha Archipelago and Rocas Atoll during spring 2015 and fall 2017. Front Mar Sci. https://doi.org/10.3389/fmars.2021.598101
Silva M, Araujo M, Geber F et al (2021b) Ocean dynamics and topographic upwelling around the Aracati seamount—North Brazilian chain from in situ observation and modelling results. Front Mar Sci. https://doi.org/10.3389/fmars.2021.609113
Silveira I, Miranda L, Brown WS (1994) On the origins of the North Brazil current. J Geophys Res Oceans. http://doi.org/10.1029/94JC01776/full
Souza MFL, Gomes VR, Freitas SS et al (2009) Net ecosystem metabolism and nonconservative fluxes of organic matter in a tropical mangrove estuary, Piauí River (NE of Brazil). Estuaries Coasts 32:111–122. https://doi.org/10.1007/s12237-008-9104-1
Souza CS, Guimarães da Luz JA, Macedo S et al (2013) Chlorophyll a and nutrient distribution around seamounts and islands of the tropical south-western Atlantic. Mar Freshw Res 64:168–184. https://doi.org/10.1071/MF12075
Stramma L, Fischer J, Reppin J (1995) The North Brazil undercurrent. Deep-Sea Res Part I Oceanogr Res Pap 42:773–795. https://doi.org/10.1016/0967-0637(95)00014-W
Stramma L, Rhein M, Brandt P et al (2005) Upper ocean circulation in the western tropical Atlantic in boreal fall 2000. Deep-Sea Res Part I Oceanogr Res Pap 52:221–240. https://doi.org/10.1016/j.dsr.2004.07.021
Takahashi T, Sutherland SC, Wanninkhof R et al (2009) Climatological mean and decadal change in surface ocean pCO_2, and net sea–air CO_2 flux over the global oceans. Deep-Sea Res II 56:554–577. https://doi.org/10.1016/j.dsr2.2008.12.009
Tchamabi CC, Araujo M, Silva M et al (2017) A study of the Brazilian Fernando de Noronha Island and Rocas Atoll wakes in the tropical Atlantic. Ocean Model 111:9–18. https://doi.org/10.1016/j.ocemod.2016.12.009
Tchamabi C, Araujo M, Silva MA et al (2018) Ichthyoplankton transport around the Brazilian Fernando de Noronha archipelago and Rocas Atoll: are there any connectivity patterns? Indian J Geo-Mar Sci 47:812–818
Travassos PEPF, Hazin FHV, Zagaglia JR et al (1999) Thermohaline structure around seamounts and islands off Northeast Brazilian coast. Arch Fish Mar Res 47:211–222
Veleda D, Araújo M, Silva M et al (2011) Seasonal and interannual variability of the southern south equatorial current bifurcation and meridional transport along the eastern Brazilian edge. Trop Oceanogr 39:27–59. https://doi.org/10.5914/to.2011.0051
Veleda D, Araujo M, Zantopp R et al (2012) Intraseasonal variability of the North Brazil undercurrent forced by remote winds. J Geophys Res Oceans. https://doi.org/10.1029/2012JC008392
Venayagamoorthy SK, Fringer OB (2012) Breaking internal waves on a shelf slope using numerical simulations. Oceanography 2:132–139. https://doi.org/10.5670/oceanog.2012.48

Yeung LY, Berelson WM, Young ED et al (2012) Impact of diatom-diazotroph associations on carbon export in the Amazon river plume. Geophys Res Lett. https://doi.org/10.1029/2012GL053356

Zhang F, Wang C, Pu Z (2019) Genesis of Tibetan plateau vortex: roles of surface diabatic and atmospheric condensational latent heating. J Appl Meteorol Climatol 58:2633–2651. https://doi.org/10.1175/JAMC-D-19-0103.1

Index

J. M. Landim Dominguez et al. (eds.), *Tropical Marine Environments of Brazil*, The Latin American Studies Book Series,
https://doi.org/10.1007/978-3-031-21329-8

SPRINGER NATURE

GPSR Compliance

The European Union's (EU) General Product Safety Regulation (GPSR) is a set of rules that requires consumer products to be safe and our obligations to ensure this.

If you have any concerns about our products, you can contact us on ProductSafety@springernature.com

In case Publisher is established outside the EU, the EU authorized representative is:

Springer Nature Customer Service Center GmbH
Europaplatz 3
69115 Heidelberg, Germany

The manufacturer's authorised representative in the EU is Springer Nature Customer Service Centre GmbH, Europaplatz 3, 69115 Heidelberg, Germany. If you have any concerns regarding our products, please contact ProductSafety@springernature.com

Printed and bound by CPI Group (UK) Ltd, Croydon, CR0 4YY

01/10/2026

02225893-0006